Biopolymers from Polysaccharides and Agroproteins

ACS SYMPOSIUM SERIES **786**

Biopolymers from Polysaccharides and Agroproteins

Richard A Gross, EDITOR
Polytechnic University

Carmen Scholz, EDITOR
University of Alabama at Huntsville

American Chemical Society, Washington, DC

Library of Congress Cataloging-in-Publication Data

Biopolymers from polysaccharides and agroproteins / Richard A. Gross, editor, Carmen Scholz, editor.

p. cm.—(ACS symposium series ; 786)

Includes bibliographical references and index.

ISBN 0-8412-3645-3

1. Polymers—Biotechnology. 2. Polysaccharides—Biotechnology. 3. Plant proteins—Biotechnology. 4. Biodegradation.

I. Gross, Richard A., 1957- II. Scholz, Carmen, 1963- III. Series.

TP248.65.P62 B547 2001
668.9—dc21 00-53117

The paper used in this publication meets the minimum requirements of American National Standard for Information Sciences—Permanence of Paper for Printed Library Materials, ANSI Z39.48-1984.

Distributed by Oxford University Press

PRINTED IN THE UNITED STATES OF AMERICA

Foreword

THE ACS SYMPOSIUM SERIES was first published in 1974 to provide a mechanism for publishing symposia quickly in book form. The purpose of the series is to publish timely, comprehensive books developed from ACS sponsored symposia based on current scientific research. Occasionally, books are developed from symposia sponsored by other organizations when the topic is of keen interest to the chemistry audience.

Before agreeing, to publish a book, the proposed table of contents is reviewed for appropriate and comprehensive coverage and for interest to the audience. Some papers may be excluded in order to better focus the book: others may be added to provide comprehensiveness. When appropriate, overview or introductory chapters are added. Drafts of chapters are peer-reviewed prior to final acceptance or rejection, and manuscripts are prepared in camera-ready format.

As a rule, only original research papers and original review papers are included in the volumes. Verbatim reproductions of previously published papers are not accepted.

ACS BOOKS DEPARTMENT

Contents

Poly(amino acids): Natural and Synthetic

Microbial and In-Vitro Synthetic Methods

Biomedical Polymers from Carbohydrates and Amino Acids

Biodegradable Polymers: Status and Testing

Preface

Nature can provide an impressive array of polymers that can be used in fibers, adhesives, coatings, gels, foams, films, thermoplastics, and thermoset resins. Polymers from renewable resources take on new importance in a world that is trying to navigate through the apparent conflict between technological development and environmental protection. It is ironic that the solutions to the problems plaguing the environment must come from the re-direction of the problem creators. Unfortunately, for too long, technological development was not linked to principles of environmental ecology. Our society has evolved so that technology lies at the core of economic development. So, we now turn to the technology innovators to create new pathways to products that are harmonious with environmental maintenance and renewal. This book is dedicated to these ideas and principles.

Biological processes are remarkable because they reside in an all-aqueous environment, function in a continuous or seamless manner, and may be conducted under ambient conditions. This is in stark contrast to most synthetic materials preparation and processing, which often require organic solvents and high temperatures and pressures. There is also generally insufficient consideration for biodegradation or recycling to replenish starting feedstocks. In the biological world, post-biosynthesis tools are used to facilitate processing. Examples worth exploring and exploiting are the formation of mesophases, enzymatic processing to control recognition events, surfactant–polymer interactions to optimize molecular recognition at interfaces, chiral control to dictate secondary and higher order structures, and cross-linking to convert water soluble precursors to water-insoluble composites and complexes. Finally, material–lifecycle concerns suggest that a great deal can be learned from biological paradigms, where all materials are returned to natural geochemical and biochemical pathways once their use is completed. Polymers from renewable resources are produced by biological processes and, therefore, benefit from the above characteristics of natural processes.

Developments in the field of polymers that exploit in-vitro enzyme-based processes continue to expand. Important new insights into novel reactions, reaction mechanisms, changes in enzyme structure in organic solvents, the use of neat reactions to generate polymers using enzymes, and molecular evolution of enzymes are all examples of this expansion. The biological world builds and modifies all of its polymers using enzymes. Enzymes provide many critical controls important to their use in polymer science including: (1) control of polymer structure (e.g., enantioselectivity and regioselectivity, (2) simple reactions over an ever-widening range of conditions (e.g., extremes of temperature, pH, pressure, and salt), (3) important options in green chemistry, and (4) rapid enzyme evolution (e.g., use of DNA shuffling and error prone PCR techniques). This book includes recent work where in-vitro enzyme-catalysis was used to synthesize oligosaccharides, star-shaped heteroarm polyesters, and new vinyl monomers from glycolipids.

The surfactant and emulsifier industry in the United States has grown nearly 300% during the past decade. In 1989, the U.S. production was estimated to be 15.5 billion pounds and the value of U.S. shipments in 1989 was approximately 3.7 billion dollars. Many of the applications of surfactants and emulsifiers (desorption of organic pollutants from soil, dispersants in cleaning formulations, facilitate emulsion formation, etc.) either involve human contact or release into the environment. Therefore, there is a critical need for a new generation of natural, low toxicity, fully biodegradable products. This book includes a chapter that describes natural polymers that are useful as surface-active agents for the control of soil erosion. In addition, powerful natural emulsifiers from microbial fermentation are the topic of one chapter.

The interface between biology and polymer science is a rapidly emerging research theology that involves the seamless convergence of concepts, ideas and research tools at the interface between biology and polymer science. Work at this interface is growing at a rapid pace and is expected to make major contributions to science in the new millennium. Scientists having core training in microbiology, genetic engineering, biophysics, agricultural science, plant physiology, food science, and other biological disciplines bring new tools, molecules, and fresh ideas to the fields of polymers and materials science. In return, polymer chemists and materials scientists bring to their colleagues in biology expertise in chemistry, materials design, structural analysis, processing, polymer solution properties, solid-state polymer physics, and an array of different characterization tools. The synergy gained by working between these fields is marvelous and exciting. What is desparately needed is a new generation of scientists who have sufficient information at the interface to begin feeling comfortable to bridge the gap between these disciplines. Many of the authors of chapters within this book describe research that has benefited by interactions at the biology–materials interface.

Water-soluble polymers are widely used in industrial products including food and beverages, cosmetics, toiletries, adhesives, coatings, pulp and paper, textiles, water treatment, detergents, and surfactants. In the United States, the quantity of these polymers used is estimated to be in excess of 13 billion pounds per year, a volume approaching that of packaging plastics. Water-soluble polymers, synthesized from readily renewable resources, can biodegrade rapidly upon disposal and produce harmless natural substances. For example, α-,β-poly(D,L-aspartate), is currently under development as a water soluble detergent additive to replace non biodegradable poly(acrylate) and related polymers. Unfortunately, cost factors are still obstacles that limit the successful implementation of this and other related new polymers that seek to replace billions of pounds of non-biodegradable polycarboxylates. The persistence in nature of down-the-drain water-soluble products is of great concern. Over time, recalcitrant water-soluble products can accumulate and may cause changes to the ecosystem. The fact that such polymers are not "seen" as they accumulate makes them less noticed by the public that might otherwise raise objections. Numerous chapters in this book describe water-soluble polymers from natural sources such as amino acids and native or modified polysaccharides.

The commercial success of product synthesis by fermentation routes depends in part on the better use of inexpensive agricultural by-products such as starch, molasses, soap-stock oils, and others. Genetic methods now exist that allow extraordinary improvements in strains. Metabolic-engineering of metabolic pathways expands the range of low/no-cost carbon sources that can be converted efficiently to products. In addition, engineered strains are leading to new options in the design and control of product structure, and, therefore, properties.

Biodegradable polymers are at an exciting stage of development and positioned to play important roles. Excellent basic and applied research in this field and increased global regulatory pressure have spawned the formation of new fledgling industries, which are creating safe biodegradable packaging materials for widespread public use, specifically designed for composting. In addition, new water-soluble degradable polymers have been developed that degrade in wastewater treatment facilities. This book highlights recent developments in biodegradable polymers. Featured topics include the establishment of regulatory guidelines, eco-labeling of products, product marketing and positioning, development of test-methods to study polymer biodegradability, and the establishment of disposal/composting infrastructure for biodegradable polymer disposal.

The papers from this book were contributed by authors that participated in the 7th Annual Meeting of the Bio/Environmentally Degradable Polymer Society, held August 19–22, 1998 at the Royal Sonesta Hotel in Cambridge, Massachusetts, as well as the symposium on *Polymers from Renewable Resources,* held at the annual American Chemical Society Meeting, August 23-27, 1998, in Boston, Massachusetts.

Of course, this book would not have been possible without the wonderful contributions by the book chapter authors. In addition, we are grateful for the help from Ann Wilson and Kelly Dennis of the ACS Books Department who made sure that this project came to a successful completion. I am also indebted to my wife Wendi, who provides the moral support and the time that is inevitably extracted from "*family hours*" that went into the editing and other tasks to complete this volume.

RICHARD A. GROSS
Professor and Herman F. Mark Chair
Department of Polymer Chemistry
Polytechnic University
Brooklyn, NY 11201

CARMEN SCHOLZ
Department of Chemistry
University of Alabama at Huntsville
John Wright Drive
MSB 333
Huntsville, AL 35899

Overview

Chapter 1

Overview: Introduction to Polysaccharides, Agroproteins, and Poly(amino acids)

Sanjay K. Singh and Richard A. Gross

Department of Polymer Chemistry, Polytechnic University, Brooklyn, NY 11201

1. General Introduction

Industrial competitiveness will increasingly depend on our ability to rapidly adapt and incorporate new technological advances. Furthermore, technological advances to reduce cost will often be associated with needs to reduce environmental contaminants and wastes from processes. This has resulted in increased global emphasis on knowledge that will enable sustainable production of materials using technologies that are safe for the environment. Cradle-to-grave evaluation of new and existing products and processes is creating new criteria for the acceptability of products. The use of natural or renewable resources is critical for this development. A wide range of naturally occurring polymers derived from renewable resources is available for material applications. Some of these, such as cellulose and starch, are actively used in products today, while many others remain underutilized. Excellent basic and applied research in this field, and increased global regulatory pressure, has spawned the formation of fledgling industries. A well-publicized example highlighting these developments is biodegradable plastics and polymers that are water-soluble. This technology represents a viable route to polymers that are designed to safely degrade in composting and wastewater treatment facilities.

Industry and government sectors continue to have a growing interest in '*Green or Environmentally Benign Chemistry and Processing*'. Multiple government agencies/organizations such as the American Chemical Society, Department of Energy, Environmental Protection Agency, National Science Foundation and others are increasingly committing greater financial support, holding workshops and creating joint-agency initiatives in '*Green Chemistry*'. Biodegradable polymers, biocatalysis, sustainable technology or renewable resource utilization are important components to environmentally benign synthesis and processing. Currently, Biodegradable Polymers is an industrial niche area. However, this may shift over time based on the following: *i*) changes in consumer values where they opt to pay a greater price for environmentally safe products, *ii*) decreased cost and increased performance of biodegradable products, *iii*) the development of an infrastructure for the disposal of bio-waste that includes biodegradable products. The existence of a network of composts that are available to households will provide a suitable outlet for biodegradable waste that can then be converted to compost. In some areas, compost is a valuable product that is crucial to the continuous replenishment of nutrients in soil. In addition, compost can be used as safe fillers for various construction and landscaping needs.

Another important consideration is that renewable resources of hydrocarbons for our petrochemical-based industries will eventually be required over the next 50 to 100 years. When one considers the combination of depleting non-renewable hydrocarbon reserves, large food surpluses and large areas of empty land, it is not surprising that there is a powerful economic and environmental incentive for the development of agrobased, value added, non-food products. This is one of the basic themes emphasized by the United States Department of Agriculture.

2. Objectives and Scope of this Book

This book on Polymers from Renewable Carbohydrates and Agroproteins and its companion on Polyesters, was assembled to provide the reader with a one-source repository for papers that survey a wide array of state-of-the-art strategies to develop and use polymers and building blocks from Renewable Resources. The following are topics that are emphasized in this Volume of the two-part series:

- Green Polymer Chemistry
- Plant and Microbial Polysaccharides
- Agroproteins and other Poly(amino acids)
- Biocatalytic Routes to value-added products
- Biodegradable Polymers: Economic and Industrial Infrastructure
- Role of Polymers in Solid Waste Management
- Evaluation of Polymer Biodegradation
- Polymers of Biomedical Importance
- "Smart" polymers
- Bio-based amphiphiles and hydrogels

3. Interface Between Biology and Materials Science

A rapidly emerging research theology involves the seamless convergence of concepts, ideas and research tools at the interface between biology and polymer science. Work at this interface is growing at a rapid pace and is expected to make major contributions to science in the new millennium. Scientists having core training in microbiology, genetic engineering, biophysics, agricultural science, plant physiology, food science and other biological disciplines bring new tools, molecules and fresh ideas to the fields of polymers and materials science. In return, polymers and materials scientists bring to their colleagues in biology expertise in chemistry, materials design, processing, polymer solution properties, solid state polymer physics, and an array of different characterization tools. The synergy gained by working between these fields is marvelous and exciting. A new generation of scientists is desperately needed that have sufficient information at the interface to begin feeling comfortable to bridge the gap between these disciplines. Examples of current research activities that were conceived and/or are developing at this interface are as follows: *in-vitro* enzyme-catalyzed monomer/polymer synthesis, metabolic engineering of polymer synthetic pathways within whole cells or plants, genetically engineered protein-based materials, and macromolecular assembly of biopolymers, and the wide area of biomedical polymers and materials science. In addition, research areas of critical importance that are thriving at the biology/materials science interface include the nanotechnology initiative, green chemistry, biobased materials or polymers from renewable resources, biodegradable polymers and studies focused on interactions between polymers and environment systems. As the tools of molecular biology become increasingly powerful at an exponential rate, the potential of work that in the past would have been an academic curiosity is being transformed into opportunities of commercial importance. Examples of this work are efforts underway to express polyesters and specialty proteins in plant production systems. Many of the Authors of Chapters within this book describe research that has benefited by interactions at the biology/materials interface.

4. Biocatalytic Routes create new opportunities

Increased reaction efficiencies based on advances in catalysis will continue to be critical to industry. There is a clear need to pave new pathways that do more than provide incremental improvements in existing conditions. This is well stated in the following quote from Professor Barry Trost "*The issues cannot simply be addressed by minor tinkering with current processes to improve their performance although, undoubtedly, that is one of many of the strategies that will be helpful. Fundamental new science derived from basic research that creates new paradigms for synthesis is also required as well as everything in between (1).*"

A new way to approach problems in polymer synthesis is beginning to appear worthy of our attention. The field of enzyme-based polymer synthesis has been growing in academic and industrial laboratories over the past ten years (*2*). This interest began in the late 1980s due to the pioneering work of Alex Klibanov and his colleagues at MIT. They were the first to demonstrate that enzyme reaction equilibria could be reversed in organic media to drive polymerization reactions instead of hydrolytic reactions. The field continues to expand since these early studies since enzymes provide critical control elements such as enantio- and regioselectivity. Furthermore, as time has progressed, much has been learned about the range of reaction conditions in which enzyme catalysis can be performed. This includes solventless or bulk reactions, biphasic and reversed micellar systems as well as reactions in supercritical fluids. Over the past five years, the Arnold and Stemmer research teams have shown how error prone PCR and gene shuffling methods along with high throughput screens can result in the rapid development of recombinant enzymes. These enzymes have enhanced performance that is based on the selected screen. The results of their work and the work of others have resulted in many new commercial enzymes, including lipases, with improved characteristics such as temperature stability and solvent tolerance.

The potential of applying the rapidly expanding arsenal of enzymes in non-traditional ways to problems in polymer synthesis is expected to provide major benefits such as milder reaction conditions, increased reaction efficiencies, processes that require less discrete steps, the avoidance of heavy metals and the potential of creating more complex but well defined products.

Curiously, the vast majority of effort on enzyme-catalysis remains directed at small molecule reactions. It is interesting to note that there is growing interest in the development of hyperbranched polymers for applications such as drug-delivery, micelle mimics and nanoscale building blocks (*3-5*). However, thus far, there has been scarce attention paid to the option of using *in-vitro* enzyme-catalysis as a means to overcome difficulties for the synthesis of these complex architectural polymers as well as to create new interesting structures.

New opportunities for the use of polymers from renewable resources have also been created by rapid advancement in understanding of fundamental biosynthetic pathways and options to modulate or tailor these pathways. Genetic control of polymer production offers many advantages in manipulation and tailoring of biopolymer structure and thus function, including molecular weight, stereochemistry, primary sequence and chemical reactivity. Tailorability is derived both at the genetic level in the control of amino acid sequence in proteins during transcription/translation, as well as at secondary levels due to the action of biosynthetic enzymes involved in the formation of other classes of biopolymers. These biopolymers are derived from a diverse set of polysaccharides, proteins, lipids, polyphenols, and specialty polymers produced by bacteria, fungi, plants and animals. These polymers have natural functions that inspire ideas for their potential use in similar or related human-required applications. For example, in addition to membrane and intracellular communication, polysaccharides form capsules or protective layers around cells, they allow cells to adhere to specific surfaces, and facilitate sequestration of nutrients including metals. Proteins not only serve as the premier natural catalysts, but also are used as structural materials, adhesives, signaling agents, create channels for transport of hydrophilic substances through membranes, and, generally, control cell function.

The remainder of this chapter will be used to highlight examples of Agroprotein and Polysaccharides from Renewable Resources. This summary is not intended as an exhaustive review but to provide a starting point for further insights in the following chapters or in other references.

5. Polysaccharides

The diversity of polysaccharides synthesized in nature may be surprising to those not familiar with this subject. Many microbial and plant polysaccharides have not yet been discovered. Their complex structures or their association with other natural systems makes them difficult to isolate and characterize. Also, the extent to which we can usurp the natural biosynthetic systems to produce new polysaccharides for targeted applications has only begun to be investigated. Figure 1 illustrates the basic structural features of polysaccharide building blocks, including the carbon numbering and configuration between monomer units.

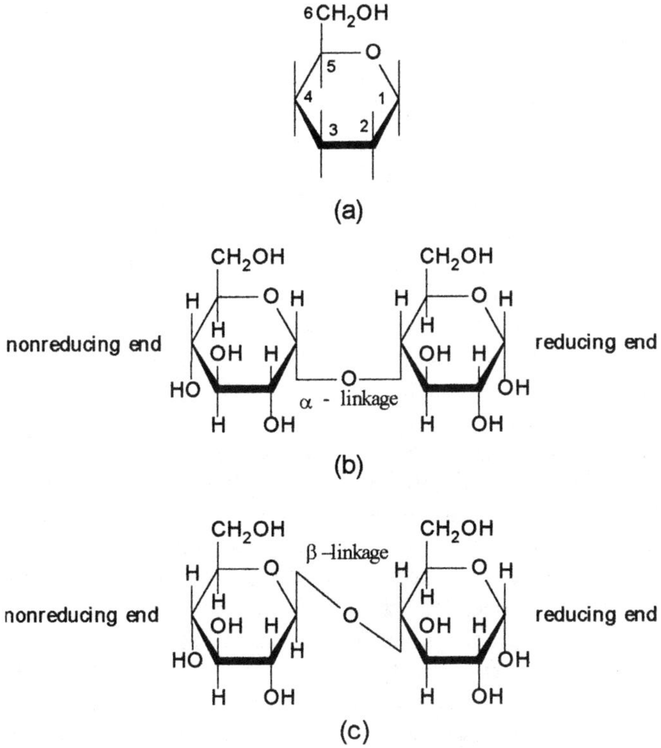

Figure 1: Nomenclature for polysaccharides - a) numbering system for sugar building blocks, b) the α-configurations of linkages between sugar repeat units, and c) β-configurations of linkages between sugar repeat units.

5.1. Polysaccharides (Plant, Algal)

5.1.1. Starch (amylose/amylopectin)

Starch is found in almost all plants and consists of a mixture of two polysaccharides, amylose (a linear α-(1→4)-glucan, DP between $1x10^2$ and $4x10^5$, Figure 2) and amylopectin (b branched α-(1→4)-glucan with α-(1→6)-glucan branches every 19 - 25 units, DP between $1x10^4$ and $4x10^7$, Figure 3) (*6-9*). The ratio of amylose and amylopectin in starch varies as a function of the source, age, etc.

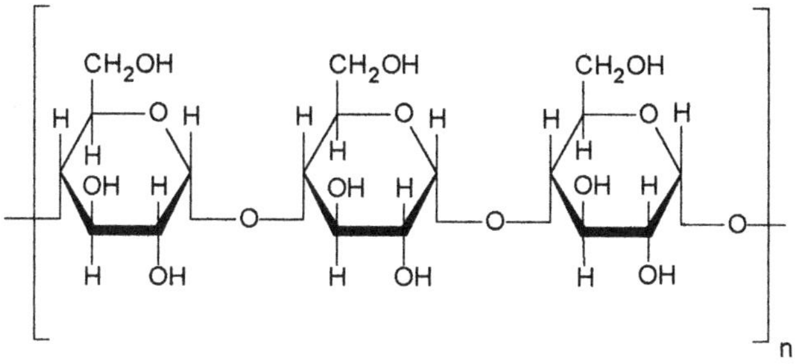

Figure 2: *Structure of Amylose*

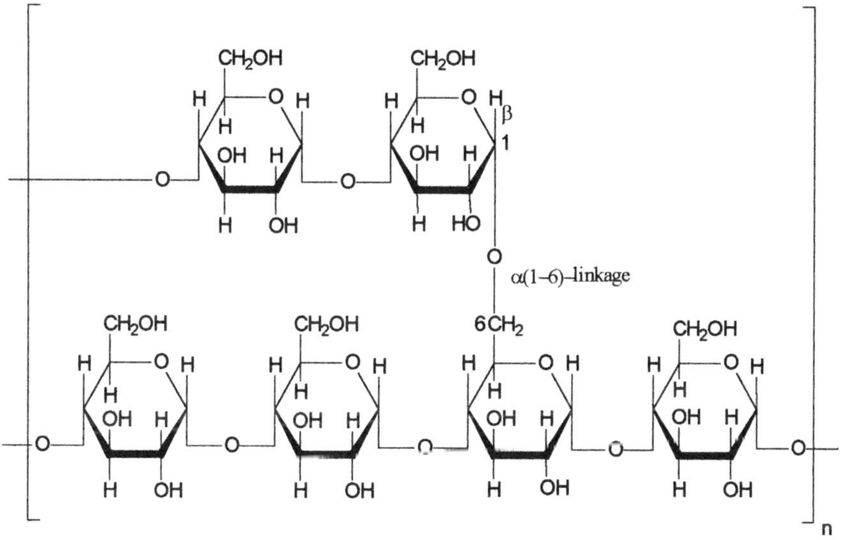

Figure 3: *Structure of Amylopactin*

Starch is the principal carbohydrate reserve of plants. It occurs as water-insoluble granules in the roots, seeds, tubers, fruits and stems of various plants, including corn, maize, wheat, rice, millet, barley, potatoes, arrowroot and cassava (*6-9*). Starch is isolated from its sources by various physical methods such as steeping, milling and sedimentation (*7-8*). Starch is not soluble in water at ambient temperature. However, in hot water, its granules gelatinize to form an opalescent dispersion (*9*). It is degraded by a number of enzymes such as amylase and amyloglucosidase. Amyloglucosidase hydrolyses starch to glucose and is used to determine the purity of starch. Like cellulose, starch can be modified by esterification, etherification, and by forming graft copolymers with synthetic materials such as polyacrylic acid. Starch is generally not suitable for plastic and fiber products due to its highly branched structure and permeability (*7*). However, starch finds extensive use in foods, as thickeners, inert diluents for drugs, adhesives, and sizing or glazing agents in the paper and textile industries (*6-9*). Amylose is used for making edible films whereas amylopectin is used for textile sizing and finishing and as a food thickener (*9*). Starch nitrate is used as explosive whereas starch acetate forms films and is used as a paper and textile-sizing agent (*9*). Starch ethers, hydroxy ethyl ether in particular, is used as an adhesive (*9*). Starch allyl ethers are used as an air-drying coating (*9*). The considerable interest in graft copolymers of starch is due to their utility as biodegradable packing materials and agricultural mulches. For example, starch cross-linked with epichlorohydrin is used in the treatment of rice granules to make them more resistant to break down in canned soups.

5.1.2. Cellulose

Cellulose is the most abundant naturally occurring linear β-(1→4)-glucan on this planet (Figure 4). This biopolymer is the principal constituent of higher plants (about 40-50%), some green algal and fungal (Valonia, a very pure form of cellulose) cell walls providing them with their structural strength (*6,10-11*). Cellulose is also the main component of a variety of natural fibers such as cotton (the most pure form of cellulose), bast fibers and leaf fibers (*6,10-11*). Wood pulp, cotton fibers and cotton linters are the most important sources of cellulose for commercial processes (*10-11*).

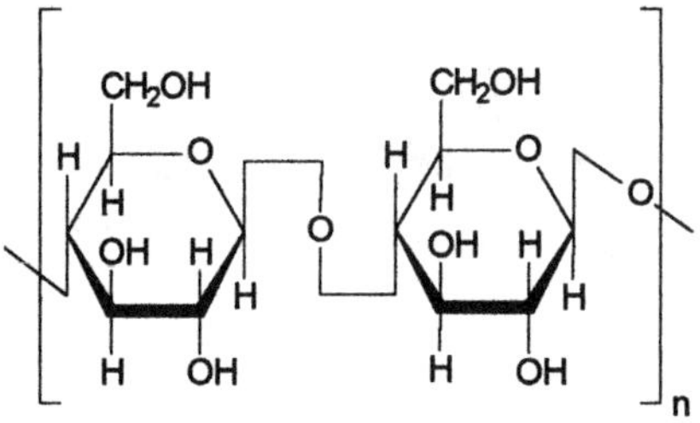

Figure 4: *Structure of Cellulose*

In nature, cellulose is generally found in close association with other polysaccharides, mainly hemicelluloses, pectin, water, wax, proteins, lignin, and mineral substances (*10-11*). It is also produced by certain types of bacteria such as *Acetobacter, Pseudomonas, Rhizobium, Agrobacterrium and Sarcina* species and can be synthesized as a continuous film by cultivating the bacteria in a glucose medium (*11*). Marine chordates also secrete cellulose as the sea squirt (*11*). Biosynthesis of cellulose by enzymatic processes is also known (*10-12*).

Cellulose is a highly crystalline polymer. Although it has a large number of hydroxyl groups, it is completely insoluble in water and most organic solvents. Many cellulose producing organisms also have depolymerase enzymes that facilitate its mineralization (*10-12*). Cellulose is soluble in strong mineral acids, sodium hydroxide solution, metal complexes solutions of copper (II)-ammonia (Schweizers reagent) and copper (II)-diamine (*13-14*). Under strong acidic conditions, it is hydrolyzed completely to D-glucose whereas very mild hydrolysis produces hydrocellulose with shorter chains, lower viscosity and tensile strength. Due to high extents of intramolecular and intermolecular hydrogen bonding, cellulose is not thermoplastic but decomposes before melting (*11*). Cellulose forms esters (e.g. cellulose nitrate, cellulose phosphate, cellulose acetates, cellulose propionates, butyrates, etc.) and ethers (e.g. carboxymethyl cellulose, methyl cellulose, hydroxy propyl cellulose, etc.) (*10-12*). Cellulose and its derivatives are widely used by humans in textiles, food, membranes, films, paper and wood products. Cellulose is also used in medicines (e.g. wound dressing), suspension agents, composites, making rope, mattress, netting, upholstery, sacking, matting, coating, cordage, twine, packaging, linoleum backing gas mantles, muslin, thermal and acoustic insulation, etc (*10-12*). Cellulose synthesized from bacteria is highly crystalline with finer diameter fibrils forming fibrous mats with greater surface area than plant-derived cellulose. Bacterial cellulose is used in the production of speaker diaphragms in headphones (*12*). Cellulose nitrate is used in gun cotton and celluloid (*10*). Cellulose esters such as acetates are used as plastic materials, fibers in textiles and coatings (*10-11*). Cellulose xanthates are employed in the viscose rayan process (*10*). Unlike cellulose, cellulose ethers e.g. carboxymethyl cellulose, methylcellulose, hydroxyl propyl cellulose, etc., are soluble in water and form films for various applications. These derivatives are also used as plastic materials. Methylcellulose is used as a food-thickening agent, and as an ingredient in certain adhesive, ink and textile-finishing formulations (*11*).

5.1.3. Hemicelluloses

Hemicelluloses are a group of polysaccharides (e.g. xylans, and glucomannoglycans) found in association with cellulose and lignin in plant cell walls (*6,15*). Their role is to provide a linkage between lignin and cellulose (*15*). They have much lower molecular weight than cellulose with very low degree of polymerization (DP, generally below 200) (*6,15*). Their primary structure depends on the plant source of the polymer. Next to cellulose, hemicelluloses are the most abundant natural polymer in the biosphere and represent about 20-30% of wood mass. They are predominantly located in the primary and secondary walls of vegetable cells (*15*). These polymers can be divided into two categories, *cellulosans* and *polyuronides* (*15*). *Cellulosans* includes the hemicelluloses made of sugar monomers, hexosans (e.g. mannan, galactan and glucan) and pentosans (e.g. xylan and arabinan). *Polyuronides* are hemicelluloses containing large amount of hexuronic acids and some methoxy, acetyl, and free carboxylic groups. The predominant constituents of hemicelluloses are 1,4-β-D-xylans, 1,4-β-D-mannans, araban (composed of arabinose units), and 1,3- and 1,4-β-D-galactans (complex mixture of monosaccharides) (*15*). Hemicelluloses exist in amorphous form. Unlike cellulose, they are generally very soluble in water. They are hydrolyzed to saccharides (e.g. glucose, mannose, arabinose, glactose, xylose), acetic acid and uronic acid.

Saccharides so obtained may be converted to useful primary chemical substances, such as acetone, butanol, ethanol, furfural, xylitol, 2-methyl furan, furan, lysine, furfuryl alcohol, glutamic acid, hydroxymethylfurfural, levulinin acid, polyols, etc. (*15*).

Hemicelluloses and their derivatives are very important in industry due to their useful chemical properties and biological activity. Isolated hemicelluloses are used as food additives, thickeners, emulsifiers, gelling agents, adhesives, and adsorbents (*15*). Some hemicelluloses have also been found to be anti-tumor agents (*15*). Hydrogenation of xylose produces xylitol that is used as a sweetener (*15*). Furfural, produced by dehydration of xylose, is a very useful precursor for the synthesis of 2-methylfuran, tetrahedrofuran, adiponitrile (hence nylon), furfuryl alcohol, tetrahedrofurfuryl alcohol, glutamic acid, polyurethane foams, resins and plastics, nitrogenated and halogenated derivatives used in drugs, etc (*15*).

5.1.4. Pectins

Pectins are a group of structural polysaccharides found in most land plant cell walls. Two major sources of pectins are apple pomace and orange peel. Pectin primarily consists primarily of α-1,4-D-galacturonic acid units and/or their methyl esters (Figure 5) (*16*).

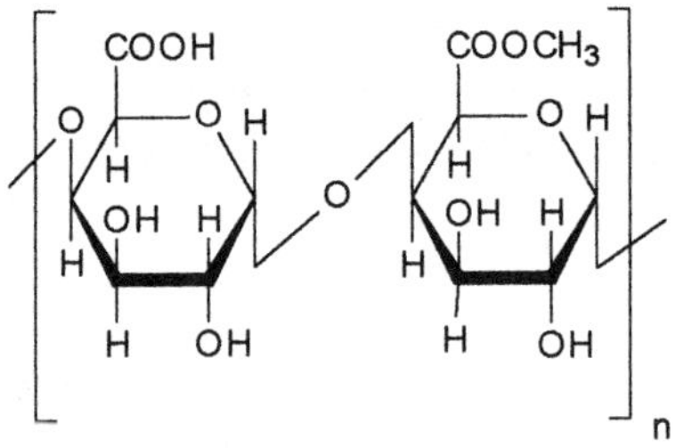

Figure 5: *Structure of pectin, poly-β-D-galacturonic acid, partially esterified main component.*

Pectin is soluble in water and forms viscous solutions. The importance of pectins in the food industry is due to its ability to form gels in the presence of Ca^{2+} ions or a solute at low pH. A number of factors that include pH, presence of other solutes, molecular size, degree of methoxylation, number and arrangement of side chains, and charge density on the molecule--influence the gelation of pectin. In the food industry, pectin is used in jams, jellies, frozen foods, and more recently in low-calorie foods as a fat and/or sugar replacement. In the pharmaceutical industry, it is used to reduce blood cholesterol levels and gastrointestinal disorders. Pectins are also used in edible films, paper substitute, foams and plasticizers, etc.

5.1.5. Konjac

Konjac is a reserve polysaccharide from *Amorphophallus konjac* tubers. It contains D-mannose and D-glucose, molar ratio of 1.6:1, with predominantly β-1,4- linkages and random acetylation (about 1 unit for every 12 to 18 monomer units) (Figure 6) (*8, 17*).

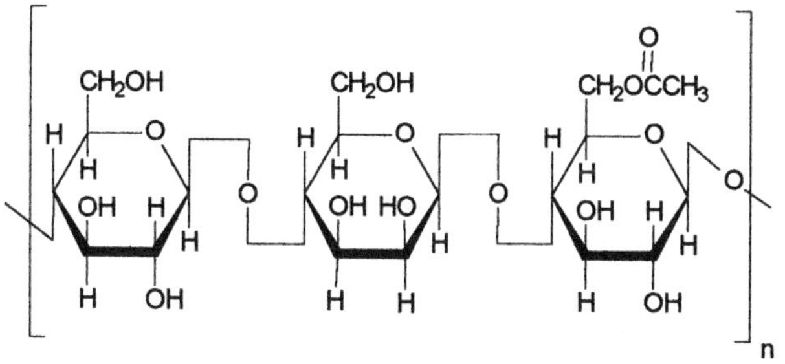

Figure 6. *Structure of konjac.*

Konjac is used as a thickener and can form mechanically strong films useful for food and pharmaceutical applications. The natural acetylation of konjac allows it to be solubilized in water. However, Deacetylation of water-cast films renders those films hot- and cold-water stable. Thus, konjac acetylation is a natural route by which an organism can disrupt chain-chain interactions leading to increased accessibility for subsequent metabolic breakdown. Enzymes are known that are capable of cleaving β-1,4-glycosidic linkages.

5.1.6. Alginate (algin)

Alginate is an extract from the cell walls of brown seaweed. These polysaccharides may differ in the composition and repeat unit sequence of linear 1,4-linked polyuronic acids. Alginates may contain segments with poly-β-D-mannuronic acid, poly-α-L-guluronic acid, and alternating D-mannuronic and L-guluronic acid (Figure 7) (*16*).

Figure 7: *Structure of alginate, D-mannuronic acid (left) and L-guluronic acid (right), blocks of either unit and segments of alternating units.*

Specific algae that form alginates include the giant kelp, Macrocystos pyrifera, as well as other species such as *Asophyllum nodosum* and *Laminaria digitata*. Alginates often comprise up to 40% by dry weight of the alginate-producing biomass. Inspection of alginates from different species and in different parts of a single species shows considerable variation in alginate composition. For example, species rich in guluronosyl residues tend to contain rigid stalks, while those rich in D-mannuronic acid are flexible (*18*). Thus far, studies on alginate biosynthesis have been more focused on bacterial rather than alga production.

Alginates are used in food and pharmaceutical industries. In foods, alginates are used as thickening agents and stabilizers. In medicine, alginates are used as wound healants, in dental impressions, and in biotechnology for cell and enzyme immobilization (*18*). Industrial uses of alginates extend into adhesive formulations and paper coatings. They are easily solution-cast into films that can be crosslinked and function as good oxygen barriers. Alginate gels are formed in the presence of calcium ions and other polyvalent cations. Calcium ions result in enhanced interchain interactions due to ionic crosslinks. Thus, useful coatings result by dipping an object into a solution of alginate followed by exposure to calcium chloride. The gels are used to coat and thereby protect foods. They prevent the foods such as meats from dehydration. Also, their oxygen barrier properties reduce deleterious effects caused by oxidative decomposition. The entrapment of live cells in bioreactors has also been accomplished by surrounding the cells with calcium alginate gels and membranes (*19*).

5.1.7. Carrageenans

Found within seaweed are a valuable polysaccharide known as carrageenan. Carrageenans found in red seaweeds are complex mixtures that consist of linear sulfated chains with D-galactose and 3,6-anhydro-D-galactose units. (Figure 8) (*6,8*).

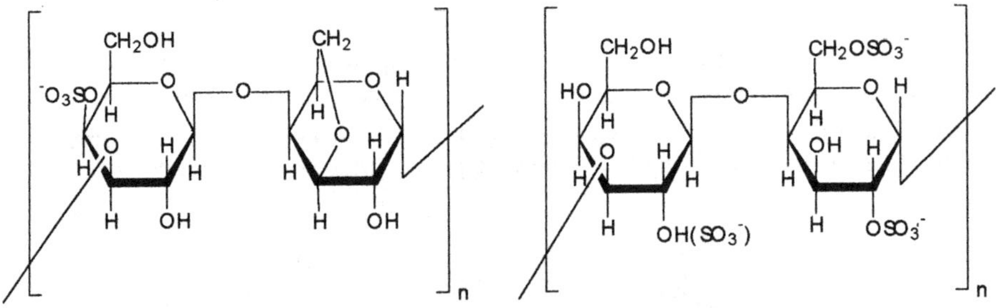

kappa form *Lambda form (most common)*

Figure 8: *Structure of carageenan (mixture of at least five polymers)*

The isolation of these polysaccharides involves the harvesting of seaweed by hand, extraction with hot water at alkaline pH, and precipitation of the carrageenans into alcohol. The molecular weights of these products range from 100,000 to one million. The gelling, thickening, and stabilizing properties of carrageenans make them useful in the food industry. Recently, they have found use in the meat industry for producing reduced fat products (*20*). In the pharmaceutical industry, they are used for the making of stable suspensions of insoluble compounds and to provide texture to products such as toothpastes. Carrageenans are also used in cosmetic formulations (*6,20*).

5.1.8. Gums (Guar Gum, Gum Arabic, Gum Karaya, Gum Tragacanth, Locust Bean Gum, etc.)

The plant gums are generally polysaccharide or polysaccharide derivatives that hydrate in hot or cold water to form viscous solutions or dispersants (*6,8,21*). Gums are generally used in foods, pharmaceuticals and material science as gels, emulsifiers, adhesives, flocculants, binders, films, and lubricants. Guar gum is a seed extract containing mannose with galactose branches every second unit. It is used in food, papermaking and petroleum production. Gum arabic is a plant exudate and consists of a highly branched polymer of galactose, rhamnose, arabinose and glucuronic acid. Gum Arabic is used in food and pharmaceutical industries and in printing inks. Gum karaya is a plant exudate and consists of galactose, rhamnose, and partially acetylated glucuronic acid repeat units. Gum karaya is used in food and pharmaceutical industries and in paper and printing. Gum tragacanth is a plant exudate and consists of a polymer of fucose, xylose, arabinose and glucuronic acid. Gum tragancanth is used in foods, cosmetics, printing, and pharmaceuticals. Locust bean gum is a seed extract that consists of mannose along the main chain with galactose branches every fourth unit. This gum is used in foods, papermaking, textile sizing and cosmetics (*6,8,21*).

5.1.9. Other Plant and Algal Polysaccharides

When red seaweed is harvested, agar can be extracted into hot water and purified by freezing. Agar contains agarose and agaropectin. Agarose is an alternating copolymer of 3-linked β-D-galactopyranose and 4-linked 3,6-anhydro-α-L-galactopyranose units. Agaropectin has the same structure as agarose but with varying amounts of the monomers replaced by 4,6-O-(1-carboxyethylidene)-D-galactopyranose or sulfated or methylated sugar residues. Agar is well known for its ability to form thermally reversible gels (agar portion). It is used extensively in dentistry, the food industry, and microbiology (*16*). Agar is not digestable in the human gut. It forms a double helix in solution, which aggregates to form a three-dimensional network. Other plant and algal polysaccharides include laminaran (β-1,3-linked glucose), inulin (β-2,1-linked fructose), xylans (β-1,4-linked xylose with glucose, arabinose and acylation), and agar.

5.2. Polysaccharides (Animal)

5.2.1. Chitin/Chitosan

Next to cellulose, chitin, is the second most abundant renewable 'celluloselike' biopolymer on the planet. This homopolymer consists of β–(1→4)–*N*-acetyl-D-glucosamine units (Figure 9) (*22*). Chitin is found in many organisms in nature where it serves a critical function for their survival. For example, it is found in a wide range of insects, crustaceans, annelids, molluscs, coelenterata, and most fungi (*23-27*). The deacetylated form of chitin is known as chitosan. Chitosan represents a group of cationic biopolymers, which includes a homopolymer of β–(1→4)-D-glucosamine units (Figure 9, R=H) and all the linear copolymers of β–(1→4)–*N*-acetyl-D-glucosamine and β–(1→4)-D-glucosamine units (Figure 10).

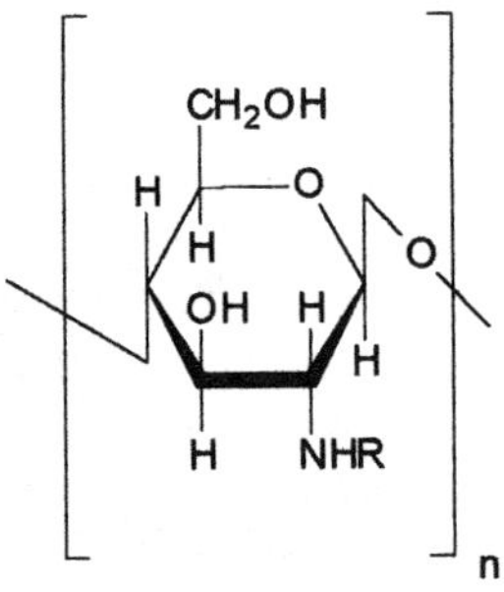

Figure 9: *Structure of Chitin (R = Ac) and Chitosan (R = H)*

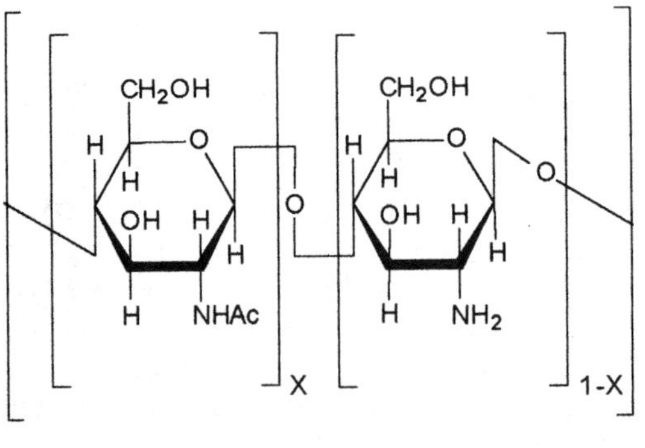

Figure 10: *Structure of Chitosan with different degree of acetylation*

Chitin and Chitosan are manufactured from crustacean (crab and crayfish) exoskeleton available as byproducts of food processing. Chitin produces chitosan by the tandem action of two enzymes, chitin synthase, and chitin deacetylase (*28*). For commercial purposes, chitosan is synthesized by base-catalyzed chemical deacetylation of chitin. Squid pens (a waste byproduct of New Zealand squid processing) are a novel, renewable source of chitin and Chitosan (*25*).

Solubility is a very important factor for different applications of chitin and chitosan. Due to the highly crystalline nature of chitin, it is not soluble in aqueous solutions and organic solvents. In contrast, chitosan is soluble in water, acidic solutions and is slightly soluble in weakly alkaline solutions. High quality chitin and chitosan are non-toxic, biocompatible and biodegradable. Chitin and chitosan can be processed into beads, gels, powders, fibers, sheets, tablets, and sponges, which are useful as base materials for cosmetics and drugs. Chitosan possesses general coagulant/flocculant characteristics towards bio-molecules and surfaces (*29-30*). Relative to other polymers such as synthetic resins, activated charcoal, and even chitin itself, chitosan is a superior metal recovery and water-purifying agent. It can be used in waste water treatment for heavy metal and radio isotope removal and valuable metal recovery, potable water purification for reduction of unwanted metals, agricultural-controlled release of trace metals essential to plant growth, and food-complex binding of iron in precooked food to reduce 'warmed-over flavor'. Furthermore, the amino group in chitosan is an effective functional group that can be exploited chemically for the production of other chitin-related derivatives with specific characteristics such as absorptive agents. Commercial chitosans may have different physicochemical characteristics, i.e., molecular weight, crystallinity, deacetylation, particle size, and hydrophilicity. These differences, for example, can be manifested in variable abilities to function as waste treatment agents. This problem of product batch-to-batch variation as well as differences in the manufacturing process between vendors must be kept in mind when developing commercial processes or products that contain chitosan.

Chitosan is currently receiving great interest in the pharmaceutical and therapeutic industries due to its very safe toxicity profile and other interesting properties. Examples of these properties include solubility in aqueous solutions, biocompatibility, nonantigenicity, ability to improve wound healing and/or clot blood, ability to absorb liquids and to form protective films and coatings. Also, chitosan selectively binds acidic lipids, thereby lowering serum cholesterol levels. Targeted applications of chitosan in medicine include wound healing and dressings; dialysis membranes; contact lenses; fibers for resorbable sutures; liposome stabilization agent; antibacterial, antiviral, and antitumer uses; serum cholesterol level reducers; hypocholesteremic and hemostatic agents; stimulants of the immune system; drug delivery and controlled release systems. In direct tablet compression, chitosan is used as a tablet disintegrant, for the production of controlled release solid dosage forms or for the improvement of drug dissolution. Chitosan has, compared to traditional excipients, been shown to have superior characteristics. In addition, chitosan is used for production of controlled release implant systems for delivery of hormones over extended periods of time. Lately, the transmucosal absorption promoting characteristics of chitosan has been exploited, especially for nasal and oral delivery of polar drugs that include peptides and proteins and for vaccine delivery (*31*). Chitosan has similar structure to heparin that is well known but an expensive anticoagulant. Chitosan is under study as a starting material for the production of heparin-like blood anticoagulants (*32*). Chitosan can immobilize whole cells, fungi, bacteria, and various enzymes (*33-35*).

Chitin and chitosan are used in paper manufacture as a wet strength additive. They are also used as chromatographic media; by the textile industry as coatings and in blends; cosmetics (shampoo, mousses, nail polish and hair spray); photographic fixing agent for acid dyes in gelatin; and in the food industry as a livestock feed, food thickener, and stabilizer for emulsions (*36*). High strength fibers have been spun from chitin and chitosan acetates and formates (*37-39*).

5.2.2. Hyaluronic Acid

Hyaluronic acid is a linear extracellular polysaccharide produced by animals. Hyaluronic acid is also produced by bacterial fermentation and subsequently modified. Its main chain is built from the 1,4-β-linked disaccharide D-glucuronosyl-1,3-β-N-acetyl-D-glucosamine. This repeating disaccharide structure allows hyaluronic acid to be described as an alternating $[A\text{-}B]_n$ copolymer (Figure 11) (*40-42*).

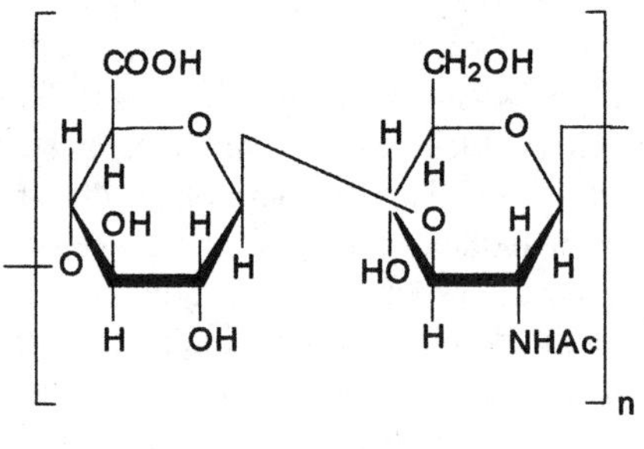

Figure 11: *Structure of hyaluronic acid.*

Hyaluronic acid exhibits high viscosity in aqueous solution at low solute concentration (*40*). In natural systems, hyaluronic acid functions to maintain mechanical properties such as compressive stiffness and swelling pressure in connective tissues, controlling hydration in soft tissue, lubrication of synovial fluid and organs, and cell surface adhesion.

Hyaluronic acid is a versatile biopolymer, which has wide application in several diverse areas. It is now extensively used in ophthalmic surgery, in the treatment of lameness in racehorses and as an ingredient in cosmetics (*42-48*). Other applications are in drug delivery, orthopedics, cardiovascular aids and in wound healing. Even so, its potential as a therapeutic agent is yet to be fulfilled. New and improved products, formed by crosslinking hyaluronic acid with itself or other chemicals are being produced that will offer therapeutic advantages (*44*).

5.3. Polysaccharides (Fungal)

5.3.1. Pullulan

Pullulan chains are neutral, linear, and normally consist of α-1→4-linked D-glucose trimers and some tetramers (maltotriose, maltotetrose) that are linked through α-(1→6) bonds (Figure 12). It is an extracellular product from the fungus *Aureobasidium pullulans* (*49*).

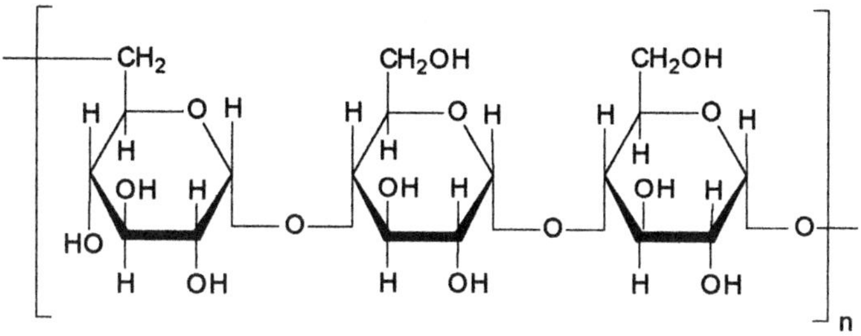

Figure 12: *Structure of pullulan.*

Pullulan can be produced from a variety of carbon sources such as glucose, sucrose and starch hydrolyzates (*49*). Product recovery and purification can be accomplished by a one-step precipitation using alcohol or acetone. Pullulan molecular weight depends on the production strain and the physiological conditions used during the fermentation (*50-53*). Pullulan has some very attractive characteristics that have made them the focus of much attention. For example, pullulans are highly water soluble, nonhygroscopic, nontoxic, tasteless, odorless, biodegradable, form excellent oxygen barriers, resistant to oils and greases as well as changes in temperature and pH. They form viscous pseudoplastic solutions in water, which are stable in the presence of cations, but do not form gels. Pullulan is not degraded by most amylases, but instead by specific pullulanases to its components (maltotriose and maltotetrose units). This enzymatic digestion provides a useful route for the preparation of these oligosaccharides. Esterification can be used to reduce their susceptibility of enzyme attack.. Pullulan has been identified as suitable polymer for a large number of industrial applications. Examples are in foods as a viscosifier, as a low calorie partial replacement for starch, and as a binder (*50*). Due to unique film forming and oxygen-barrier properties, pullulan has been used in protective and adhesive edible coatings. Other proposed applications of pullulan are in degradable films and fibers, paper coatings and binders, cosmetics, pharmaceutical tablet coatings, and contact lenses that contain slow release bioactive medicines (*54*). Pullulans produced in narrow polydispersity are commonly used as standards in gel-permeation chromatography. Chemically modified pullulans have been prepared that have enhanced or 'tailored' physical properties relative to its native form. Like Pullulan, its derivatives are useful as adhesives, films, molded products, and as a coating for products such as fertilizers, pharmaceuticals, paper, tobacco, shatterproof glass, and lithographic plates (*55-59*). Pullulan has also shown promise in vaccine production and as a plasma extender (*58-59*).

5.3.2. Elsinan

Elsinan is a neutral linear α-D-glucan that conists of linkages, 1→4- and 1→3 (2.0:1 to 2.5:1, mol/mol, Figure 13). Approximately one in 140 linkages are α-1→6. Elsinan is an extracellular product of the fungus *Elsinoe*. Its molecular weight and polydispersity may to some extent be controlled based on regulation of physiological variables during the fermentation process (*60*). Films and coatings of elsinan have low oxygen permeability. The fact that salivary α-amylases can depolymerize elsinan, points to the use of elsinan in coatings and in other forms that are edible and provide nutritional value.

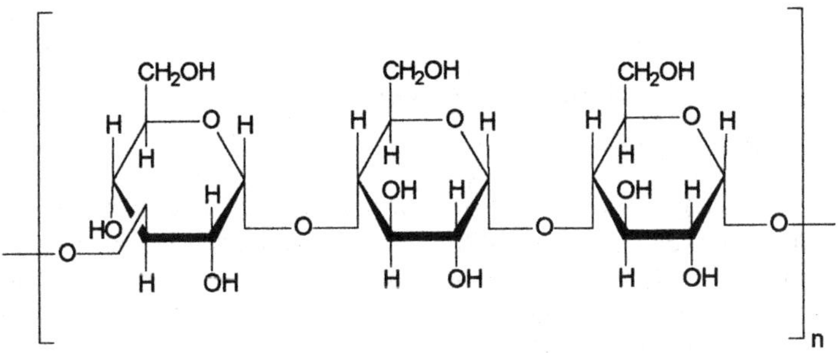

Figure 13: Structure of elsinan.

5.3.3. Scleroglucan

Scleroglucan represents a group of neutral extracellular homopolysaccharides produced by several fungal species such as *Sclerotium rolfsii, Sclerotium glucanicum* and *Schizophyllum commune*. The main chain of scleroglucan is composed of β-1,3-D-glucose units bearing β-1,6-D-glucose units as side chains, regularly or randomly with varying degree of frequency (Figure 14) (*61*). This high molecular weight polymer forms highly viscous solutions with water and exists in a stable triple helix conformation (*62*). Solutions of Scleroglucan are pseudoplastic and their viscosity remains virtually unaffected over the temperature range 10° to 90°C. This behavior of scleroglucan is also not affected by various salts that can be a critical characteristic for certain applications. Scleroglucan is insoluble in alcohol that is used to isolate and concentrate the material. The most important application for Scleroglucan is as a mobility control agent for enhanced oil recovery. This polymer is also used in drilling muds, ceramic glazes, paints and inks, and in agricultural sprays. Like curdlan and other β-D-1,3-glucans, Scleroglucan can also stimulate an immune response and repress the development of some forms of cancer (*63*).

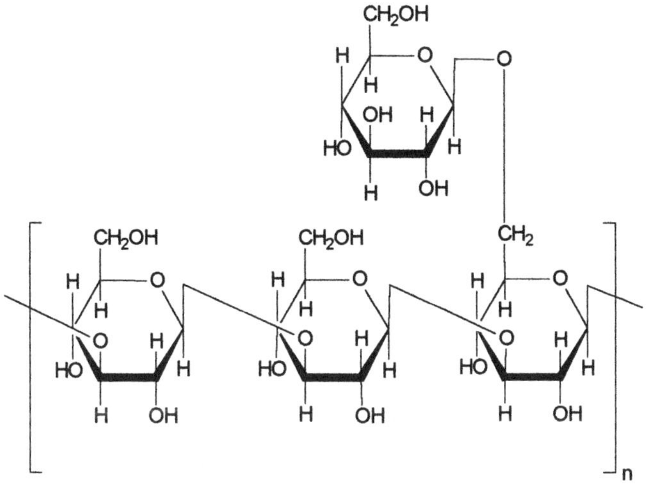

Figure 14: Structure of Scleroglucan.

5.3.4. Other Fungal Polysaccharides

Examples of other prominent fungal polysaccharides include chitin/chitosan (described earlier under animal polysaccharides), lichenan (β-1,3 and 1,4-linked glucose) and nigeran (α-1,3 and 1,4-linked glucose).

5.4. Polysaccharides (Bacterial)

5.4.1. Xanthan gum

Xanthan gum is an anionic extracellular exopolysaccharide produced by several bacterial species of the genus *Xanthomonas,* such as *X. campestris* (Figure 15) (*16,21,64-65*). This high molecular weight polymer is composed of repeating pentasaccharide units consisting of a main chain of linear β-1,4-D-glucose units with trisacharide side chains composed of D-mannose and D-glucuronic acid units with variable proportions of O-acetyl and pyruvyl residues.

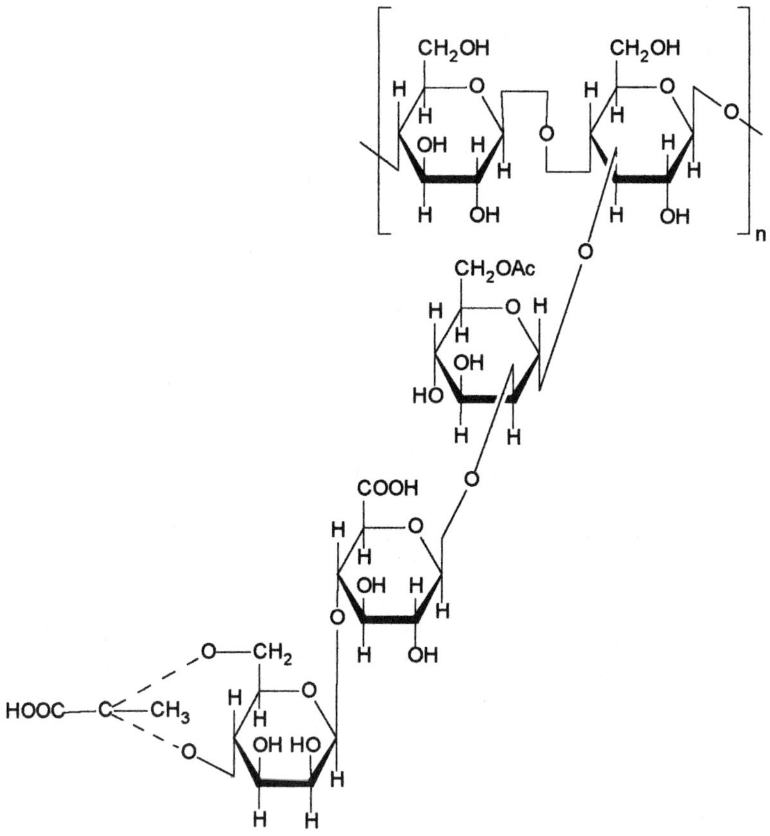

Figure 15: *Structure of xanthan.*

Xanthan gum solutions are highly pseudoplastic, rapidly regaining viscosity on removal of shear stress. Due to their unusual rheological properties, xanthan gum is widely used as a thickener or viscosifier in both food and non-food industries. Xanthan gum is also used as a stabilizer for a wide variety of suspensions (food, agrochemical pesticides and sprays), emulsions (foods and thixotropic paints), and foams (beer, fire-fighting fluids). It is used in petroleum production to increase the viscosity of drilling fluids, and extensively in food and pharmaceutical industries to stabilize emulsions and improve texture.

A 15 kilobase DNA region containing a cluster of 12 genes is required for xanthan biosynthesis. Recent work has focused on genetic engineering involving the production of mutants defective in various parts of the biosynthetic pathway. The result of these manipulations has been the production of a series of modified xanthans with different rheological properties produced by fermentation. The degree of acetylation and pyruvylation influence these viscometric properties, as does the elimination of sugar residues from the side chains (*66-67*).

5.4.2. Polygalactosamine

This linear polysaccharide, consisting of α-1,4-linked units, is produced by *Paecilomyces* sp. (Figure 16) (*68*). Along with chitin, this is one of the few naturally occurring cationic polysaccharides. The molecular weight of the polymer is around 300,000. Polygalactosamine is insoluble in water, alkali and ordinary organic solvents, and it is soluble in dilute organic acids. It is commercially available and currently being evaluated for applications in foods, cosmetics and in pharmaceutical industries.

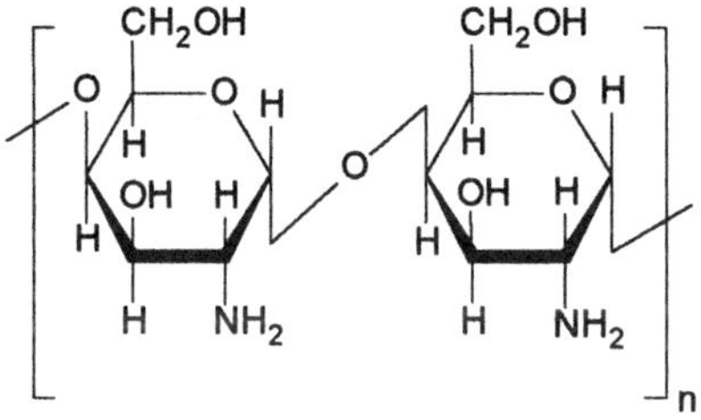

Figure 16: Structure of polygalactosamine.

5.4.3. Curdlan

Curdlan is a neutral, extracellular, linear β(1→3)-D-glucan produced by several bacteria including *Alcaligenes faecalis* var. *myxogenes, Agrobacterium radiobacter, Rhizobium meliloti,* and *R. trifolii* (Figure 17) (*50,69*). Curdlan is insoluble but swells in water. It dissolves in alkaline solution, formic acid, dimethylsulphoxide, aqueous saturated urea or thiourea, and 25% potassium iodide. This polymer is considered by many as a model-system for the study of gel formation mechanisms. An aqueous suspension of curdlan turns clear when heated to 55°-60°C and forms a weak, low set gel upon cooling. At higher temperatures (80-100°C), it forms a firm, resilient, and irreversible high set gel unlike agar. Autoclaving at 120°C changes the molecular structure to a triple helix (melting temperatures above 140°C). These gels are very susceptible to shrinkage and syneresis but are resistant to degradation by most β-D-glucanases. Dialyzing alkaline curdlan solutions against distilled water or neutralization also forms gels. The property of such gels on acidification can be used as a method for *in situ* gelation in oilfields (*70-71*).

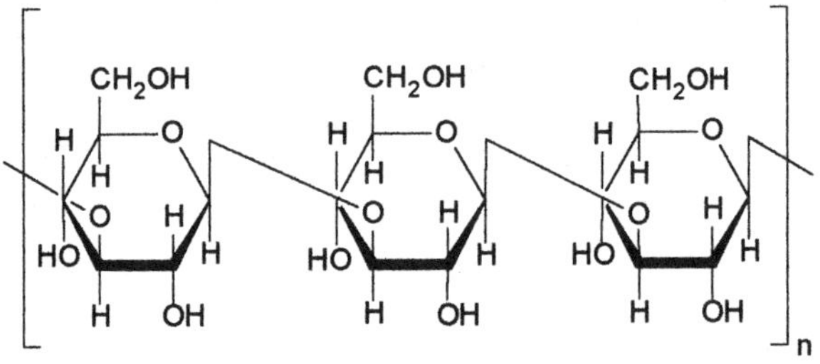

Figure 17: *Structure of curdlan.*

In food industries, Curdlan is used as an additive to improve the texture of food products such as jellies, jams, puddings, noodles, ice cream, fish paste and meat products. It is also used in developing new food products such as tofu, low fat sausages, and non-fat whipped cream substitutes (*72*).

Biomedical applications of curdlan and various modified forms have also been explored. Curdlan and its carboxymethylated derivative exhibit antitumor activity (*73-74*). Curdlan has also been evaluated for use in controlled drug delivery systems (*75*). Sulfated curdlan exhibits antithrombotic activity (*76*). By increasing the degree of curdlan sulfonation, the resulting products have shown strong anti-HIV activity (*77-78*).

The water-retention, adsorbing, and calcium gel forming properties of curdlans make them useful as a segregation reducing agent in making super-workable concrete form cement and small stones (*79*). It is also used as a binding agent for ceramics, active carbon, and reconstituted tobacco sheets. Treatment of curdlans with cyanogen halides forms curdlan derivatives that have shown promise as carriers for enzyme immobilization as well as for affinity chromatography (*80*). Curdlans ability to cause precipitation of an alkaline suspension makes it useful for water purification. Since it remains in the form of a gel at high temperatures, it can be used in cultures of microorganisms.[81] A film of acetyl curdlan can be used to separate sugars or organic acids from complex mixtures.

5.4.4. Gellan

Gellan is an anionic, extracellular, linear polysaccharide that consists of D-glucopyranosyl, L-rhamnopyranosyl and D-glucopyranosyluronic acid units (Figure 18) (*8,44,49-50*). It also carries some O-acetyl and L-glyceric acid substituents on D-glucopyranosyl units (*44*). Gallan is synthesized by a group of bacteria now named *Sphingomonas paucimobilis* (*44*). Native gellan forms thermoreversible gels. The rigidity and brittleness of these gels is generally increased by deacetylation. This polysaccharide is considered as an excellent substitute for agar due to its higher purity, good clarity, and the ability to obtain strong gels at low concentrations (*49-50*). In the food industry, gellan is used as an additive that stabilizes, thickens, and gels food products. Gellan can be used in icings, frostings, jams, jellies, fillings, meat products, pet foods, candies, cheeses, yogurt, dressings, sauces, ice cream and other frozen desserts (*49-50*). Other industrial applications of gellans are for air-freshener gels, dental and personal-care toiletries, deodorants and paper manufacturing (*8,49-50*).

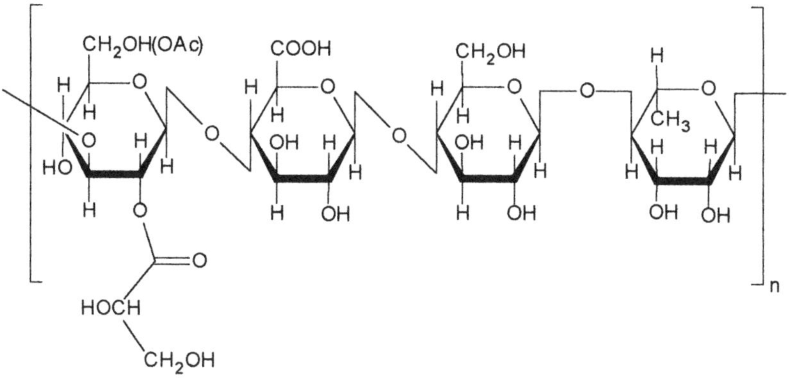

Figure 18: Structure of gellan.

5.4.5. *Dextrans*

Dextrans are extracellular bacterial α-D-glucans containing mainly α-(1→6)-D-glucopyranosyl repeat units that make up the chains. These polymers are formed with variable degrees of branching that occurs through α(1→2), α(1→3), or α(1→4) linkages (Figure 19) (*16,49,82-83*). These high molecular weight, polydisperse, and uncharged polymers are synthesized from sucrose by a number of lactic acid bacteria such as *Leuconostoc mesenteroides* and certain *Streptococcus* species. They are also formed by enzymatic methods with dextran sucrase. The dextran from *L. mesenteriodes* contains 95% α-(1→6) main-chain linkages and 5% α(1→3) branch linkages. Dextran was the first extracellular microbial polysaccharide to find uses in industry. Primary uses of dextran have been as a blood plasma extender and in separation technology. Dextrans exhibit low solution viscosity. They complex some metal ions and are compatible with salts, acids, and bases over a broad temperature range. Crosslinked dextran gels are used in chromatography/separation applications.

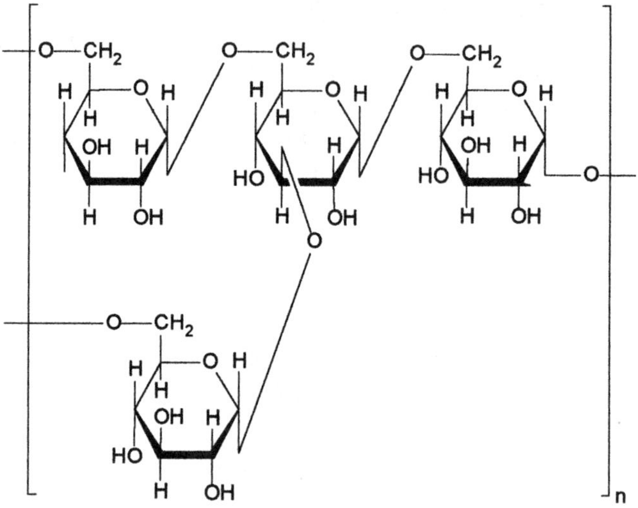

Figure 19: Structure of dextran.

5.4.6. Other bacterial polysaccharides (succinoglycan, welan and levan)

Succinoglycan, an acidic extracellular polymer with octasacharide units, is produced by several bacteria, including *Alcaligenes facalis*, *Agrobacterium radiobacter*, *Rhizobium meliloti*, and *R. trifolii*. This polymer has properties that make useful as a thickening agent in foods and for industrial processes (*83*). Welan is structurally similar to gellan with additional side groups of α-L-rhamnopyranosyl or an α-L-mannopyranosyl unit linked (1→3) to β-D-glucopyranosyl unit in the polymer backbone. Welan is produced by an *Alcaligenes* species by aerobic fermentation, and marketed under the trade name BIOZAN (Merck and Co., Inc.) (*8,49,83*). Thus far, welan has properties that make it useful for oil recovery, oil-field drilling operations and hydraulic fracturing projects. The compatibility of ethylene glycol and welan has led to the formation of a viscous welan-ethylene glycol composition that is useful as an insulating material (*84*). Levan is a water-soluble homopolysaccharide composed of β-(2→6)-linked D-fructofuranoside units, with branch chains linked to the main chain *via* β-(2→1)-linkages. This polymer is produced by several bacteria (*49*). Due to the low inherent viscosity and sensitivity to hydrolysis of levan, it is not a suitable thickening or viscosifying agent. Many of the applications proposed for levan appear to be better served by dextran. Food applications proposed for levan include their use as fillers, bulking agents and as a substitute for gum arabic. In addition, levan is of interest as a fructooligosaccharide analog of Neosugar, produced by Meiji Seika Kaisha, Ltd (*49*).

6. Lignin

Lignins are a major byproduct from the pulp and paper industry (20 million tons per year in the United States). They are complex heteropolymers that contain aromatic moieties and ether linkages. Lignins derive their basic structural elements from the enzyme-catalyzed free radical polymerization of p-hydroxy cinnamyl alcohols. They are synthesized by plants and are mineralized by a variety of organisms, particularly fungi (*85-87*). Due to the abundance of lignins and their structural characteristics, they are under study as a base material for composites and in polymer grafts (*85*). Thus far, lignocellulosic composites for rigid structural materials have suffered from dimensional instability due to moisture absorption, are flammable, and are subject to degradation by ultraviolet light. To address these limitations, a major research objective is to chemically modify lignins and thereby overcome these problems (*88*). Recent work on graft copolymers such as with polystyrene may help address some of these needs, although the issue of biodegradability of these grafts remains a question (*89-91*). In addition, studies are underway in a number of laboratories to genetically engineer lignin biosynthesis, particularly in rapid growth trees such as poplars. Some of the goals are to enhance pulping operations and optimize cellulose recovery by tailoring lignin composition and content.

7. Agroproteins

7.1. Soy

Soy protein is derived from soybean that consists of about 40% proteins and 20% oil on a dry-weight basis. Once the oil is extracted, the protein content of the remaining solid, called 'meal', rises to about 44%. Among the plant proteins, soy protein contains relatively high contents of glutamic and aspartic acid residues. These anionic repeat units are responsible for its solubility in alkali and precipitation at pH 4.5. Compared to other plant protein preparations, soy-proteins have high value (*92*). Soy proteins are globular, reactive and often water-soluble. In addition, they are biodegradable, renewable, widely available and competitively priced. Much work is ongoing to develop methods and compositions to better use soy-proteins in foods and for other applications. Soy protein is one of the common ingredients in many foodstuffs. Various kinds of soy protein, including soy flour, soy-protein concentrate, and isolated soy protein, are currently used in human foods. Soy proteins are used in infant formulas and enteral nutrition products, as ingredients in meat products, and as protein supplements. Soy protein films are formed on the surface of heated soymilk and plasticizers are required to maintain their elasticity. These films are used in the food industry as coatings. Experiments on animals have shown that soybean protein has hypocholesterolemic and antiatherogenic properties. In human beings, substitution of soy protein for dietary animal protein or addition of soy protein to the diet lowers total and low-density-lipoprotein cholesterol levels in individuals with hypercholestero-lemia (*93*). Soy protein-based plastics are biodegradable, non-electrostatic and non-flammable. They can be designed with a moisture barrier layer. Thus, there is considerable interest in developing soy-based engineering biodegradable plastics as one approach to easing current problems of solid-waste disposal and toxic chemical generation associated with some traditional polymer synthetic methods (*94*). It is claimed that molded plastics made of soy protein alone, or mixed with starch, have adequate mechanical and water resistance properties for many one-time use consumer products such as disposable containers, utensils, toys, and outdoor sporting goods. Soy protein films are good oxygen barriers and UV-blockers. Thus, they may be used as packaging materials to protect products from UV-light induced oxidation and deterioration. Also, these materials may be used as agricultural mulch films and for other agricultural or horticultural uses. Biodegradable foams from soy-protein are also being evaluated for their potential to replace Styrofoam for numerous one-time use products such as picnic plates, clamshell containers, packaging material, and non-flammable thermal-insulation (for construction).

7.2. Zein

Zein protein, a prolamine, is found in amounts of 2.5 to 10 % in corn (*Zea mays L., gramineae*). The greater fraction of zein has a molecular weight of 38,000. It does not contain lysine or tryptophan. Zein is insoluble in water or acetone alone. However, zein is soluble in water/acetone mixtures, aqueous alcohol, the glycols, mono-ethyl ether of ethylene glycol, furfuryl alcohol, tetrahedrofuryl alcohol and aqueous alkaline solutions with pH $\geq$ 11.5. For commercial purposes, it is extracted from gluten meal with dilute isopropenol (*95-96*). The zein proteins are used in manufacturing plastics, as paper coatings, adhesives, substitutes for shellac, laminated board, in solid color printing, films, as edible coatings, and in confectionery products such as nuts, fruit and candies.

7.3. Gluten

Gluten is a wheat protein intermixed with the starchy endosperm of the grain. This protein causes the dough to retain the carbon dioxide produced during dough fermentation, thus providing the porous and spongy structure of bread. Gluten is insoluble in water but partly soluble in alcohol, dilute acids and soluble in alkalies. It is used as an adhesive and as a substitute for flour.

7.4. Casein

Casein is a mixture of related phosphoproteins found in milk and cheese. In bovine milk, casein is found as calcium caseinate (up to 3%) whereas in human milk it is present as potassium caseinate. Casein is one of the most nutritive milk proteins that contains all the common amino acids and is rich in essential ones. The major component of casein are α-, β-, γ-and κ-caseins which can be distinguished by electrophoresis (*97*). Casein is only sparingly soluble in water and nonpolar solvents but is soluble in aqueous alkaline solution. Hydrolysis of caseins gives casamino acids (*98-99*). Casein can be obtained from milk by removing the cream followed by acidification of the skimmed milk. This causes the precipitation of casein. In cheese manufacture, casein is precipitated by the lactic acid formed from the same milk by fermentation.

Casein is used in the manufacture of molded plastics, adhesive, paints, glues, textile finishes, paper coatings, and man-made fibers. Vitamin-free casein is used in diets of animals employed for the biological assay of vitamins (*100*). Medicinal grade caseins are used in dietetic preparations. Also, medicinal grade caseins provide information on the effectiveness of digestive enzyme preparations such as pepsin, trypsin, and papain. Casamino acids are used in microbial assays and in the preparation of microbiological media.

8. Non-Ribosomal Poly(amino Acids)

These polymers represent a broad family of poly(amino acids) that may be produced by microbes or by chemical synthesis (Figure 20). Some of these polymers are referred to as psuedopoly(amino acids) since they are not linked between α-amino and α-carboxyl groups (α-linked). Instead, esters or other functionalities within amino acids may link amino acids together. Also, amide bonds that are not α-linked may be used to join amino acid repeat units. Non-ribosomal poly(amino acids) are also formed by the ring-opening of N-carboxyanhydride monomers. This provides a route to model α-linked poly(amino acids) that, for example, may consist of only one type of amino acid along the chain.

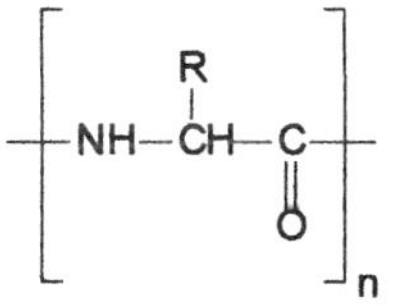

$$\left[-NH-CH(R)-C(=O)- \right]_n$$

Figure 20: *Structure of polyamino acid, linkages may be between N, C or R groups.*

L-arginyl-poly(aspartic acid) is a branched polypeptide synthesized by cyanobacteria such as *Anabena cylindrica* (*101*). This high molecular weight (Mr=25 000-125 000) polymer is composed of a poly(aspartic acid) core with L-arginyl residues that are amide bonded to each side-chain carboxyl group of the poly(aspartic acid) (*102*). Polymers with carboxylic acid side chains, such as those consisting of poly(acrylic acid) and acrylic acid copolymers, are used as builders in the detergents, dispersants in paints, additives to control scaling in piping, and absorbents in diapers and medical products. However, with the exception of low molecular weight oligomers, poly(acrylic acid) is highly recalcitrant and, when disposed, will accumulate instead of degrading within the environment. Thermal polyaspartic acid (TPA), a poly(amino acid) with carboxylic side chains, is synthesized by the thermal polycondensation of aspartic acid to form poly(succinimide) (PSI) that is subsequently hydrolyzed. (Scheme 1) (*103-107*).

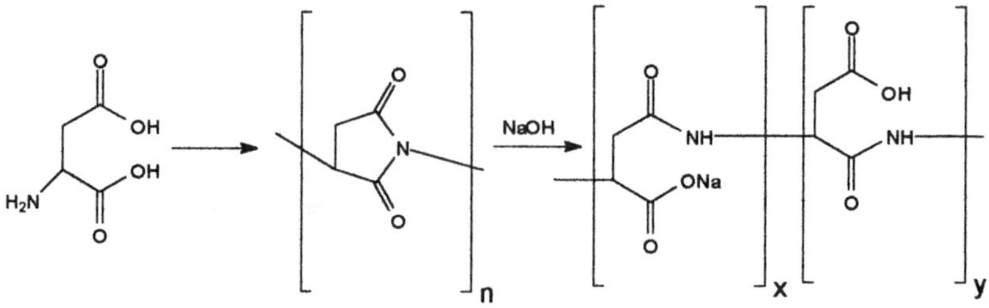

L-Aspartic Acid *Poly(succinimide) (PSI)* *Thermal polyaspartate (TPA)*

Scheme 1: *Synthesis of thermal polyaspartate (TPA)*

The polymer chain of TPA is connected by a mixture of α- and β-bonds. In addition, aspartic acid is isomerized during the condensation to form PSI so, the aspartic residues of TPA are racemic. Despite the fact that TPA is racemic and consists of a mixture of α- and β-linked aspartate residues, linear TPA is highly biodegradable in wastewater environments. In addition, TPA and polyacrylic acid perform similarly as chelators and dispersants in detergent formulations. Also, F. Littlejohn *et al.* has reported the use of TPA for the removal of hydroxyapatite/brushite deposits from stainless steel tubing (*106*).

γ–Poly(glutamic acid), γ–PGA, is a natural extracellular product of several species of *Bacillus* genus; *B. anthracis, B. licheniformis, B. megaterium,* and *Bacillus subtilis* (NATTO). This water-soluble polyanion consists of glutamate repeat units linked between α-amino and γ-carboxyl functional groups (Figure 21) (*108-112*).

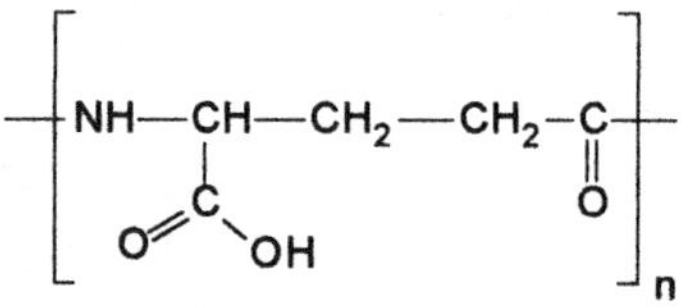

Figure 21: *Structure of poly-γ-glutamic acid.*

The nutritional requirements for γ–PGA production varies according to the strain used. In addition, a combination of the strain used as well as the physiological conditions during microbial production will ultimately dictate γ–PGA molecular weight as well as the stereochemical composition of the main chain repeats (*113*). For example, in some microbial production strains, γ–PGA stereochemistry can be regulated by the concentration of divalent metals, such as zinc and manganese in the culture media. γ–PGA is a component of NATTO, a traditional fermented food produced in Japan from soy beans. Recent studies have explored possible enhancement of yields of polymer under controlled fermentation conditions (*114*). Films can be formed from the polymer subsequent to purification from culture filtrates. Applications are targeted for controlled release and medical implants, and potentially, in the food industry. Poly(L-glutamic acid)-paclitaxel (PG-TXL) is a new water-soluble paclitaxel derivative that has shown remarkable antitumor activity against both ovarian and breast tumors (*115*).

9. Overview on Biosurfactants

Biosurfactants are natural surface-active molecules with complex chemical structures and physical properties. They are derived from microbial, plant or animal sources (*116-120*). In general, their structure includes a hydrophobic moiety (e.g. a long-chain saturated, unsaturated or hydroxy fatty acid) and the hydrophilic moiety (carbohydrate, carboxylic acid, phosphate, amino acid, cyclic peptide, or alcohol). To the best of our knowledge, biosurfactants with cationic head groups are not known. These biomolecules have the ability to reduce surface and interfacial tensions in both aqueous solutions and hydrocarbon mixtures making them candidates for enhanced oil recovered and deemulsification processes (*120-122*).

Mainly, their chemical composition and their origin are used to classify biosurfactants. The major classes of biosurfactants are glycolipids, lipopeptides/lipoproteins, phospholipids, fatty acids, polymeric and particulate surfactants (*116-120*). Glycolipids, which are made of a long-chain saturated, unsaturated or hydroxy fatty acid and a carbohydrate (e.g. rhamnolipids, trehalolipids, sophorolipids, cellibioselipids, mannosylerythritol lipids, and trisaccharide lipids with varying substitutions), are prominent examples in the literature (*116-120*).

Certain industry sectors are considering replacing synthetic surfactants with their natural counterparts. Motivation for such replacements are their perceived properties of safety and mildness, better physical properties, low toxicity, biodegradability, environmental compatibility, higher foaming ability, high selectivity and specific activity at extreme temperatures, pH, and salinity and the ability to be synthesized from renewable feedstock (*120-123*). Research on the applications of biosurfactants have mainly been focussed on enhanced biodegradation of emulsified crude oil and in oil recovery (*120-123*). Other areas of interest have been to use biosurfactants for the removal of hazardous chemicals, stabilization of coal slurries for pipeline transportation, cosmetics, detergent, foods, drugs and drug delivery, etc., see Table I) (*119-138*).

Table I: Some biosurfactants and their applications

Type of biosurfactants	Sources	Applications and References
A. Glycolipids:		
1. Rhamnolipids	Bacterial (e.g.*Pseudomona,Cornyebacterium, Arthobacter* and *Nocardia* sp.)	Bioremediation, emulsifying agent, enhanced oil recovery, removal of hazardous chemicals (such as 2,4-trichlorophenyloxyacetic acid, chlorophenols), wound healing, in treatment of burn shock, organ transplants, arteriosclerosis, depression, dermatological diseases, papilloma viral infections, schizophrenia, cosmetics and detergents (*119-128*).
2. Sophorolipids	Yeast (*Turulopsis* or *Candida* sp.)	In oil recovery, cosmetics (especially for antidandruff, bacteriostatic agents, and deodorants), detergent, macrophage activator, fibrinolytic agent, skin healing agent, desquamating agent, depigmenting agent, and in germicidal compositions (suitable for cleaning fruits, vegetables, skin and hair (*119-120,127-133*).
3. Trehalose lipids	Bacterial (*Rodococcus, Mycobacterium, Arthrobacter, Brevibacterium, Cornybacterium* sp.)	Bioremediation, emulsifying agent, enhanced oil recovery, removal of hazardous chemicals and metals, inducer for leukemic cell strain and surface activities, cell differentiation inducer, emulsifying dispersant for cosmetics or foods, gelling agent, raw material for liposomes, additive to antiviral agent (*119-120,135-138*).
4. Cellobiose lipids	Bacterial (*Ustilago* sp.)	In detergent (*120,128*).
B. Lipopeptides and lipoproteins:		
1. Surfactin	Bacterial (*Bacillus* sp.)	A powerful biosurfactant, mechanical dewatering of peat, treating or preventing hypercholesterolemia, good inhibitor of eukaryotic protein kinase activity (119-120,139-140).
2. Mycosubtilin	Bacterial (*Bacillus*	As antibiotic (141)

Table I. *Continued*

	sp.)	
3. Viscosin	Bacterial (*Pseudomonas* sp.)	As antibacterial, antiviral and antitrypan-osomal therapeutic agent, effective inhibitors of eukaryotic protein kinase activity (142-143).
4. Viscosinamide	Bacterial (*Pseudomonas* sp.)	Antifungal agent (144).
5. Germicidins	Bacterial (*Bacillus* sp.)	As antibiotics (120).
6. Polymyxins	Bacterial (*Bacillus* sp.)	As antibiotics (120).
7. Cerlipin (ornithine- and taurine-containg lipid)	Bcterial (*Gluconobactor* sp.)	Excellent biosurfactant (120).
8. Lysine containg lipids	Bacterial (*Acinetobacter* sp.)	Excellent biosurfactant (120).
9. Ornithine-containg Lipids	Bacterial (*Pseudomonas* and *Thiobacillus* sp.)	Excellent biosurfactant (120).
D. Polymeric biosurfactants:		
1. Biodispersan	Bacterial (*Acinetobacter* sp.)	As a dispersing agent, mining and paper industries (prevents flocculation of minerals while stabilizing aqueous suspensions of minerals) (119-120,145).
2. Emulsan	Bacterial (*Acinetobacter sp.*)	A very potent emulsifying agent used in cleaning oil-contaminated vessels, control of dental plaque and caries, treating chlamydial infection, patients having Gram-positive bacterial infections such as septicemia and associated pathophysiolo-gical states such as septic shock (119-120,146-147).
3. Liposan	Bacterial (*Candida* sp.)	Emulsifying agent (120).

Continued on next page

Table I. *Continued*

E. Phospholipids	Bacterial and fungal (*Aspergillus, Acinetobacter, Arthobacter, Pseudomonas, Thiobacillus sp.*)	As a cytotoxic agent, treatment of HIV-1, hepatitis B and herpes viruses, as auxiliary agents in the production and application of spray mixtures containing plant protectants, as additive in anti-inflammatory, antiulcer and healing agents, etc. (*119-120,148*).

9.1. Emulsan

Emulsans are a family of extracellular lipoheteropolysaccharide surfactants formed by the bacterium, *Acinetobacter calcoaceticus,* a hydrocarbon-degrading organism (*149-151*). The main chain of emulsans is believed to consist of amino sugars, including galactosamine (2-amino-2-deoxy-galactose) and a 2-amino-2-deoxy-glucuronic acid. Appended to the main chain by ester and amide bonds are fatty acids that vary in chain length from C10 to C18, are saturated and monounsaturated, and include 2-hydroxyl and 3-hydroxyl C12 fatty acids (Figure 22).

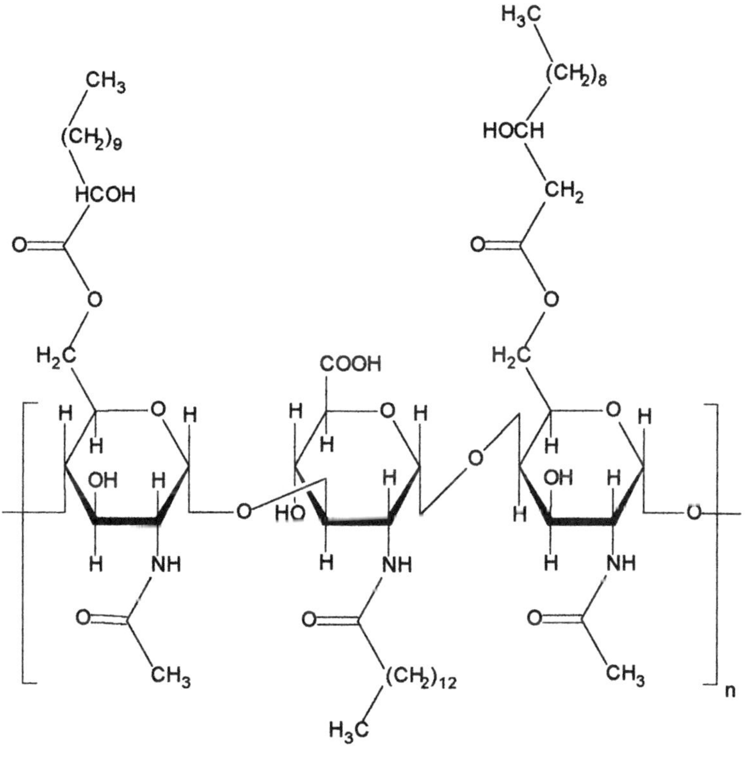

Figure 22: *Structure of emulsan.*

Emulsan is polyanionic, high molecular weight (approximately one million), and contains varying amounts of fatty acids (from about 10 to 25% by weight). The degree of substitution of fatty acids as well as the composition of fatty acids can be regulated by selected feeding experiments (*150*). Fermentations to produce emulsans are conducted in submerged aerobic culture, and the produce can be recovered with ammonium sulfate precipitation or solvent extraction. Large-scale fermentation has also been developed with ethanol as the carbon source, and more recently with triglycerides and fatty acids (*152*). Emulsan forms stable oil/water emulsions by forming a film around oil droplets to prevent coalescence. This surfactant promotes low viscosity emulsions and is used in oil transport though pipes and in enhanced oil recovery. Unlike many bioemulsifiers, emulsan biosynthesis does not require prior growth on hydrocarbon substrates. In contrast to many synthetic surfactants, emulsan is only found at the oil/water interface and not in the water or oil phases, thus only small amounts are required even in excess water. In addition, emulsans are nontoxic and biodegradable. Emulsan-stabilized emulsions resist inversion even when water/oil ratios fall below 1/4. Some studies have been carried out where genes involved in emulsan production have been isolated and cloned into *E. coli*, including esterase genes and alcohol dehydrogenase complexes (*153*).

10. Conclusions

All living organisms produce polysaccharides and poly(amino acids). Thus, these naturally occurring polymers are ubiquitous in the world around us. The different classes of polymers presented here represent a wide range of structural and functional features. However, only selected, better-characterized polymers have been described herein. Many partially characterized natural polymers, or polymers that are found only at low levels in nature, were not included. Furthermore, it is not possible to estimate the number of natural polymers that have yet to be discovered and what role they may have in providing important new material or biological functions. Aside from cellulose and starch, most of the natural polymers reviewed remain relatively high in cost. This cost barrier is normal for materials that are not currently manufactured in large quantity. As environmental concerns force us to develop polymers that can be manufactured from renewable resources by natural or integrated natural/chemical processes, increased production scale and developments in process engineering will result in products from natural polymers that are of competitive and acceptable in cost. Furthermore, product development from readily renewable feedstocks will be driven by the volatility of petroleum prices that have been especially apparent in the year that this book chapter was written. One of the authors of this chapter (R.G.) observed during the years 1999 to 2000 that the price per gallon of gasoline (87 octane level) has varied from about $1.05 to $1.80 per gallon.

There is not question that, in the future, we will see increasing use of renewable materials in polymers. A significant portion of these developments will involve natural products that serve as building blocks for polymers synthesized by otherwise conventional methods. Recent examples of this are the development of a commercial processe for the synthesis of 1,3-propanediol and lactic acid that are both precursors for the synthesis of valuable polyesters.

As we move beyond the year 2000, it is apparent that developments in biotechnology will begin to either take center stage, or at least share co-billing with other processes to make polymeric products. Insights into biosynthetic mechanisms, coupled with advanced methods for genetic improvements, will lead to a new list of possibilities that were only recently on the list of the impractical or even impossible. Surely the goal to incorporate chemical synthesis into plant systems will take on increased importance and positive outcomes. Our continuously advancing ability to regulate biosynthetic pathways will lead to clean, multi-step syntheses in biological factories such as photosynthetic microbes where energy is derived from the sun.

New learning in genetics and biosynthetic pathways will allow unprecedented capabilities to control non-ribosomal polymer syntheses within organisms. For example, the ability to regulate the relative rates of chain initiation, propagation, chain transfer and termination reactions will lead to natural polymers with high levels of control of polymer molecular weight, dispersity, main and side chain composition, repeat unit sequence distribution, and incorporation of specific groups at chain ends.

Literature Cited

1. Trost, B. M. In *Green Chemistry*, Anastas, P. T.; Williamson, T. C., Eds.; Oxford University Press, New York, **1998**.
2. *Enzymes in polymer Synthesis, ACS Symposium series 684*; Gross, R. A.; Kaplan, D. L.; Swift, G., Eds; American Chemical society, Wasington, DC, **1998**.
3. Uhrich, K. E.; Cannizzoaro, S. M.; Langer, R. S.; Shakesheff, K. M. *Chem. Rev.* **1999**, *99*, 3181.
4. Lloyd, A. W.; Hunter, A. C. *Pharmaceutical Science and Technology Today* **2000**, *3*, 210.
5. WTEC Panel Report on *Nanostructure Science and Technology*, September **1999** (web site: http://itri.loyola.edu/nano/toc.htm).
6. Collins, P.; Ferrier, R. *Monosaccharides*; John Wiley & Sons., New York, **1995**, 463.
7. Stevens, M. P. *Polymer Chemistry;* Oxford University press, New York, **1999**, 489.
8. Kaplan, D. L. In *Biopolymers from Renewable Resources;* Kaplan D. L. (Ed), Springer, New York, **1998**, 1.
9. Shogren, R. L. In *Biopolymers from Renewable sources;* Kaplan D. L., (Ed), Springer, New York, **1998**, 30.
10. Gilbert, R. D.; Kadla, J. F. In *Biopolymers from Renewable sources;* Kaplan D. L., Ed.; Springer, New York, **1998**, 47.
11. Stevens, M. P. *Polymer Chemistry* Oxford University press, New York, **1999**, 484.
12. Hon, D. N. -S. In *Polysaccharides in Medicinal Applications;* Dumitriu, S. (Ed), Marcel Dekker, Inc., New York, **1996**, 87.
13. Hudson, S. M.; Cuculo, J. A. *J. Macromol. Sci. Rev. Mocromol. Chem. Phys.* **1980**, *C18*, 1.
14. Gadd, K. F. *Polymer* **1982**, *23*, 1867.
15. Popa, V. I. In *Polysaccharides in Medicinal Applications;* Dumitriu, S.,Ed.; Marcel Dekker, Inc., New York, **1996**, 107.
16. Teot, A. S. *Encyclopedia of Chemical Technology* John Wiley & Sons, New York **1982**, *20*, 207.
17. Maeda, M.; Shimahara, H.; Sugiyama, N. *Agric. Biol. Chem.* **1980**, *44*, 245.
18. Sutherland, I. W. In *Biomaterials* Byrom, D., Ed.; Stockton Press, New York, **1991**, 307.

19. Smidsrod, O.; Skjak-Braek, G. *Trends in Biotechnol.* **1990**, 8.
20. Trius, A.; Sebranek, J. G. *Crit. Rev. Food Sci. Nutr.* **1996**, *36*, 69.
21. Cottrell, I. W.; Baird, J. K *Encyclopedia of Chemical Technology* John Wiley & Sons, New York, **1980**,*12*, 45.
22. Rinaudo, M.; Domard, A. In *Chitin and Chitosan* Skjak-Brak, G.; Anthonsen, T.; Sandford, P., Eds.; Elsevier, London, **1989**, 71.
23. Austin, P. et al. *Science* **1981**, *212*, 749.
24. Cohen-Kupiec, R.; Chet, I. *Curr. Open. Biotech.* **1998**, 9, 270.
25. Shepherd, R.; Reader, S.; Falshaw, A. *Glycoconj. J.* **1997**, *14*, 535.
26. Arcidiacono, S.; Kaplan, D. L. *Biotechnol. Bioeng.* **1992**, *39*, 281.
27. Hudson, S. M.; Smith, C. In *Biopolymers from Renewable sources*; Kaplan D. L., Ed.; Springer, New York, **1998**, 96.
28. Skjak-Braek, G.; Anthonsen, T.; Sandford P. *Chitin and Chitosan;* Elsevier, London, **1989**.
29. Onsoyen, E.; Skaugrud, O. *J. Chem. Technol. Biotechnol.* **1990**, *49*, 395.
30. No, H. K.; Meyers, S. P. *Rev. Environ. Contam. Toxicol* **2000**, *163*, 1.
31. Illum, L. *Pharm. Res.* **1998**, *15*, 1326.
32. Whistler, R. J.; Koski, M. *Arch. Bioch. Biophys.* **1971**,*142*, 106.
33. Muzzarelli, R. A. A. *Chitin;* Pergamon Press, New York, **1977**.
34. Synowiecki, J.; Sikorski, Z. E.; Naczk, M. *Biotechnol. Bioeng.* **1981**, 23, 2311.
35. Stanley, W. L.; Watters, G. G.; Kelly, S. H.; Chan, B. G.; Garibaldi, J. A.; Schade, J. E. *Biotechnol. Bioeng.* **1976**, *18*, 439.
36. Mathur, N. K.; Narang, C. K. *J. Chem. Education* **1990**, *67*, 938.
37. DeLucca, G. V.; Kezar, H. S.; O'Brien, J. P., U.S. Patent 4,833,238, **1989**.
38. DeLucca, G. V.; Kezar, H. S.; O'Brien., J. P. U.S. Patent 4,857,403, **1989**.
39. DeLucca, G. V.; Pelosi, L. F.; O'Brien, J. P. U.S. Patent 4,861,527, **1989**.
40. Meyer, K. *Fed. Proc.* **1958**, 17, 1075.
41. Flowers, H. M.; Jeanloz, R. W. *Biochemistry* **1964**, 3, 123.
42. Swann, D. A.; Kuo, J. In *Biomaterials - Novel Materials from Biological Sources;* Byrom, D., Ed.; Stockton Press, New York, **1991**, 285.
43. Goa, K. L.; Benfield, P. *Drugs* **1994**, 47, 536.
44. Sutherland, I .W. *Trends in Biotechnology* **1998**, *16*, 41.
45. Burns, J. W.; Cox, S.; Waits, A. E. U.S. Patent 5,017,229, **1991**.
46. Goldberg, E. P.; Yaacobi, Y. U.S. Patent 5,080,893, **1992**.
47. Goldberg, E. P.; Yaacobi, Y. U.S. Patent 5,140,016, **1992**.
48. Bender, H.; Lehmann, J.; Wallenfels, K. *Biochim. Biophys. Acta* **1959**, *36*, 309.
49. Cote, G. L.; Ahlgren, J. A. In *Kirk-Orthmer Encyclopedia of Chemical Technology;* John Wiley & Sons, Inc., Fourth Ed., **1995**, *16*, 577.
50. Gibbs, P. A.; Robert, J. S. In *Polysaccharides in Medicinal Applications*; Dumitriu, S. (Ed), Marcel Dekker, Inc., New York, **1996**, 59.
51. Kaplan, D. L.; Mayer, J. M.; Lombardi, S. J.; Wiley, B.; Arcidiacono, S. *Polymer Preprint, American Chem. Sec. Div. Polymer Chem.* **1989**, 30, 509.
52. Mayer, J. M.; Greenberger, M.; Ball, D. H.; Kaplan, D. L. *Preprints Proceed. American Chem. Sec. Div. Polymeric Materials. Sci. Eng.* **1990**, 63, 732.

53. Wiley, B. J.; Ball, D. H.; Arcidiacono, S.; Sousa, S.; Mayer, J. M.; Kaplan, D. L. *J. Environ. Polym. Degradation* **1992**, 1, 3.
54. Tsujisaka, Y., Mitsuhashi, M. In *Polysaccharides and their Derivatives;* Whistler, R. L.; BeMillerand, J. N. (eds), Acedemic Press, Inc., San Diego, California, **1993**, 446.
55. Sugimoto, K. *J. Ferment. Assoc. Jpn.* **1979**, *36*, 98.
56. LeDuy, A.; Choplin, L.; Zajic, J. E.; Luong, J. H. T. In *Encyclopedia of polymer Science and Engineering*; H. F. Mark et. al., Eds.; 2nd Ed., Wiley & sons, New York, **1988**, 13, 650.
57. Ball, D. H.; Wiley, B. J.; Reese, E. T. *Canadian J. Microbiol.* **1992**, *38*, 324.
58. Sugimoto, K. *J. Ferment. Assoc. Jpn.* **1979**, *36*, 98.
59. Uchida, T.; Ikegmi, H.; Ando, S.; Kurimoto, M.; Mitsuhashi, M.; Naito, S.; Usui, M.; Matuhasi, T. *Int. Arch. Allergy Immunol.* **1993**, *102*, 276.
60. Wiley, B. J.; Arcidiacono, S.; Ball, D. H.; Mayer, J. M.; Kaplan, D. L. *Report 89/035, U. S. Army Natick Research Development and Engineering Center;* Natick, Massachusetts, **1989**.
61. Rinaudo, M.; Vincendo, M. *Carbohydr. polym.* **1982**, *2*, 135.
62. Bluhm, T. L.; Deslandes, Y.; Marchessault, R. H.; Perez, S.; Rinaudo, M. *Carbohydr. Res.* **1982**, *100*, 117.
63. Pretus, H. A. et. al. *J. Pharm. Exp. Therapeut.* **1991**, *257*, 500.
64. Becker, A.; Katzen, F.; Puhler, A.; Ielpi, L. *Appl. Microbiol. Biotechnol.* **1998**, *50*, 145.
65. Marchessault, R. H. *Topics Polym. Sci.* **1984**, *5*, 15.
66. Betlach, M. R.; Capage, M. A.; Doherty, D. H.; Hassler, R. A.; Henderson, N. M.; Vanderslice, R. W.; Marelli, J. D.; Ward, M. B. In *Prog. in Biotechnol.;* Yalpani M., Ed; **1987**, *3*, 35.
67. Hassler, R. A.; Doherty, D. H. *Biotech. Progress* **1990**, *6*, 182.
68. Takagi, H.; Kadowaki, K. *Agric. Biol. Chem.* **1985**, *49*, 3159.
69. Harada, T. *Biochem. Soc. Symp.* **1986**, *48*, 97.
70. Nakao, Y. *Agro. Food Ind. Hi Tech.* **1997**, *8*, 12.
71. Vossoughl, S.; Buller, C. S. *SPE J. Reservoir Eng.* **1991**, 485.
72. Harada, T.; Harada, A. In *Polysaccharides in Medicinal Applications;* Dumitriu, S., Ed.; Marcel Dekker, Inc., New York, **1996**, 21.
73. Kunimoto, T.; Baba, H.; Nitta, K. *J. Biol. Response Modifiers* **1986**, *5*, 160.
74. Aketagawa, J.; Tanaka, S.; Tamura, H.; Shibata, Y.; Saito, H. *J. Biochem.* **1993**, *113*, 683.
75. Kanke, M.; Koda, K.; Koda, Y.; Katayama, H. *Pharm. Res.* **1992**, 9, 414.
76. Franz, G.; Alban, S. *Int. J. Biol. Macromol.* **1995**, *17*, 311.
77. Yamamoto, I.; Takayama, K.; Konma, K.; Gonda, T.; Yamazaki, K.; Matsuzaki, K. Hatanaka, K.; Uryu, T.; Yoshida, O.; Nakashima, H.; Yamamoto, N.; Kaneko, Y.; Mimura, T. *Carbohydr. Polym.* **1991**, *14*, 53.
78. Yoshida, T.; Hatanaka, K.; Uryu, T.; Kaneko, Y.; Suzuki, E.; Miyano, H.; Mimura, T., Yoshida, O.; Yamamoto, N. *Micromolecules* **1990**, *23*, 3717.
79. Nara, K.; Schindo, T.; Yada, H.; Miwa, S. *Concrete Technol. Rep.* **1994**, 5, 23.

80. Murooka, Y.; Yim, M.; Yamada, T.; Harada, T. *Bichim. Biophys. Acta* **1977**, *485*, 134.
81. Hasegawa, T.; Takizawa, M.; Tanida, S. *Actinomycertes J. Gen. Appl. Microbial.* **1983**, *29*, 319.
82. Yalpani, M. *Prog. In Biotechnol.* Elsevier, Amsterdam, **1987**, 3.
83. Yalpani, M. *Polysaccharides: Synthesis, modifications and structure/property relations. Studies in Organic Chem.* Elsevier. Amsterdam, The Netherlands, **1988**, 36.
84. Ramsay, A. M.; Trimble, G.; Seheult, J. M.; O'Brien, M. S. U.S. Patent 5,290,768, **1991**.
85. Glasser, W. G. *Assessment of Biobased Materials, Report SERI/T~R-234-3610*, Solar Chum, H. L. (Ed.), Energy Research Institute, Colorado, **1989,** 41.
86. Kaplan, D. L.; Hartenstein, R. *Soil. Biol. Biochem.* **1980**, *12*, 65.
87. Srinivasan, V. R.; Cary, J. W. *Wood and Cellulosics: Industrial Utilisation, Biotechnology, Structure and Properties* Ellis Horwood Ltd., Chichester, England **1987**, 267.
88. Rowell, R. M.; Young, R A. *Assessment of Biobased Materials*, Chum, H. L., Ed.; Report SERI/TR-234-3610, Solar Energy Research Institute, Colorado, **1989**, 11.
89. Meister, J. J.; Patil, D. R., *Macromolecules* **1985**, *18*, 1559.
90. Meister, J. J.; Patil, D. R.; Channell, H., *J. Appl. Polymer Sci.* **1984**, *29*, 3457.
91. Milstein, O.; Gersonde, R.; Huttermann, A.; Chen, M. J.; Meister, J. J. *Appl. Environ. Microbiol.* **1992**, *58*, 3225.
92. Slavin J. *J. Am. Deit. Assoc.* **1991**, *91*, 816.
93. Corroll, K. K. *J. Am. Diet. Assoc.* **1991**, *91*, 820.
94. Wang, S.; Sue, H-J; and Jane, J. *JMS Pure Appl. Chem.* **1996**, *33*, 557.
95. Swallen *Ind. Eng. Chem.* **1941**, *33*, 394.
96. Carter, R. Ger. pat. 2,002,337, **1971**.
97. McKenzie *Advan. Protein Chem.*, **1967**, *22*, 75.
98. Mueller, M. *J. Immunol.* **1941**, *40*, 21.
99. Mueller, M. *J. Immunol.* **1941**, *40*, 33.
100. Kissel US Patent 2,853,479, **1958**.
101. Mackerras, A. H.; de Chazal, N. M.; Smith, G. D. *J. Gen. Microbiol.* **1990**, *136*, 2057.
102. Simon, R. D. *Biochim. Biophys. Acta* **1976**, *422*, 407.
103. Matsubara, K.; Nakato, T.; Tomida, M. *Macromolecules* **1997**, *30*, 2305.
104. Matsubara, K.; Nakato, T.; Tomida, M. *Macromolecules* **1998**, *31*, 1466.
105. Kakuchi, T.; Shibata, M.; Matsunami, S.; Nakato, T.; Tominda, M. *J. Polymer Science: Part A: Polymer Chemistry* **1997**, 35, 285.
106. Littlejohn, F.; Saez, A. E.; Grant, C. S. *Int. Eng. Chem. Res.* **1998**, *37*, 2691.
107. Koskan, L. *Industrial Bioprocessing* May 1-2, **1992**.
108. Gross, R. A. In *Biopolymers from renewable sources;* Kaplan, D.L., Ed.; Springer, New York, **1998**, 195.
109. Byrom, D. In *Biomaterials - Novel Materials from Biological Sources;* Byrom, D. Ed.; Stockton Press, New York, **1991**, 263.

110. Housewright, R. D. In *The Bacteria: A Treatise on Structure and Function;* Gunsalus, L. C.; Stanier, R Y., Ed.; Academic Press, New York, **1962**, *III*, 389.
111. Kunioka, M.; Goto, A. *Appl. Microbiol. Biotechnol.* **1994**, *40*, 867.
112. Troy, F. A. In *Peptide Antibiotics: Biosynthesis and Functions* Kleinkauf, H., Dohren, H. V., Eds.; Waiter de Gruyter, Berlin, **1982**, 49.
113. Cromwick, A. M.; Birrer G. A.; Gross, R. A. *Biotechnol. Bioeng.* **1996**, 50, 222.
114. Giannos, S.; Shah, D.; Gross, R. A.; Kaplan, D. L.; Arcidiacono, S.; Mayer, J. *Preprints Am. Chem. Sec. Div. Polymeric Materials Sci. and Eng.* **1990**, *62*, 236.
115. Li, C.; Price, J. E.; Milas, L.; Hunter, N. R.; Ke, S.; Yu, D. F.; Charnsangavej, C.; Wallace, S. *Clin. Cancer Res* **1999**, *5*, 891.
116. *Biosurfactants and Biotechnology;* Kosaric, N.; Cairns, W. L.; Grey, N. C. C., Eds.; New York, Marcel Dekker, **1987**.
117. *Biosurfactants;* Kosaric, N., Ed.; New York, Marcel Dekker, **1993**.
118. Ishigami, Y. *Informa* **1993**, *4*, 1156.
119. Finnerty, W. R. *Curr. Opin. Biotechnol.* **1994**, *5*, 291.
120. Desai, J. D.; Banat, I. M. *Microbiology and Molecular Biology Reviews* **1997**, *61*, 47.
121. Muller-Hurtig, R.; Wargner, F.; Blaszczyk, R.; Kosaric, N. *Biosurfactants;* New York, Marcel Dekker, **1993**, 447.
122. Jack, T. R. *Curr. Opin. Biotechnol.* **1991**, *2*, 444.
123. Falatko, D. M.; Novak, J. T. *Water Env. Res.* **1992**, *64*, 163.
124. *Biosurfactants and Biotechnology;* Kosaric, N.; Cairns, W. L.; Grey, N. C. C., Eds.; New York, Marcel Dekker, **1987**, pp 183.
125. *Biosurfactants and Biotechnology;* Kosaric, N.; Cairns, W. L.; Grey, N. C. C., Eds.; New York, Marcel Dekker, **1987**, pp 26.
126. Berg, G.; Seech, A. F.; Lee, H.; Trevors, J. T. *J. Env. Sci. Health,* **1990**, *7*, 753.
127. Benerjee, S.; Duttagupta, S.; Chakrabarty, A M. *Arch. Microbiol.* **1983**, *135*, 110.
128. Oberbremer, A.; Muller-Hurtig, R.; Wagner, F. *Appl. Microbiol. Biotechnol.* **1990**, *32*, 485
129. Hall, P. J.; Haverkamp, J.; van Kralingen, C. G.; Schmidt, M. US Patent 5,520,839, **1996**.
130. *Biosurfactants and Biotechnology*; Kosaric, N.; Cairns, W. L.; Grey, N. C. C., Eds.; New York, Marcel Dekker, **1987**, pp 34.
131. Mager, H.; Roethlisberger, R.; Wagner, F. Ger. Offen. DE 3526417 A1, **1987**.
132. Maingault, M. U.S. Patent 5,981,497, **1999**.
133. Goclic, E.; Muller-Hurtig, R; Wagner, F. *Appl. Microbiol. Biotechnol.* **1990**, *34*, 120.
134. Pierce, D.; Heilman, T. J. WO 9816192 A1, **1998**.
135. Finnerty W. R.; Singer, M. E. *Dev. Ind. Microbiol.* **1984**, *25*, 31.
136. Miller, R. M. *Environ. Health Perspect.* **1995**, *103*, 59.
137. Tadaatsu, N.; Hiroko, I.; Toshiaki, N. Japanese Patent 10072478A, **1998**.
138. Kawai, A.; Kayano, M.; Funada, T.; Hirano, J. Japanese Patent 63126493A, **1988**.
139. Arima, K.; Tamura, G.; Kakinuma, A. U.S. Patent 3,687,926, **1972**.
140. Davies, J. E.; Waters, B.; Saxena, G. WO 9920792A1, **1998**.

141. Peypoux, F.; Michel, G.; Delcambe, L. *Eur. J. Biochem.* **1976**, *63*, 391.
142. Burke Jr., T.; Chandrasekhar, B.; Knight, M. U.S. Patent 5,965,524, **1999**.
143. Nielson, T. H.; Christophersen, C.; Anthoni, U.; Sorensen, J. *J. Appl. Microbiol.* **1999**, *87*, 80.
144. Rosenberg, E.; Rubinovitz, C.; Gottlieb, A.; Rosenhak, S.; Ron, E. Z. *Appl. Env. Microbiol.* **1988**, *54*, 317.
145. Rosenberg, E.; Rubinovitz, C.; Legmann; R.; Ron, E. Z. *Appl. Env. Microbiol.* **1988**, *54*, 323.
146. Gutnick, D. L.; Rosenberg, E. U.S. Patent 4,276,094, **1981**.
147. Eigen, E.; Simone, A. U.S. Patent 4,737,359, **1988**.
148. Istrate, N.; Muni, G.; Brauner, E.; Raheman, F. U.S. Patent 5,853,738, **1998**.
149. Gutnick, D. L. *Biopolymers* **1987**, *S223*, 26.
150. Rosenburg, E. *Appl. Environ. Microbio.* **1979**, *37*, 402.
151. Gorkovenko, A.; Zhang, J.; Gross, R. A.; Kaplan, D.; Allen, A. *Carbohydrate Polymers* **1999**, *39*, 79.
152. Shabtai, Y.; Wang, D. I. C. *Biotechnol. Bioeng.* **1990**, 35, 753.
153. Gutnick, D. L.; Allon, R.; Levy, C.; Petter, R.; Minas, W. *The Biology of Acinetobacter* Towner, K. J. (Ed.), Plenum Press, New York, **1991**, 411.

Polysaccharide–Based Materials

Chapter 2

Mechanical and Physical Properties of Microcellular Starch-Based Foams Formed from Gels

G. M. Glenn, W. J. Orts, R. Buttery, and D. Stern

Western Regional Research Center, Agricultural Research Service, United States Department of Agriculture, 800 Buchanan Street, Albany, CA 94710

Microcellular foams can be formed from rigid aqueous gels of starch. Foams were made from starch gels by exchanging the aqueous phase with ethanol before air-drying. The foam product had an extremely fine microstructure comprised of starch granule remnants embedded in a fibrous network. The starch foams had a moderate density (0.14-0.37 g/cm^3), low thermal conductivity (0.024-0.040 W/mK) and high compressive strength. The foams exhibited the capacity to adsorb polar compounds and alkylpyrazines. When the foams were compression molded, a starch plastic was formed with a tensile strength of more than 12 MPa. Tensile strength and the elongation to break more than doubled in the starch plastics when 33% cellulose fiber was incorporated in the formulation. X-ray diffractograms revealed only small amounts of recrystallization in starch foams and plastics after three years of aging.

Introduction

Starch is the least expensive and most abundant worldwide food commodity (*1*). Its low cost and wide availability from year to year have made starch attractive as an industrial raw material. More than 4.5 billion pounds of starch are used in the United States for industrial applications (*2*). Industrial products made from starch include chemicals derived from fermentation processes, adhesives, sizing products, and soil conditioners (*1-7*). Starch is used on a limited scale for single-use disposable products that replace some petroleum-based products. For instance, various starch-based resins have been developed for making films or injection molded articles (*8*). Starch has also been used in making polymeric foams (*9-11*). The best example is extruded starch-based foams that have gained a considerable portion of the loose fill packaging market.

Polymeric foams consist of a solid polymer matrix formed into discrete elements or empty cells and a gas phase that fills the void space (*12-13*). Polymeric foams may be categorized based on cell size as macrocellular or microcellular. Low density, macrocellular starch-based foams, such as those used for loose-fill packaging, are made by extrusion using water/steam as a blowing agent (*9-11*). These foams typically have cell diameters ranging from 100 to 1000 μm (*9-10)*.

Microcellular foams have cell diameters under 10μm and have unique properties that are of commercial interest (*14-17*). For instance, they may provide remarkable sound and thermal insulation and can reduce material density without compromising strength. Microcellular foams have been successfully made from a wide array of polymers using replication of removable pore formers, polymerization of inverse monomer emulsions, blowing/nucleation methods, and phase separation of polymer solutions (*16,18*).

A blowing/nucleation method was reported for making microcellular plastic foams. The technique used CO_2 as a blowing agent in an extrusion process where the CO_2 was injected into the extruder barrel (*15,17,19-20*). A similar process was reported for making CO_2 expanded starch foams (*21*). An aqueous starch slurry was processed just below 100 °C in an extruder to prevent steam from forming and functioning as a blowing agent as the melt exited the die. Carbon dioxide was injected into the starch melt within the extruder barrel. The starch melt expanded and formed a solid foam as it exited the extruder die. The cells of the CO_2 blown starch foam were quite uniform but were much larger than 10 μm in diameter and had thick cell walls (*21*).

A second method of particular interest for making microcellular plastic foams involves solubilizing a high molecular weight, semicrystalline polymer in a solvent that is then cooled to induce phase separation and produce a gel (*16*). A foam is formed if the gel is rigid enough to withstand the compressive forces created by surface tension as the solvent evaporates. Starch, a high molecular weight, semicrystalline polymer, swells and hydrates in excess water at temperatures above the gelatinization temperature. As a gelatinized starch solution is cooled, a gelation process occurs in which the starch molecules reassociate. The starch gel is not rigid enough to withstand the surface tension created by water evaporation in the pores of the starch gel matrix and simply collapses into a film if air-dried. However, Glenn and Irving (*22*) demonstrated that the gel structure of wheat and corn starches could be at least partially preserved by exchanging the water within the gel matrix with a solvent having a lower surface tension before proceeding with solvent evaporation. This chapter describes methods of making starch-based microcellular foams and describes their properties and potential applications.

Experimental

Foam Forming Process

Unmodified starches of wheat (Midsol 50, Midwest Grain Products, Atchison, KS) and two from corn, a regular Dent corn starch (Melogel) and a high-amylose starch (Hylon VII) (National Starch and Chemical Company, Bridgewater, NJ) were used to prepare foams. The wheat and Dent corn starches were composed of ~28% amylose and 72% amylopectin. The high-amylose corn starch contained ~70% amylose and 30% amylopectin.

Starch Gelatinization.
Aqueous suspensions (7 l) of wheat or Dent corn starches (8% w/w) were vigorously mixed while being heated in a boiling water bath. Viscosity was monitored (Brookfield, Model RVT, Stoughton, MA) to determine when peak viscosity was reached. High-amylose corn starch had a much higher gelatinization temperature than wheat or Dent corn starch and was prepared in a pressure reactor (Paar Instrument Co., Moline, IL) equipped with a mixer and controller (model 4843). The aqueous suspension of high-amylose corn starch (8% w/w) was heated at 2 °C min^{-1} up to 140 °C. The starch melt was then cooled to 90 °C using an internal cooling coil.

Slab Foams.
The gelatinized starch melts were poured into slab molds and chilled (5 °C) overnight to promote gelation. The starch gels were removed from the molds and equilibrated 48 hr each in successive baths of ethanol (40, 60, 70, 90% and three changes of 100% ethanol). During the ethanol equilibration process, the gels shrank and became very rigid. The changes in the gel strength enabled them to withstand the compressive force created by surface tension at the air/solvent interface within the pores of the starch foam matrix. The slabs were dried in air filtered through desiccant until the ethanol odor was no longer detected.

A variation of the wheat starch foam was created by adding a bleached softwood pulp fiber (Leafwood, Georgia Pacific, 4% w/w) to the aqueous starch (8% w/w) suspension. The starch-fiber suspension was heated and processed identically to the gelatinized wheat starch melts containing no fiber. The dry weight concentration of fiber in the foam was approximately 33%.

A modified procedure was used to make foam from gels of high amylose corn starch. The gels were first equilibrated in ethanol as previously described. However, only a small amount of shrinkage occurred in the high-amylose starch gels and they did not gain sufficient compressive strength to withstand the compressive forces of surface tension created during air-drying. Consequently, the gels were equilibrated in a pressurized autoclave containing a second solvent (liquid CO_2) of even lower surface tension. After several changes of liquid CO_2, the high amylose corn starch gels were air-dried by gradually evaporating the liquid CO_2 and storing the foams in a desiccator.

Foam Beads.
A second processing technique was used to make beaded foam from wheat starch with a processing time much shorter than that of the foam slabs. An aqueous suspension of wheat starch (8%, w/w) was gelatinized as previously described. The hot starch melt (90 °C) was injected under pressure (0.069 to 0.14 MPa) through nozzles with a 0.5 mm diameter into a stream of chilled (10 °C) vegetable cooking oil. The jet stream of starch melt quickly formed into small (0.25 mm to 1.0 mm) spherical beads at a distance of 6-8 mm from the nozzle. The beads were collected in a chilled oil stream and carried to a settling tank composed of two liquid phases approximately 15 cm each in depth. The upper phase consisted of chilled vegetable oil and the lower phase consisted of a mixture of ethanol in water (35% w/w). The beads were collected from the bottom of the settling tank and equilibrated 24 hrs each in two changes of two volumes of 100% ethanol. The beads were dried in air filtered through desiccant until the ethanol odor was no longer detected. The density, thermal conductivity and mechanical properties of the foams were determined as described earlier (*22*). All starch samples were equilibrated at 50% relative humidity.

Starch Plastics.
Starch foams and fiber/starch foams were prepared as previously described. Starch plastics and fiber/starch plastics were formed by pressing starch foams with approximately 69 MPa of force using an hydraulic press (*23*).

Characterization of Starch Products.

Physical and Mechanical Properties.
Thermal conductivity was measured at a mean temperature of 22.7 °C according to standard methods (ASTM C 177-85). Density was determined from measurements of sample weight and volume. The samples were tested in compression and tension using a universal testing machine (model 4500, Instron Corp., Canton, MA). Two methods, porosimetry analysis using nitrogen gas and Horvat-Kawazoe plots using argon, were used to determine pore size distribution in the submicron and Angstrom ranges, respectively. Pore size distribution measurements were performed by an independent laboratory (Porous Materials, Inc. Ithaca, NY).

The adsorbent properties of the foams were studied to determine their potential in encapsulating chemicals. Experiments were performed in which 0.5 g of starch foam beads were sealed in a flask with a 5 μl sample of volatile compounds having different chemical properties. The adsorption properties of the starch foams were compared to those of two commercial adsorbents; charcoal and Tenax. The amount of compound adsorbed was determined by the vapor pressure depression measured in the headspace of the sealed flask using gas chromatography.

X-ray Diffractometry.
X-ray diffractograms were obtained using a Phillips X'Pert MPD diffractometer employing Cu **K**α radiation. Diffractograms were obtained from raw wheat starch powder, aqueous gels (8%, w/w), gels that were oven dried (80 °C) and oven dried gels that had aged for four weeks at 50% relative humidity. Diffractograms were also obtained from starch foam slabs and starch plastics that were freshly made or aged for three years at room temperature and 50% relative humidity.

Results and Discussion

The densities of the foams made of wheat, corn and high amylose corn starch gels ranged from approximately 0.14 g/cm^3 to 0.37 g/cm^3 (Figure 1). The foams made of high amylose corn starch had the lowest density (0.14- 0.20 g/cm^3) while foams made of unmodified Dent corn starch had the highest range in density (0.25-0.37 g/cm^3). The range in density (0.24-0.30 g/cm^3) of wheat starch-based foams was intermediate. The density of wheat starch/fiber foams was nearly half the density of wheat starch foams containing no fiber (Table I). The lower density of the fiber/starch foam compared to the starch foam was attributable to the difference in the amount of shrinkage during the ethanol equilibration process. The fiber/starch gels exhibited only a small degree of shrinkage compared to the starch foams.

The thermal conductivity of the wheat starch and fiber/starch foams was similar to that of a commercial beaded polystyrene (PS) foam (Table I) even though the densities of the starch and fiber/starch foams were one order of magnitude higher than the density of the PS foam. The low thermal conductivity observed in the starch foams in spite of their relatively high density in comparison to PS foam may be explained by their microstructure. The PS foam had a closed cell structure with cells approximately 100 μm in diameter. The matrix of the starch and fiber/starch foams is composed of remnants of starch granules and many pores that are smaller than 2 μm (Figure 2). A small pore size in the foam matrix effectively reduces the amount of heat transfer by convection and conduction compared to larger foam pore sizes (*24*). The data support that relative to the PS foam, the effect of higher densities of the starch and fiber/starch foams on thermal conductivity was offset by a smaller pore size.

The mechanical properties of the fiber/wheat starch foams were more similar to the properties of PS foams than the wheat starch foams (Figure 3). The compressive strength and compressive modulus for the fiber/starch and PS foams were both approximately one order of magnitude smaller than that of the wheat starch foam containing no fiber (Table I). The wheat starch foams typically had a yield point in the range of 0.4 to 0.6 MPa. Beyond the yield point, the stress

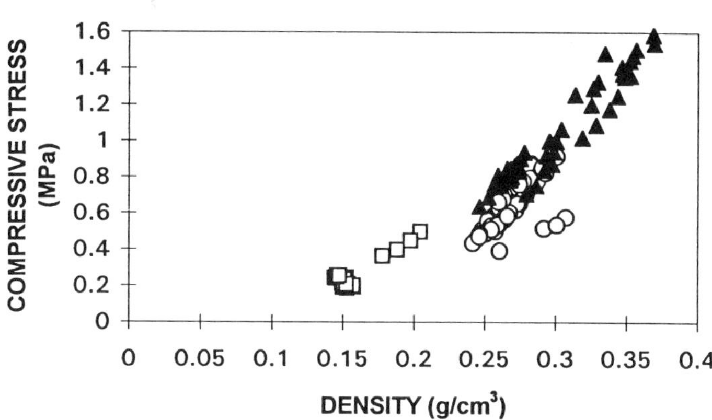

Figure 1. The compressive strength of starch foams at 10% deformation was related to foam density. Starch foams made of high amylose corn starch (□), had the lowest relative densities. The range in density of foams made of regular corn starch (▲) was greater than the density range for the wheat starch foams (O).

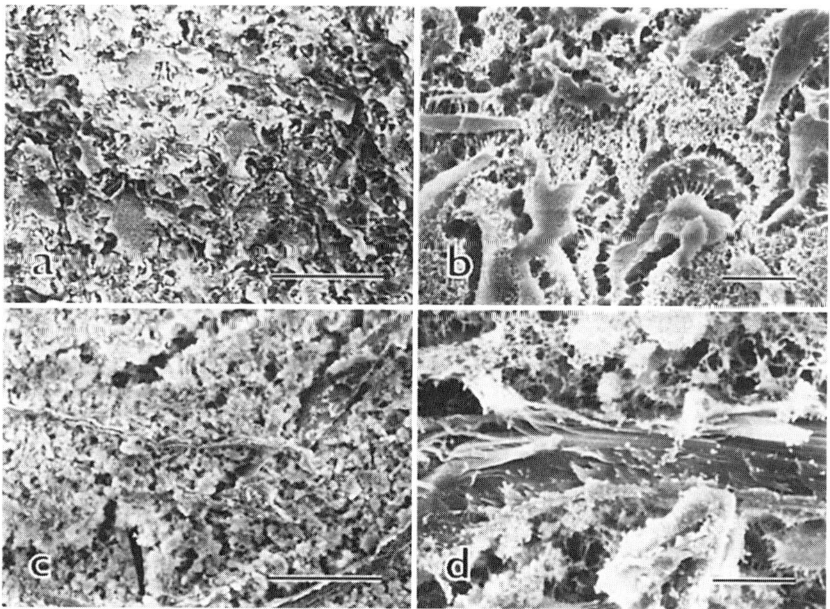

Figure 2. Scanning electron micrographs of wheat starch foams (a, b) and fiber/starch foams (c, d). The wheat starch foam (a, b) contained numerous starch granule remnants interspersed in a fibrous matrix. Most of the pore diameters were smaller than 2 μm. The fiber/starch foam (c, d) had a lower density and a more porous matrix. Notice the cellulose fiber embedded in the starch matrix (c. d). Magnification bars: a, c = 100 μm; b, d = 5 μm.

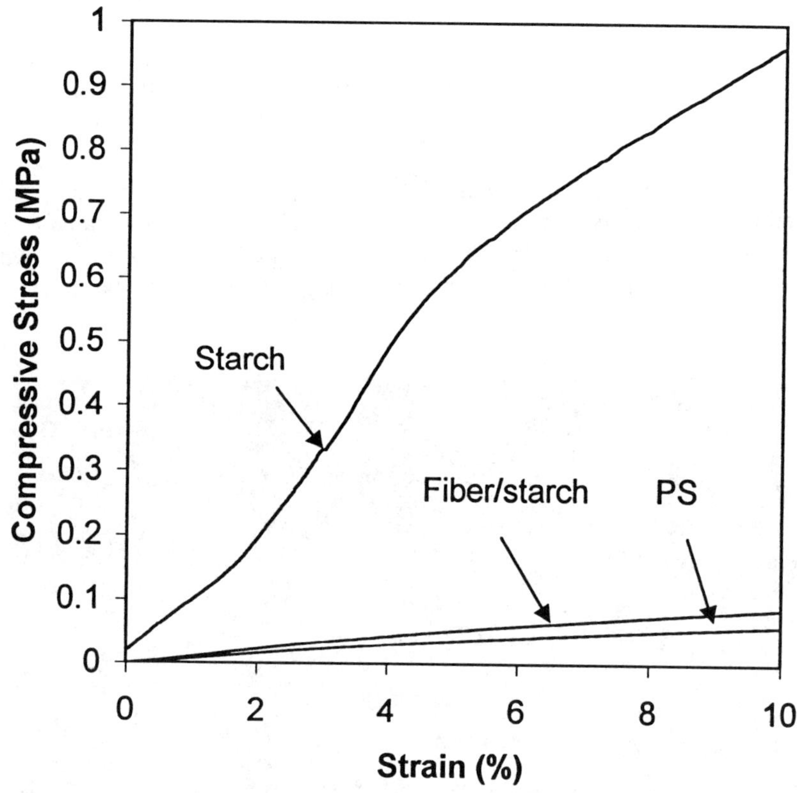

Figure 3. Compressive stress/strain curves for wheat starch, fiber/starch and polystyrene (PS) foams. The foams were tested to 10% strain. Note that the starch foams had a much higher modulus and compressive strength than either of the other two foams.

increased linearly with strain in the range tested. The stiffness of the starch foam was much higher than that of the fiber/starch and PS foams as indicated by the compressive moduli (Table I).

Table I. Density, compressive strength and modulus at 10% deformation, and thermal conductivity of wheat starch foams both with and without 33% (dry weight) softwood fiber. Data for polystyrene (PS) foam was included for comparison.

Sample	*Starch Foams*	*Fiber/Starch Foams*	*PS Foam*
Density (g/cm^3)	0.28a[a]	0.16b	0.019c
Compressive Strength (MPa)	0.64a	0.086b	0.098b
Compressive Modulus (MPa)	11.2a	1.3b	3.6b
Thermal Conductivity (W/mk)	0.04a	0.039a	0.036a

[a]Values within rows followed by a different letter are significantly different at the 95% confidence level.

Tensile tests of the wheat starch, fiber/starch and PS foams revealed that the wheat starch foam was much stronger in compression than tension (Compare Tables I & II). The tensile strength of the fiber/starch foams was similar in both compression and tension. The tensile strength of the PS foam was more than one order of magnitude greater in tension than compression and more than three times greater than that of the starch and fiber/starch foams (Table II). The comparatively high tensile strength of the PS foams is attributable to the structure of its matrix as well as the mechanical strength of PS polymer. Micrographs of the PS foams show a matrix of uniform cell size with thin sheet-like cell walls that form an interconnected matrix (*22*). In contrast, the matrix of the starch and fiber/starch foams was fibrous and less interconnected (Figure 2). It was not surprising to find that the fiber component did not markedly increase the tensile strength of the fiber/starch foams compared to the starch foams because the fibers appeared loosely dispersed throughout the matrix and because the fiber/starch foams were less dense. The fiber did, however, increase the elongation to break and toughness of the fiber/starch compared to the wheat starch foams without fiber (Table II). The PS foams had a higher elongation to break than either the starch or fiber/starch foams but a lower tensile modulus than the starch foams (Table II).

Table II. Tensile strength, modulus, elongation and toughness of wheat starch, fiber/wheat starch and polystyrene (PS) foams.

Sample	*Starch Foams*	*Fiber/Starch Foams*	*PS Foams*
Tensile Strength (MPa)	0.094a[a]	0.092a	0.32b
Tensile Modulus (MPa)	20.2a	6.73b	9.14b
Elongation (%)	0.77a	2.4b	4.8c
Toughness (J)	0.006a	0.028b	0.010a

[a]Values within rows followed by a different letter are significantly different at the 95% confidence level.

One concern with pregelatinized starch products is that the mechanical properties may change with age (*25*). Dried films of gelatinized starch become hard and brittle over time (*24*). Shogren (*26*) reported that starch embrittlement over time was due to a combination of water loss and free volume relaxation. Starch may slowly recrystallize if moisture levels are high enough to drop the glass transition temperature closer to the temperature of the starch sample (*27-28*).

X-ray diffractometry has been used to help characterize the crystalline properties of starch powders (*28*). X-ray diffractometry was used in the present study, to determine the relative changes in the crystallinity of wheat starch over time. The X-ray diffractogram for raw wheat starch (Figure 4) was typical for A-type starch (*29*). The diffractogram for aqueous starch gels consisted of a broad curve with no distinct peaks, indicative of a completely amorphous structure (Figure 4). The so-called "amorphous halo" or amorphous scattering occurred ~10-12 degrees higher compared to the curve for raw starch due to the high moisture content of the gel. The dried gel sample was nearly completely amorphous except for three small peaks (Figure 4).

During a four-week period in which the dried gel samples were stored at 50% relative humidity, the area of each of the three peaks increased (Figure 4). The data show that dried starch gels retrograded to some extent but the degree of crystallinity remained much lower than that of raw starch. In addition, the diffractogram of the dried gel was no longer an A-type pattern but was more similar to the B-type pattern found in potato and high amylose corn starches (*29-30*). The B-type structure is

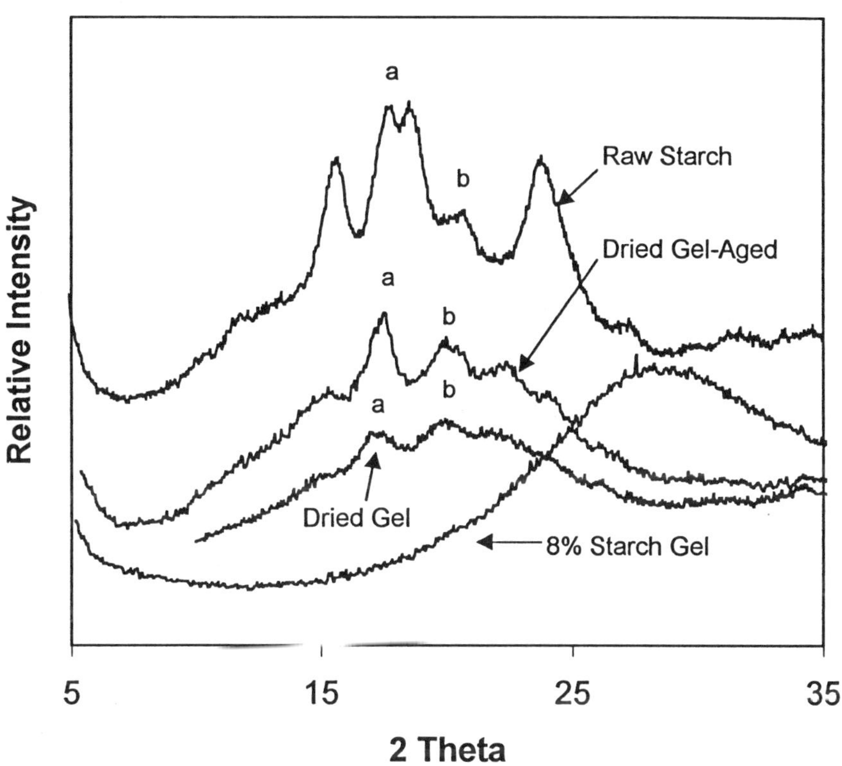

Figure 4. X-Ray diffractograms of raw wheat starch powder, an aqueous starch gel, dried gel, and a dried gel that had aged for 4 weeks at room temperature and 50% relative humidity. The dried gel and aged, dried gel had two peaks that occurred as similar angles (a=17.6, b=20.2) as two peaks found in the starch powder. The curves were purposely offset to facilitate visual comparisons.

described as a more loosely packed assembly of helices with more inherent water than the A-type (*31*).

The X-ray diffractograms of the starch foam slabs, which were prepared with ethanol, were compared to those of aged, dried gel that had not been exposed to ethanol (Figure 5). Although both the foams and the dried gel were made from starch gels, there was the possibility that the ethanol dehydration step used in the process for making the foams altered the diffraction pattern. Starches generally contain 1% lipid or less and these lipids can form complexes with amylose with a unique X-ray diffraction pattern (*32*). However, the results of this study show that the X-ray diffraction pattern obtained from non-aged foams was similar to the X-ray diffraction pattern obtained from dried gels that were not exposed to ethanol (Figure 5). The peak heights were slightly greater in the diffractogram of the starch foams that had been aged for three years (50% relative humidity, ~22 °C) but the change over time was only minor. There was no embrittlement or any other detectable change in the mechanical properties of the foams as they aged (data not shown). However, it should be noted that an increase in embrittlement would have been difficult to detect since the foams were brittle (low percent elongation to break) from the outset.

The potential of the starch-based foams for commercial products based on their mechanical properties was considered. The low thermal conductivity and high compressive strength of the starch foams make them attractive for several commercial products. A vapor barrier could be applied for applications requiring moisture resistance.

A second product of interest made from the starch foam or fiber/starch foam was a starch plastic (*23*). The tensile properties of the starch plastics were studied and compared to starch plastics that had been aged three years and a fiber/starch plastic (Table III). The tensile strength of the starch plastic was approximately one half that of the fiber/starch plastic (Figure 6). The tensile strength of the starch plastics did not change significantly during a 3-year storage period. The tensile modulus and toughness were higher for the fiber/starch plastic than the starch plastic (Table III). The tensile modulus of the starch plastic increased over time while toughness decreased. The elongation to break was nearly three times higher for the fiber/starch plastic compared to the starch plastic. The starch plastics embrittled slightly during storage as was evident by the lower elongation to break values (Table III).

A comparison of the mechanical properties of fiber/starch plastic and five commercial plastics revealed that the fiber/starch plastic had tensile strength in the range of cellulose acetate, high-density polyethylene and polypropylene (Table IV). Elongation to break, which was low for the fiber/starch plastic, was comparable only to polystyrene. Tensile modulus was comparable for the fiber/starch plastic and cellulose acetate (Table IV).

As mentioned earlier, embrittlement of pregelatinized starch products could be due to moisture loss, structural relaxation, or crystallization *(26)*. The X-ray diffraction pattern of newly formed starch plastics was similar to that of dried gels that had not been aged (compare Figures 4&7). However, the aged starch plastics

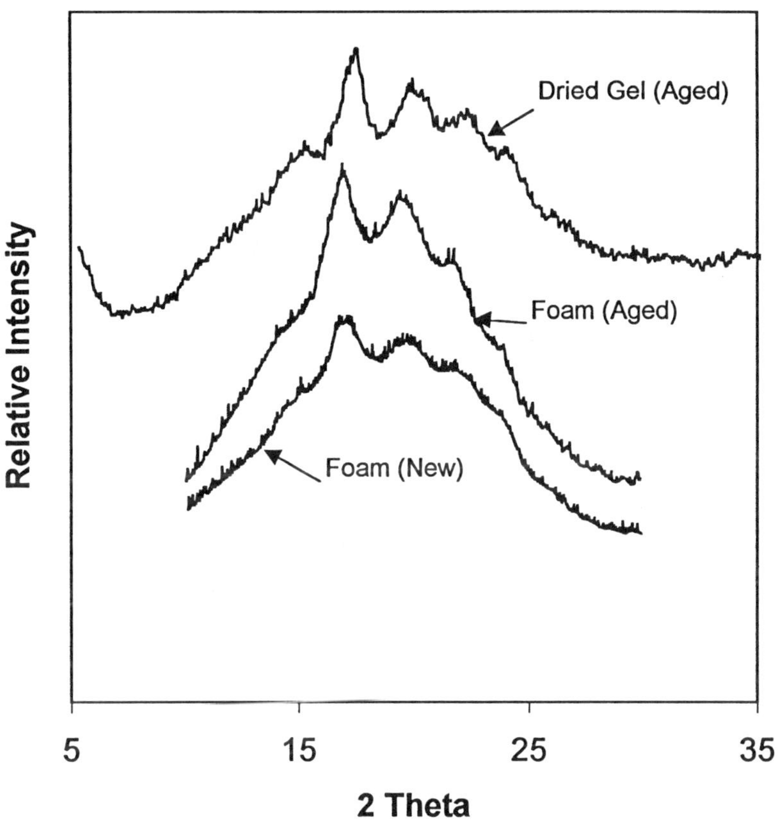

Figure 5. X-Ray diffractograms of a dried wheat starch gel and a new and aged (3 years) wheat starch foam. The curves were purposely offset to facilitate visual comparisons.

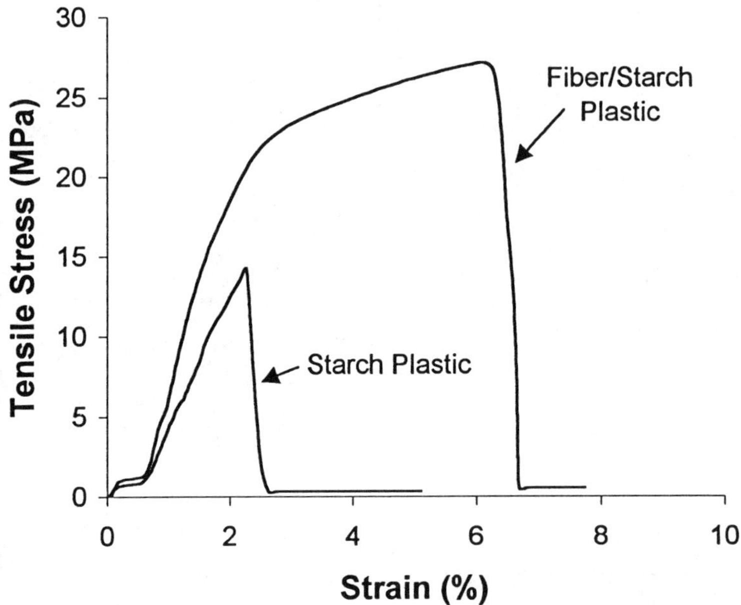

Figure 6. Typical tensile stress/strain curves for wheat starch plastics with and without softwood fiber reinforcement.

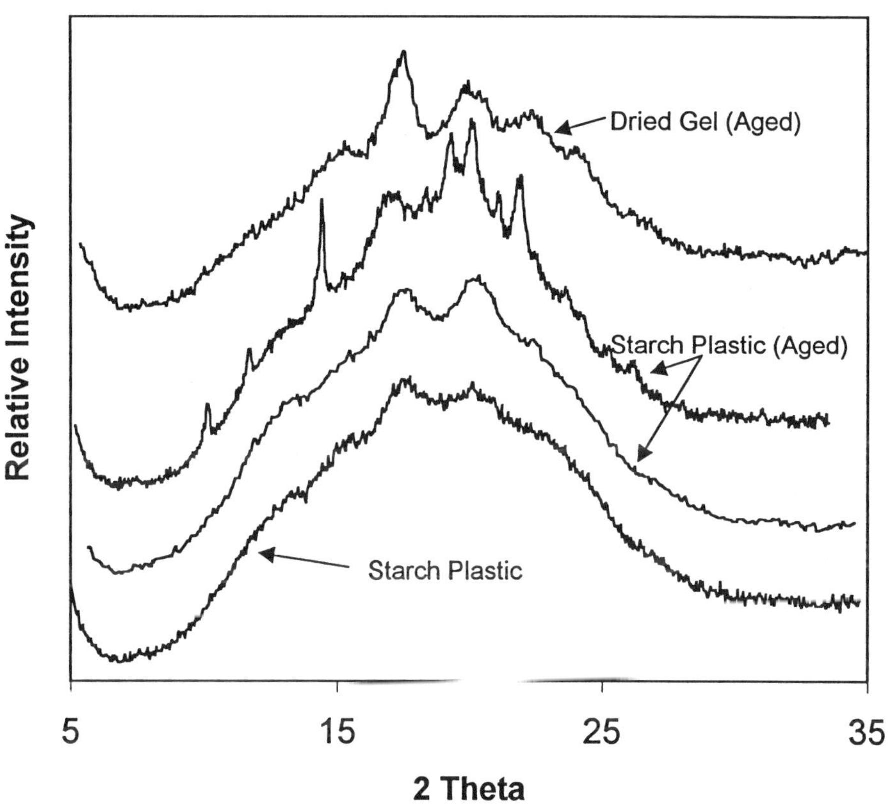

Figure 7. X-Ray diffractograms of a dried wheat starch gel, a new wheat starch plastic and two aged (3 years) wheat starch plastics. The aged starch plastics gave two distinct X-ray patterns. The curves were purposely offset to facilitate visual comparisons.

had a diffraction pattern that deviated from the typical B-type pattern observed for the starch foams and aged, dried gel (Figure 7). The diffraction pattern for the aged starch plastic was variable. One pattern observed was similar to a B-type pattern except that the peak near 20.1 degrees was larger than the peak at 17.5 degrees. A second pattern was observed for the aged starch plastic that had several new peaks not apparent in other starch diffractograms. The increase in the crystallinity of the starch plastics over time could partly account for the changes observed in tensile properties. More research is needed to determine why two different diffraction patterns developed and whether the crystalline regions observed in the starch plastics are unique.

Table III. Tensile strength, modulus, elongation to break and toughness of plastics made of wheat starch or wheat starch plus 33% softwood fiber. The samples consisted of a newly made starch plastic, a starch plastic that had been aged for 3 years at room temperature and 50% relative humidity, and a newly made starch plastic containing fiber.

Sample	*Tensile Strength (MPa)*	*Elongation To Break (%)*	*Tensile Modulus (MPa)*	*Toughness (MPa)*
Starch Plastic	12.4a[a]	1.6a	900a	0.36a
Starch Plastic (Aged)	11.6a	0.92b	1230b	0.16b
Fiber/Starch Plastic	27.5b	4.7b	1383c	2.3c

[a]Values within columns followed by a different letter are significantly different at the 95% confidence level.

The commercial potential of starch plastics is of wide interest. One application of particular interest is for microwavable food trays. There are concerns that plasticizers, unreacted monomers and other contaminants of plastics currently used for microwavable food trays can be absorbed by foods during baking creating a safety issue (*33-35*). The starch plastics were tested in microwave cooking studies using low moisture, high moisture, and oily foods (*23*). The starch plastics had some properties such as high initial tensile strength, good appearance and desirable food safety qualities. However, the functional properties of the starch plastics were inadequate for cooking applications. The starch plastic had a tendency to blister and deform during microwave baking. A starch copolymer with less hydrophilic properties would be better suited for this application.

Table IV. Typical values for tensile strength, modulus and elongation to break of fiber/starch plastics and various commercial plastics.

Sample	*Tensile Strength (MPa)*	*Elongation To Break (%)*	*Tensile Modulus (MPa)*
Fiber/Starch	27.5	4.7	1383
Cellulose Acetate	21	40	1300
Polyethylene - LD	8.3	550	180
Polyethylene -HD	29	60	830
Polystyrene	52	2.0	3450
Polypropylene	36	300	916

The commercial potential of the starch plastics is largely determined by the cost of the final product. There may be niche applications where the starch plastics could function well. However, unless the product costs are comparable with the cost of commercial plastics, the starch plastics will not succeed in the marketplace. The raw material costs for making starch foams is relatively small but the processing costs are considerable. A more cost efficient way of making the starch foams is by making small beads as described earlier. The beads can be made using a continuous process and have a much shorter processing time than the slab gels. The beads could function as loose-fill insulation or be molded into foamed articles.

One method used commercially for molding foamed beads into cups or bowls requires that the foam bead be expandable when it is heated. Polystyrene cups and bowls are made from pre-expanded polystyrene foam beads. The beads are pneumatically transported into a mold in which they are further expanded with heat. The heat also makes the beads fuse together as they expand and fill the void space in the mold. The starch foam beads have an appearance very similar to that of expanded polystyrene beads. However, when starch foam beads are heated in a cup molder they do not expand and fuse together as do the polystyrene foam beads. The reason for the failure of the starch beads to expand is readily apparent from SEM pictures of beads (Figure 2). The starch foams have an open cell structure with fibrous cell walls. In addition, the starch foam does not melt in the temperature range of PS. Consequently, the starch foam beads do not expand or fuse when heated in a cup molder. Further research into starch blends or starch copolymers could result in the production of foam beads that can be processed similar to PS foam beads.

The starch foam beads had other unique physical and mechanical properties that could be utilized for other applications. For instance, as mentioned earlier, the starch foams had a very fine microstructure. Scanning electron micrographs (Figure 2b) showed that the wheat starch foam had a high percentage of pores smaller than 1 μm with some as large as 2-3 μm. The lower range in pore size was too small to determine from SEM micrographs and required porosimetry analysis to determine the pore size range. The total pore volume of the foams was 3.23 ml/g. The median pore diameter was 0.616 μm and the mean pore diameter was 0.065 μm based on pore volume. Measurement of pore sizes in the 5-14 Angstrom range revealed that the foams had a cumulative pore volume of 0.004 ml/g attributable to pores of 5 Angstroms in diameter and a cumulative pore volume of 0.06 ml/g attributable to pores smaller 14 Angstroms or less in diameter.

The volume of extremely small pores in the starch foams seems to confer chemical adsorption properties that could have commercial interest. The results indicated that the adsorption of non-polar compounds such as decane was low compared to charcoal and Tenax (Table V). The adsorption of 2-octanone, a polar compound, was higher for the starch foam beads than for charcoal. The starch foam beads were also very effective in reducing the vapor pressure of alkylpyrazines such as 2-ethyl-3-methylpyrazine (Table V). There is also some evidence the foams may adsorb and help stabilize chemically unstable compounds.

Table V. Vapor pressure depression of three volatile chemicals (5 μl) sealed in a flask with 0.5 g of wheat starch foam, charcoal or Tenax.

Compound	*Starch*	*Charcoal*	*Tenax*
Decane	1.4	152	36
2-octanone	77	42	140
2-ethyl-3-methylpyrazine	150	–	–

The small pore size of the foams appears to be a critical factor in chemical adsorption since starch powder alone did not depress the vapor pressure. Popped corn, a starch foam with large pore size, was also ineffective in lowering the vapor pressure. The volatile compounds were released from the starch foam beads when the structure was disrupted by the addition of water. The results indicate that the starch beads could have commercial value as a chemical absorbent. Applications in the food industry may include using the starch foam beads, which are food grade, as a carrier of flavor compounds in dry food products that are reconstituted in water.

Conclusion

Starch can be processed into microcellular foams by air-drying starch gels that have been equilibrated in ethanol. The foams have low thermal conductivity, high compressive strength and have useful chemical absorbency properties. The foams can also be pressed into a plastic or a fiber/starch plastic if fiber is incorporated in the formulation. Small spheres made in a continuous process may minimize the fabrication cost of the foams and be commercially viable.

References

1. Whistler, R.L. In *Starch Chemistry and Technology*; Whistler, R.L.; Bemiller, J. N.; Paschall, E. F., Eds.; Academic Press: New York, NY., 1984, pp 1-9.
2. U.S. Congress, Office of Technology Assessment, *Biopolymers: Making Materials Nature's Way*, OTA-BP-E-102; U.S. Government Printing Office: Washington DC, 1993; pp19-50.
3. Kirby, K. W. In *Developments in Carbohydrate Chemistry*; Alexander, R. J.; Zoebel, H. F., Eds.; American Association of Cereal Chemists: St. Paul, MN, 1992, pp 371-386.
4. Otey, F.H.; Doane, W. M. In *Starch Chemistry and Technology*; Whistler, R.L.; Bemiller, J. N.; Paschall, E. F., Eds.; Academic Press: New York, NY., 1984, pp 389-414.
5. Koch, H.; Roper, H. *Starch* **1988**, *40*, 121-131.
6. Roper, H.; Koch, H. *Starch* **1990**, *42*, 123-130.
7. Bushuk, W. *Cereal Foods World*, **1986**, *31*, 218-226.
8. Griffin, G. J. L. In *Chemistry and Technology of Biodegradable Polymers*, Griffin, G. J. L., Ed.; Blackie Academic & Professional: New York, NY., 1994, pp 135-150.
9. Chinnaswamy, R. Hanna, M. A. *Journal of Food Science* **1988**, *53*(3), 834-836, 840.
10. Chinnaswamy, R. Hanna, M. A. *Cereal Chemistry* **1988**, *65*(2), 138-143.
11. Tiefenbacher, K. F. *J.M.S.-Pure Appl. Chem.* **1993**, *A30(9 & 10)*, 727-731.
12. Hilyard, N. C.; Young, J. In *Mechanics of Cellular Plastics*, Hilyard, N.C., Ed.; Macmillan Publishing Co.: New York, NY., 1982, 1-26.

13. Saunder, J. H. In *Handbook of Polymeric Foams and Foam Technology*; Klempner, D.; Frisch, K.C., Eds.; Oxford university Press: New York, NY., 1991, 1-25.
14. Williams, J. M.; Wrobleski, D. A. *J. Mater. Sci. Letters*, **1989**, *24*, 4062-4067.
15. Baldwin, D. F.; Park, C. B.; Suh, N. P. In:*Cellular and Microcellular Materials*; Kumar, V.; Seeler, K. A., Eds.; ASME: New York, NY, 1994, Vol. 53; pp 85-107.
16. LeMay, J. D.; Hopper, R. W.; Hrubesh, L. W.; Pekala, R. W. *ater. Res. Soc. Bull.* **1990**, *194*, 19-45.
17. Park, C. B.; Baldwin, D. F.; Suh, N. P. In:*Cellular and Microcellular Materials*; Kumar, V.; Seeler, K. A., Eds.; ASME: New York, NY, 1994, Vol. 53; pp 109-124.
18. Pekala, R. W.; Alviso, C. T.; Hulsey, S. S.; Kong, F. M. *Cellular Polymers*, **1992**, *38*, 129-135.
19. Martini, J. E.; Suh, N. P.; Waldman, F. A. *Microcellular Closed Cell Foams and Their Method of Manufacture*; US Patent, #4473665, 1984.
20. Martini, J. E.; Waldman, F. A.; Suh, N. P. *Soc. Plastics Eng. Technol.* **1982**, *20*, 1094-1097.
21. Ferdinand, J. M.; Lai-Fook, R. A.; Ollett, A. L.; Smith, A. C.; Clark, S. A. J. *Food Engin.* **1990**, *11*, 209-224.
22. Glenn, G. M.; Irving, D. W. *Cereal Chem.* **1995**, *72*, 155-161.
23. Glenn, G. M.; Hsu, J. *Industrial Crops and Products* **1997**, *7*, 37-44.
24. Buttner, D.; Hummer, E.; Fricke, J.; In:Aerogels; Fricke, J., Ed.; Springer-Verlag: New York, N.Y., 1985, Vol. 6; pp 116-120.
25. Zoebel, H. F. *Starch* **1988**, *40*, 44-47.
26. Shogren, R. L. *Carbohydrate Polymers* **1992**, *19*, 83-90.
27. Slade, L.; Levine, H. In *Food Structure-Its Creation and Evaluation*; Blanshard, J. M. V.; Mitchell, J. R., Eds.; Butterworths, London, pp 115.
28. Chinnaswamy, R.; Hanna, M. A.; Zobel, H. F. *Cereal Foods World* **1989**, 34, 415-422.
29. Zobel, H. F. *Starch*, **1988**, *40*, 1-7.
30. van Soest, J. J. G.; Hulleman, S. H. D.; de Wit, D.; Vliegenthart, J. F. G. *Industrial Crops and Products* **1996**, *5*, 11-22.
31. Imberty, A.; Buleon, A.; Tran, V.; Perez, S. *Starch*. **1991**, *43*, 375-384.
32. Kugimiya, M.; Donovan, J. W.; Wong, R. Y. *Starch* **1980**, *32*, 265-270.
33. Booker, J. L.; Friese, M. A. *Food Technology* **1989**, *43*, 110-117.
34. Castle, L.; Nichol, J.; Gilbert, J. *Food Addit. Contam.* **1992**, *9*, 315-330.
35. McNeal, T. P.; Hollifield, H. C. *J. Assoc. Off. Anal. Chem.* **1993**. *76*, 1268-1275.

Chapter 3

Cellulose Color Effects Copied from Nature with Natural Materials: Solid Opalescent Films Originated from Cellulose Derivatives

Georg Maxein, Manfred Müller, and Rudolf Zentel*

Department of Chemistry and Institute of Materials Science, Gauss Strasse 20, D–42097 Wuppertal, Germany

Solid opalescent films, which owe their color to Bragg reflection of visible light, can be prepared from cholesteric cellulose derivatives. Both thermotropic and lyotropic systems can be used. They are accessible from commercial products by simple reactions and a subsequent photo polymerization (crosslinking). We found cellulose carbanilates and hydroxypropylcellulose esters most promising. By careful selection of the substitutents, the degree of substitution and the molecular weight, systems with brilliant reflection colors are available.

The cholesteric phase (chiral nematic phase) of liquid crystals shows selective reflection of light, if the pitch of the cholesteric helix coincides with the wavelength of light within the material ($\lambda = n * p$) (see figure 1). Since the reflection conditions vary with the angle between the cholesteric helix and the incident light, different reflection colors are seen depending on the observation angle. Recently highly crosslinked cholesteric pigments have found a lot of interest as dye pigments for cars or as "copy safe" colors for documents or money *(1)*. These cholesteric pigments have so far been prepared from cholesteric monomers or oligomers by a photocrosslinking process *(2,3)*.

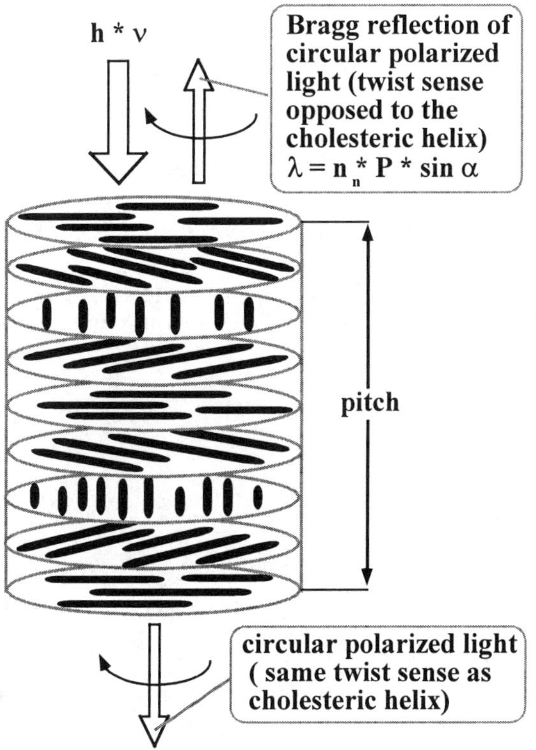

Figure 1: Schematic representation of the cholesteric phase and it's optical properties. . (λ: wavelength of the reflected light; n_n: average refractive indicex of a nematic layer; P: pitch; α angle of reflection)

Helical biopolymers offer the same potential and a great deal is known about thermotropic or lyotropic phases of polypeptides or cellulose derivatives. In this case crosslinking is possible either by crosslinking a suitable thermotropic system *(4)* or by the crosslinking of vinyl monomers used as solvents for a lyotropic phase, thus producing a semi-interpenetrating network. The last example has been successfully demonstrated for polyglutamates *(5)*.

Among cellulose derivatives many chiral nematic phases are known, both lyotropic *(6-8)* and thermotropic *(9-11)*. The mesogens of these liquid crystals are usually commercial products or accessible by simple reactions. If the conditions are right, a great number of these liquid crystals exhibit selective reflection between 400 and 800 nm. In addition, only moderately polar solvents such as ketones, glycol ethers, and glycol acetates are necessary in order to obtain colored mesophases from the urethanes of cellulose *(12,13)* (many technologically used derivatives of acrylic and methacrylic acid are of similar moderate polarity and of similar structure).

In this work we describe both lyotropic and thermotropic systems based on cellulose derivatives. These systems can be prepared from commercial products by simple reactions. They show excellent optical properties and can be photo patterned. Lyotropic mesophases were prepared from aryl urethanes of cellulose in commercially available mono- or bifunctional derivatives of acrylic and methacrylic acids (see Scheme 1). To obtain solid films and to conserve the selective reflection the solvents were polymerized photochemically, thus yielding a semi-interpenetrating network of cellulose urethanes in polyacrylates. Thermotropic systems were prepared from esters of hydroxypropylcellulose (HPC) with propionic acid and acrylic acid (see Scheme 2).

Lyotropic Cholesteric System

In the search for lyotropic cholesteric systems (described in detail in ref. 14) cellulose carbanilates in mono- and bisacrylates of oligoethylen glycols gave the best results (see Scheme 1). Lyotropic mesophases formed by these compounds orient well and show a sharp absorbance of light due to selective reflection (see figure 3). Photo polymerization transforms this lyotropic cholesteric phase into a solid film. However, the sample remains clear and no signs of phase separation are seen. The selective reflection (absorption) after crosslinking is as sharp as before, but it is shifted to a shorter wavelength, presumably because of a volume shrinkage during polymerization.

A crucial point in the preparation of cholesteric films is the orientation of the cholesteric phase in the planar texture (helical axis perpendicular to the film surface). Such an orientation takes a very long time, if the DP (degree of polymerization) of the cellulose derivatives is high. A misalignment leads to broad reflection bands and a great deal of scattering in the sample. In order to overcome these problems cellulose with a DP of 100 or 50 was used, which is accessible by saponification of cellulose acetate or propoinate. It was later transformed into the cellulose carbanilates under investigation (see Scheme 1).

Especially cellulose carbanilates prepared from 3-trifluoromethyl-phenyliso-cyanate proved to be advantages, as their viscosity in the about 45 weight % solutions, necessary to prepare a cholesteric phase with selective reflection in the visible range, is rather low. Nevertheless, orientation times of half an hour are necessary to obtain a perfect orientation of the cholesteric phase and narrow reflection bands. To reduce the viscosity and the polydispersity of the cholesteric solutions further, a sole gel fractionation of the cellulose carbanilates (system acetone/ water) was performed. It reduced the polydispersity to 1.5 and produced samples ranging in molecular weight

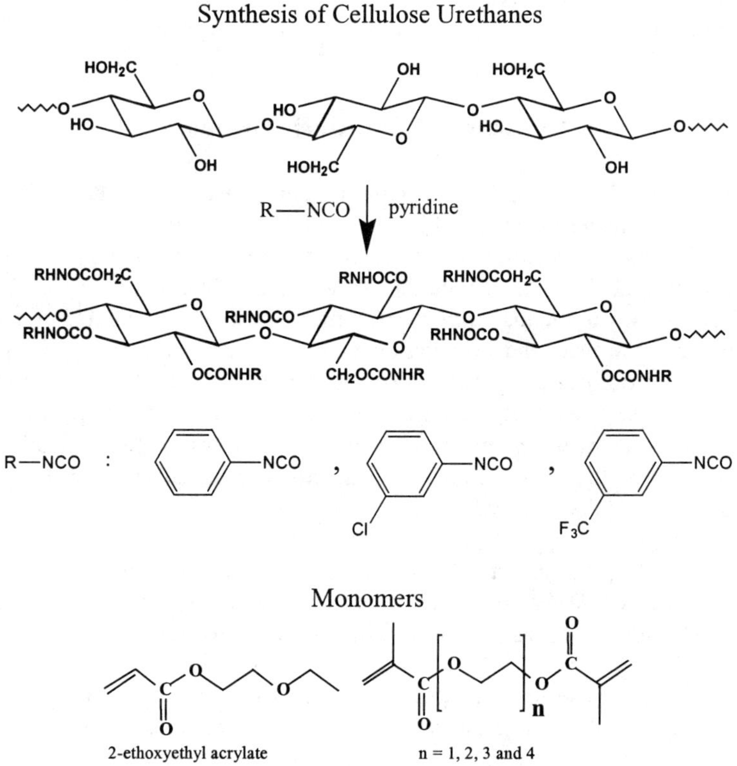

Synthesis of Cellulose Urethanes

R—NCO pyridine

R—NCO :

Monomers

2-ethoxyethyl acrylate

n = 1, 2, 3 and 4

Scheme 1

(M_{PS}: peak maxima, GPC in THF against polystyrene) from M_{PS} = 21,000 to 134,000 (see figure 2). All samples gave lyotropic cholesteric samples, which showed the reflection wavelengths displayed in figure 2 for crosslinked samples. Two results are noteworthy: 1.) A decrease of the molecular weight leads to a blue shift (shrinkage of the cholesteric helix) of the reflected light: 2.) Even the sample with the lowest molecular weight (M_{PS} = 21,000) produces a 100% cholesteric solution. This behavior is different from that of cellulose carbanilates samples, for which the low molecular weights had been prepared by acidic hydrolysis of high molar mass polymers. For these systems with a broad polydispersity (about 2.7) the samples with a molecular weight below M_{PS} = 50,000 do - no longer - give rise to lyotropic cholesteric phases.

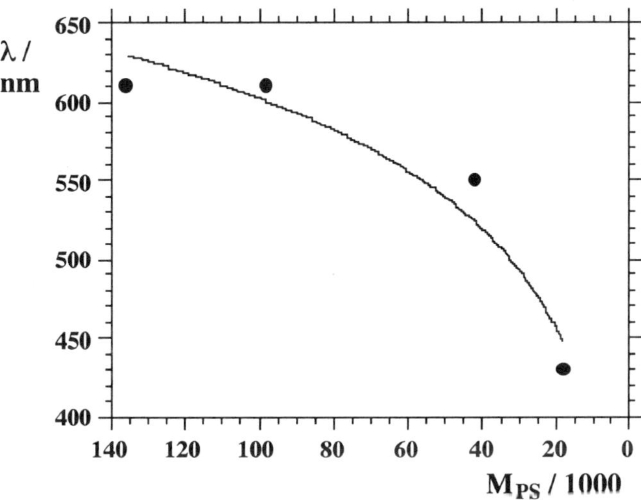

Figure 2: Influence of the molecular weight on the maxima of the selective reflections. The cholesteric mixtures contained 45 wt. % of trifluormethyl phenyl urethanes in diethylene glycol dimethacrylate. The measurements were carried out at room temperature after cross linking.

Lyotropic solutions prepared from the low molar mass system obtained by fractionation (M_{PS} = 21,000) orient perfectly within a few minutes. By changing the temperature it is - in addition - possible to observe selective reflection throughout the whole visible range. Figure 3 shows some films - sometimes as big as 100 cm^2 - obtained by crosslinking these mixtures at temperatures ranging from 24 to 53^{0}C. By performing the photochemical crosslinking at various temperatures it is possible to pattern films as described in ref. 14.

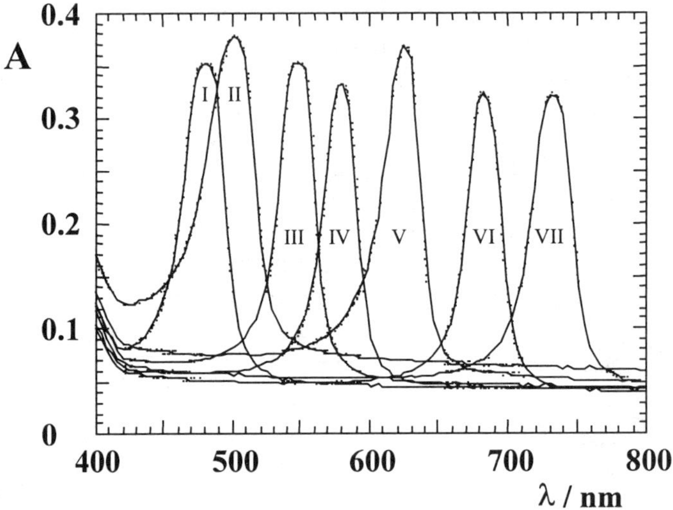

Figure 3: Influence of the temperature on the maxima of the selective reflections. The cholesteric mixtures contained 45 wt. % of trifluormethyl phenyl urethanes in diethylene glycol dimethacrylate. The cholesteric mixture was irradiated at: 24^0C (I), 27^0C (II), 31^0C (III), 38^0C (IV), 41^0C (V), 46^0C (VI) and 53^0C (VII). The Spectra were recorded at room temperature.

Thermotropic Cholesteric Systems

Based on the well-known liquid crystalline properties of 2-Hydroxypropyl cellulose (HPC) derivatives, a thermotropic system could be realized as described in detail in ref. 15. There are two major advantages with this concept: a) cellulose represents a renewable raw material with almost unlimited accessibility b) the resulting polymeric liquid crystal is a single component cholesteric system, containing no evaporatable or toxic low molecular weight compounds.

The "2-step-single-pot" polymer analogous reaction of HPC with acrylic and propionic acid chloride in acetone yields statistically substituted HPC mixed esters, which represent highly crosslinkable thermotropic liquid crystalline macromers. The naturally chiral cellulose backbone induces the desired selective reflecting cholesteric structure that can be permanently frozen in by curing films of the material with a photo initiator and UV-light (see Scheme 2).

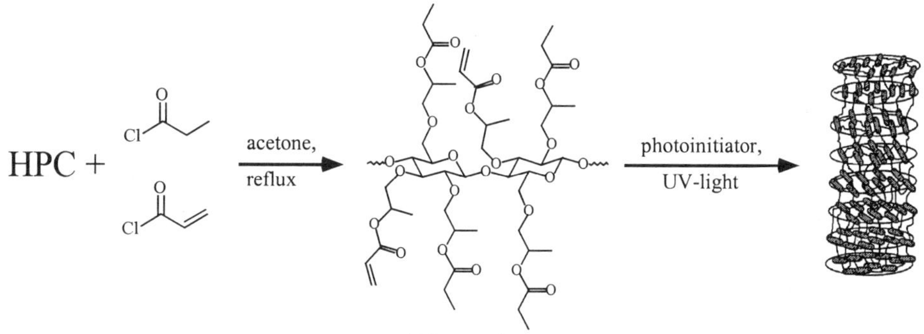

HPC +

acetone,
reflux

photoinitiator,
UV-light

Scheme 2

UV/VIS spectra on thin uncrosslinked films of these derivatives show that the selective reflection wavelength and thus the cholesteric pitch (see Figure 2), increases linearly with temperature. This is analogous to the lyotropic systems. The reflection wavelength is -however- more strongly determined by polymer specific parameters. So it was found that a smaller degree of substitution (DS) as well as a smaller degree of polymerization (DP) of the HPC result in a significant "red-shift" of the reflection wavelength (see Figures 4 and 5).

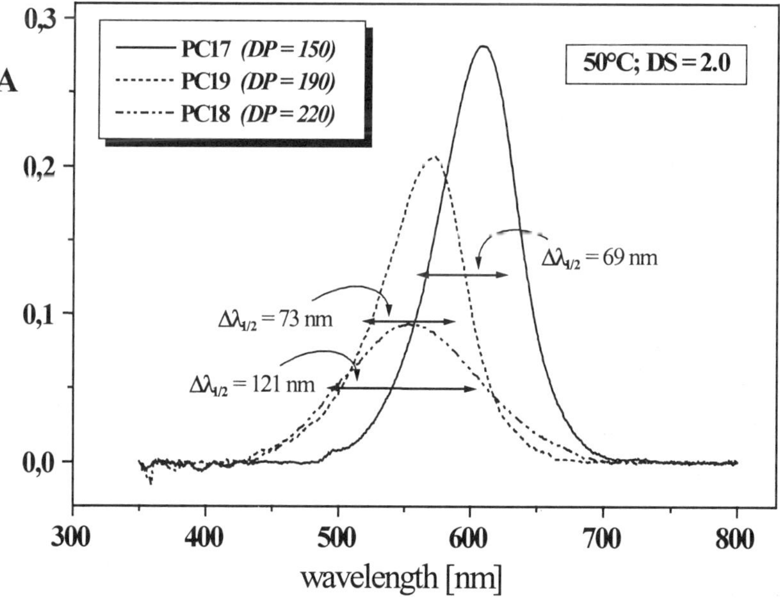

Figure 4: Dependence of the selective reflection λ_0 on the degree of polymerization DP: A smaller DP leads to a more intensive reflection and a smaller half width of the reflection peak.

Figure 4 shows the high sensibility of the cholesteric phase towards small changes in the DS, the increase by 0.5 in DS shifts the samples reflection color from red to blue (see Figure 4 at 70^{0}C). Furthermore, a smaller DP exhibits more intense and more brilliant reflection colors (high and narrow peaks) (see Figure 5). This is most probably due to a more perfect alignment of the cholesteric phase during annealing. For bulky substituents one finds red-shifted, but flat and broadened, reflection signals.

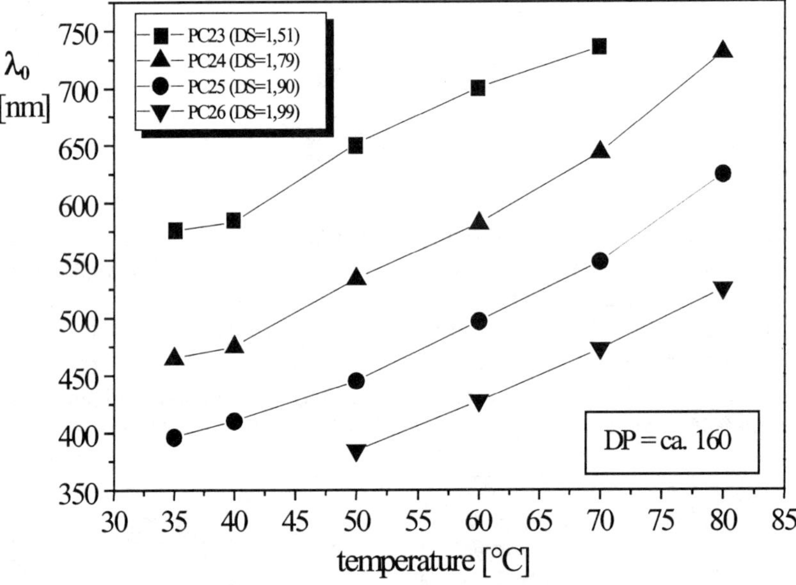

Figure 5: Influence of the degree of substitution DS on λ_0: Significant "blue-shift" for the higher substituted HPC-derivatives.

Whereas the DP is more or less controlled by the initial HPC, the DS can be controlled by the reaction conditions and time respectively. Finally only the use of HPC with a low degree of polymerization in combination with a mere partial substitution of the HPC-OH - groups in a esterification reaction free of a proton trap (in situ: acid catalyzed depolymerization possible) yields cholesteric phases, which orient well to give brilliant reflection colors.

Due to the temperature dependence of the cholesteric phase an exact adjustment of the color effect prior to the crosslinking process is possible. This is illustrated in figure 6. Here the circles represent the linear relation between temperature and reflection color for an uncrosslinked film (photo initiator already added. Photo crosslinking at 60^{0}C or 90^{0}C respectively yields in permanently locked in cholesteric phases (filled squares) that reflect blue or red light respectively. As can be seen from figure 6, the crosslinking process causes almost no change in the cholesteric phase. The previously adjusted reflection color only suffers a small blue shift (5 nm) due to volume shrinkage. This effect is much smaller than in the lyotropic systems. The

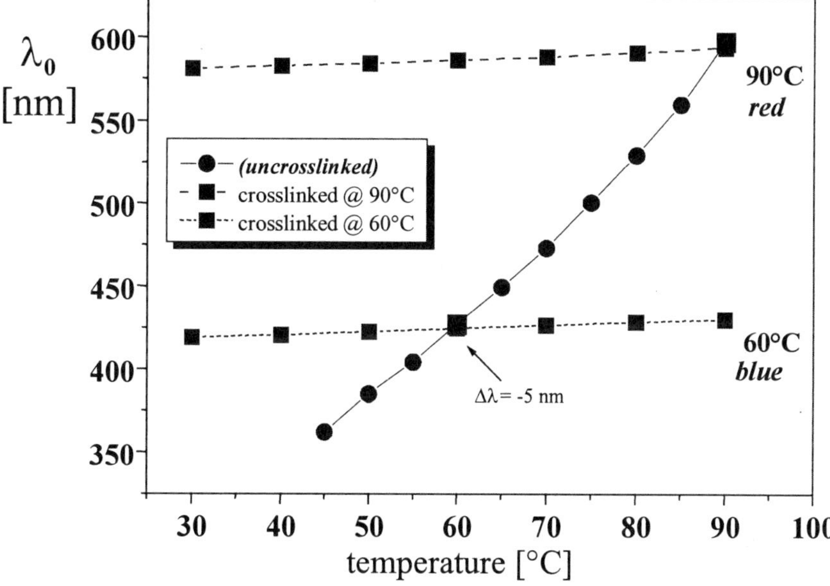

*Figure 6: Fixation of the cholesteric phase by crosslinking: * represents the thermotropic phase that allows precise color adjustment.*

resulting solid phases exhibit a excellent thermal stability of their reflection colors over the whole temperature range (filled squares in figure 6).

The original aim for the development of cholesteric systems is their angle dependent selective reflection. This effect makes a coated surface change their color appearance to a spectator, who varies his viewing angle towards the surface. Angle dependent UV/VIS-spectroscopy proves this desired optical effect (see figure 7). The samples color changes from almost red for the nearly orthogonal view ($\beta = 10^0$) on the surface to blue for a flat angle ($\beta = 70^0$).

In conclusion we have shown that also the cellulose based systems are suitable for achieving the desired optical properties of crosslinked cholesteric films.

Acknowledgment

Financial support from the DFG (Schwerpunkt : Cellulose) is highly appreciated.

References

[1] *Spiegel* 1995, 49, 256.
[2] Daimler-Benz AG, Wacker-Chemie GmbH, DE 44 18 076 A1, 1995
[3] BASF-AG, DE 4342280 A1, 1995.

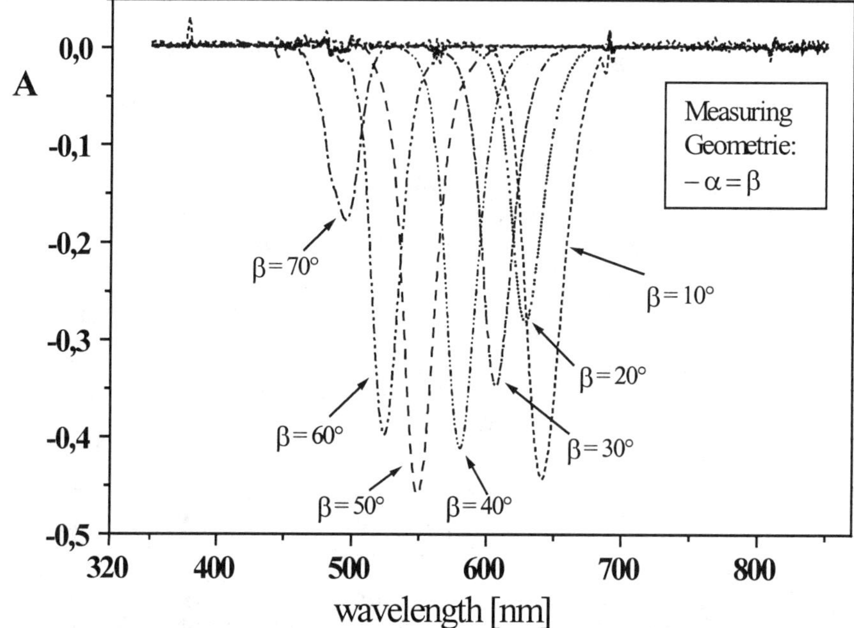

Fig. 7: Angular dependence of the reflections; The "negative absorption" correspond to the intensity of the reflected light. The measurement was performed in a UV-spectrometer versus a black reference.

[4] Bhadani, S. N.; Gray, D. C., *Mol. Cryst., Liq. Cryst.* **1984**, 102(Letters), 255.

[5] Tsutsui, T.; Tanaka, R., *Polymer* **1980**, 21, 1351.

[6] Gray, D. G.; *Appl. Polym. Symp.* **1983**, 37, 179.

[7] Nishio, Y.; Yamane, T.; Takahashi, T., *J.Polym. Sci., Polym. Phys. Ed.* **1985**, 23, 1043.

[8] Zhao, C.; Cai, B., *Makromol. Chem., Rapid Commun.* **1995**, 16, 323.

[9] Bhadani, S. N.; Gray, D. G., *Macromol. Chem. Rapid Commun.* **1982**, 3, 449.

[10] Tseng, S.; Laivins, G. V.; Gray, D. G., *Macromolecules* **1982**, 15, 1262.

[11] Laivins, G. V.; Gray, D. G., *Polym. Papers* **1985**, 26, 1435

[12] Siekmeyer, M.; Zugenmayer, P., *Makromol. Chem.* **1990**, 191, 1177

[13] Zugenmeier, P. *Cellulose Polymers Blends and Composites*; Gilbert, R. D., Ed.; Carl Hanser Verlag: Munich, Germany, 1994; p.71.

[14] Müller, M.; Zentel, R.; Keller, M., *Adv. Materials* **1997**, 9, 159.

[15] Maxein, G.; Ph. D. thesis; University Mainz: Mainz, Germany, 1998.

Surface Active Polymers and Solution Properties

Chapter 4

Solution Properties of Some Amphiphilic Polysaccharide Derivatives

J. Desbrieres and M. Rinaudo

Centre de Recherches sur les Macromolécules Végétales, CERMAV-CNRS, affiliated with the University Joseph Fourier, BP 53, 38041 Grenoble Cedex 9, France

Amphiphilic polymers give very interesting properties in aqueous solution; some of them give a large increase of the viscosity in semi-dilute regime or some form a noncovalent gel. The hydrophilic-hydrophobic balance plays a large role in the interchain interactions, depending on charge density of the polymer (electrostatic character), the temperature, the nature of hydrophobic substitution or external salt concentrations.

In our work we have investigated two systems :

* methylcelluloses which form gel in given thermodynamic conditions when highly substituted hydrophobic zones are present in the molecules (block-like copolymers),

* alkylchitosans obtained by grafting alkyl chains on a poly-D-glucosamine backbone.

The relation between the chemical structure of the modified polysaccharides and the physical properties of the solution will be discussed.

Introduction

For many years, our work has concerned the extension of the knowledge of natural polymers and especially polysaccharides. Our goal is to develop new ways to extend the use of these important sources of polymers such as starch, cellulose or chitin. Due to the difficulty to process these polymers in their native form, it is useful to perform chemical or enzymic modifications to open new applications. Specifically in the field of water soluble polymers, the cellulose derivatives found a lot of developments. For this purpose we have proposed to perform homogeneous modifications of these polysaccharides such as to be able to relate the chemical modifications with the physical properties in solution or solid state (1). One of the chemical modifications we have recently investigated is the introduction of hydrophobic character in these usually hydrophilic polysaccharides.

Hydrophobic associating water soluble polymers represent a new class of industrially important macromolecules. They possess unusual rheological characteristics which are thought to arise from the intermolecular association of neighboring hydrophobic substituants (2) which are incorporated into the polymer molecule through chemical grafting (3,4) or suitable copolymerisation procedures (5). The hydrophobic associations give rise to a three-dimensional polymer network.

For synthetic polymers, the large variety of behaviours observed are now qualitatively understood in terms of nature, number and length of the hydrophobic substituents and also in terms of distribution along the backbone. In contrast natural polysaccharides suffer from a lack of fundamental studies. The difficulty arises mainly in the absence of model polymers with a precise distribution of substituents. The hydrophilic-hydrophobic balance plays a large role in the interchain interactions, depending on charge density of the polymer (electrostatic character) or the solubility for the neutral polymers, the temperature, the nature of hydrophobic substitution or external salt concentrations.

In this work we have investigated two types of polymers, one being neutral and block-like copolymer (methylcellulose) and the other one ionic and having grafted hydrophobic chains (alkylchitosan). The hydrophilic-hydrophobic balance may be adjusted under controlled chemical modifications using numerous parameters such as the charge density, the length and the number of hydrophobic groups. The related properties will be discussed.

Experimental

Products

Commercial samples of methylcelluloses were kindly supplied by Dow Chemical company under the trade name Methocel A4C. Laboratory made methylcellulose samples (M12, M18) were prepared according to an original procedure (6) based on a previously proposed methylation method (7). The cellulose was dissolved in DMAc (dimethylacetamide) / 6 wt% LiCl from a swelling procedure followed by solvent exchange. Then the dimsyl solution (NaH – DMSO) was added to the cellulose solution for activation of the reactive sites of the cellulose and finally iodomethane was added to the mixture, stirred at room temperature for different times to reach different substitution degrees (8).

The original chitosan samples were from Aber Technologies (Plouguerneau, France) and Protan (Norway). The alkylated derivatives were obtained by reductive amination following the procedure described by Yalpani (9). It is a versatile and specific method for creating a covalent bond between a substrate and the amine function of the chitosan. They are called CCx, x being the number of C atoms of the hydrophobic chain. The degree of substitution is defined as τ.

In any case we have developed procedures for performing chemical reactions in homogeneous phase. This allows a better accessibility of the reactive sites to the reagent and hence a regular repartition of substituents compared with the commercial samples which have a heterogeneous repartition of substituents leading to block-like copolymers.

The physicochemical characteristics of the samples are given in table I.

Methods

The rheological measurements were performed with a Couette type rheometer (Contraves Low Shear 40) or a stress-controlled rheometer (CarriMed CS50) according to the concentration and the temperature of the sample.

The calorimetric experiments were carried out with a Micro DSC III calorimeter from Setaram (France). The temperature rate was 0.5 deg/min.

The fluorescence spectra were obtained using a LS50B luminescence spectrometer from Perkin-Elmer. The experiments were carried out in the temperature and concentration domains where no turbidity was observed. The pyrene concentration was 10^{-7} M due to its low solubility in water and the excitation wavelength was 334 nm. The studied parameter is the I_1/I_3 ratio of the intensities of first and third peaks of fluorescence spectrum of pyrene in chitosan derivative solutions.

Gelation of methylcelluloses

Commercial methylcellulose (MC) is a heterogeneous polymer consisting of highly substituted zones called "hydrophobic zones" and less substituted ones called "hydrophilic zones" (10) as block-like copolymers. It has the ability to form a gel on heating which melts again on cooling. Due to these unusual properties, most of the experimental work reported in the literature was dedicated to the evolution of the viscosity and turbidity of MC solutions during heating and cooling cycles in a small polymer concentration range (10-25 g/L) (11-14). The behaviour of this polymer is not monotonous when the temperature varies : the viscosity of a semi-dilute solution decreases when temperature is increased up to a critical value over which the viscosity increases. Then, the formation of a gel may be observed and this

Table Ia : Characterization of methylcelluloses

	A4C	M12	M18
DS[a]	1.7	1.5	2.2
% Non S	10	5	9
% MonoS	29	51	12
% DiS	39	29	36
% TriS	22	15	43
M_w	149,000		

[a] Determination by ^{13}C n.m.r. in DMSO-d_6 (353K)
DS is the average degree of substitution per glucose residue

Table Ib : Characterization of chitosan and alkyl derivatives

	Chitosan	CC12	CC8	CC10
DA	12	12	2	2
M_v	190,000			
τ		0.04	0.12	0.12
[η] (mL/g)	765**	1990**	1250*	1200*
k_H	0.35	1.52	1.50	1.92

DA : acetylation degree
τ : degree of substitution
* : solvent AcOH 0.3M / AcOH 0.05M, T=20^0C
** : chlorhydrate form in water, T=5^0C

phenomenon is associated with a turbidity, indicating phase separation (11). All these phenomena are reversible.

We have studied the evolution of MC solutions with the temperature using different techniques such as n.m.r., rheometry (15), calorimetry (8) and fluorescence spectroscopy (16). In-depth studies were carried out on the commercial sample A4C. The phase diagram of A4C sample is very complex (figure 1). A cloud point is observed indicating that the A4C aqueous solution presents a LCST-type phase separation and this curve can be considered as the binodal curve with a minimum corresponding to the critical point (17). Below this curve a clear homogeneous phase was present, while above, the phase separation was incomplete due the superposition of a gelation process. The gelation was all the more slowed down because the gap to the binodal curve was small. It was demonstrated that there is a competition between the phase separation leading to polymer-rich regions, and the gelation which prevents the mobility of chains, hence the growth of dense zones. Moreover, the homogeneous domain appeared also to be very complex. It is constituted by two gel phases separated by a sol phase. In the dilute regime there was no gel formation, or for very high temperatures there was direct phase separation. For low concentration solutions ($c<20$ g/L) rheological experiments were carried out on MC solutions at equilibrium and within the linear regime in order to determine the sol-gel transition temperature. For a given concentration it was found (17) that the sol-gel transition observed as temperature increased was well described by a percolation process. The gel I - sol transition would be due to the appearance of the heterogeneities which, generating high local concentrations, lead the system below the percolation threshold of the former aggregates. The transition was completely reversible in temperature as well as in concentration at a fixed temperature.

To obtain more information these rheological experiments were completed by oscillatory tests at a fixed frequency (1 Hz) during a temperature sweep. For the A4C sample (figure 2), for a concentration larger than 2.5 g/L the evolution of the storage modulus G' showed two distinct waves. At low temperature the solution was clear, and in the range 30-50°C a weak gel appeared. In this temperature range, the solution became progressively turbid (depending upon the polymer concentration). Then, an elastic, turbid gel with a large increase of viscosity was observed over 60°C within the experiment time scale. The second wave, at high temperature, was more pronounced with the highest polymer concentration solutions. The 2.5 g/L concentration (which corresponds to $c[\eta] = 1.27$ at 25°C) may be considered for the A4C sample as the critical polymer concentration needed for gelation. The gelation phenomenon was time dependent (15) and reversible. After gelling upon heating, aqueous methylcellulose solutions were recovered upon subsequent cooling and the values of the moduli were the same at 20°C before and after the thermal treatment.

To determine the role of the structure (i.e. the repartition of the substituents along the chain) we have prepared MC samples using a homogeneous process to allow a better accessibility of the hydroxyl groups to the reagent and hence a statistical distribution of the substituents along the backbone (18). Using the same rheological procedure we have studied the behaviour of homogeneously prepared MC samples. With the samples of low degree of substitution (DS<1.5) only one wave was

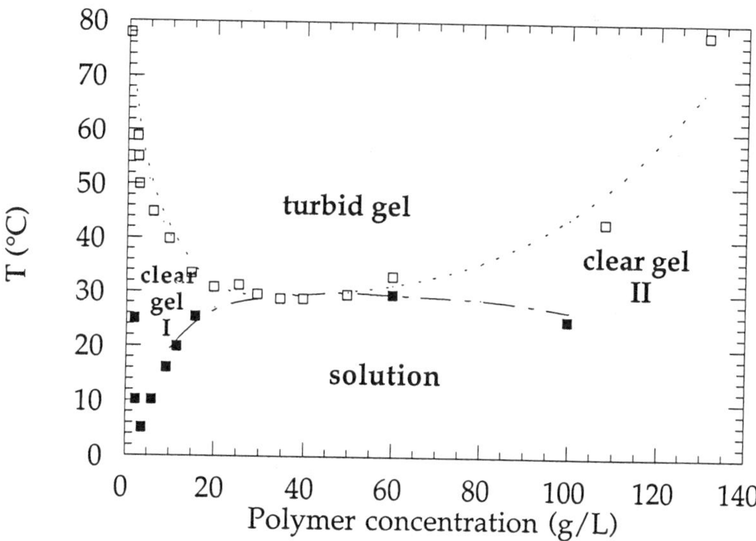

Figure 1. Phase diagram of an aqueous A4C methylcellulose (DS=1.7) solution including the cloud point curve (□) and the sol-gel line (■) determined from oscillatory experiments

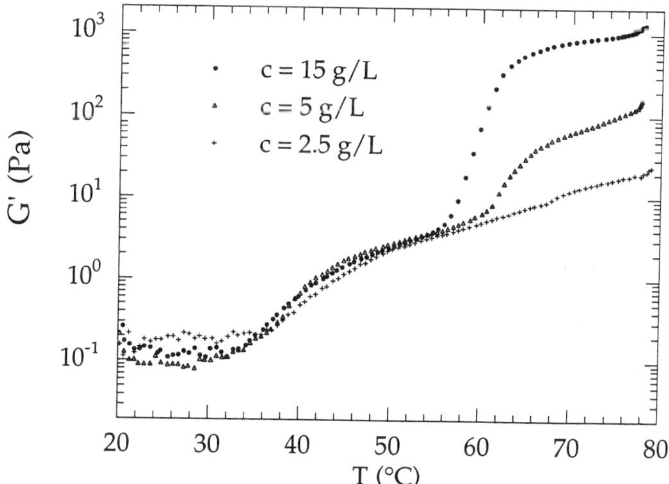

Figure 2. Influence of polymer concentration and the temperature on the elastic modulus of A4C solutions (G' modulus at 1 Hz, DS=1.7, solvent : water)

observed, with a slight increase of G' with temperature. Furthermore the temperature for which G' is equal to G" is shifted towards higher temperatures with a temperature displacement of around 15°C. We had the same observation for a homogeneously prepared MC sample with a similar degree of substitution than A4C sample (figure 3). But for the sample with the highest DS value (M18, DS=2.2) a similar behaviour to that of A4C was observed with the presence of two waves on the evolution of G' in the studied temperature domain. With this degree of substitution, the proportion of trisubstituted units was relatively high and the presence of blocks of such units may be suspected. Same conclusions would be made from calorimetric experiments for which a similar temperature displacement is observed at equivalent DS, this phenomenon being related to the heterogeneity of the substituent distribution (8). When temperature cycles of A4C (DS = 1.7) and M18 (DS = 2.2) were compared, hysteresis was still present, but the width of the cycle was much larger with A4C meaning stronger interactions. The hydrophobic nature of the interactions was well demonstrated either with n.m.r., fluorescence spectroscopy or the influence of the salts on the gelation characteristics (19).

From all these results we may go deeper into the mechanism of gelation (19). At low concentration and low temperature the methylcellulose is well dissolved in water and solutions without aggregates were obtained. When the temperature and concentration increase, for all of the methylcellulose samples, hydrophobic interactions due to the presence of methyl groups appear. According to the structure (repartition of substituents and presence or not of highly substituted units) the interactions lead to a viscosity increase and sometimes (in the case of the presence of highly substituted zones) to the formation of percolation aggregates leading to a clear gel. From calorimetric experiments (19) low substituted chains do not participate in junction zones, but favour their connection leading to a three-dimensional network. At high concentration pseudo-crystalline zones form slowly within junction zones. From X-ray diffraction (20) the structure of these 'crystallites' is very close to trimethylcellulose crystals and confirms that the highly substituted units are involved in these junctions. This was previously observed by Kato et al. (21) and it was necessary to have from 4 to 8 methylcellulose units to form these junction zones. When the temperature is increased above the binodal curve the concentration fluctuations generated by the phase separation lead to the formation of dense aggregates. The growth of these aggregates is limited by the concomitant gelation which decreases the mobility of the chains and hence forms rigid gels. The temperature at which the phase separation occurs (or the turbid gel is observed) depends on polymer concentration and chemical structure but also molecular weight (22). Whatever the samples, the same steps occur, but physicochemical observations depend upon the substitution characteristics. Critical temperatures for the changes in physical states are displaced towards higher values when the repartition of the substituents is more uniform, or the degree of substitution is lower due to a smaller quantity of highly substituted zones. In any case, the presence of trisubstituted units is compulsory for observable gelation.

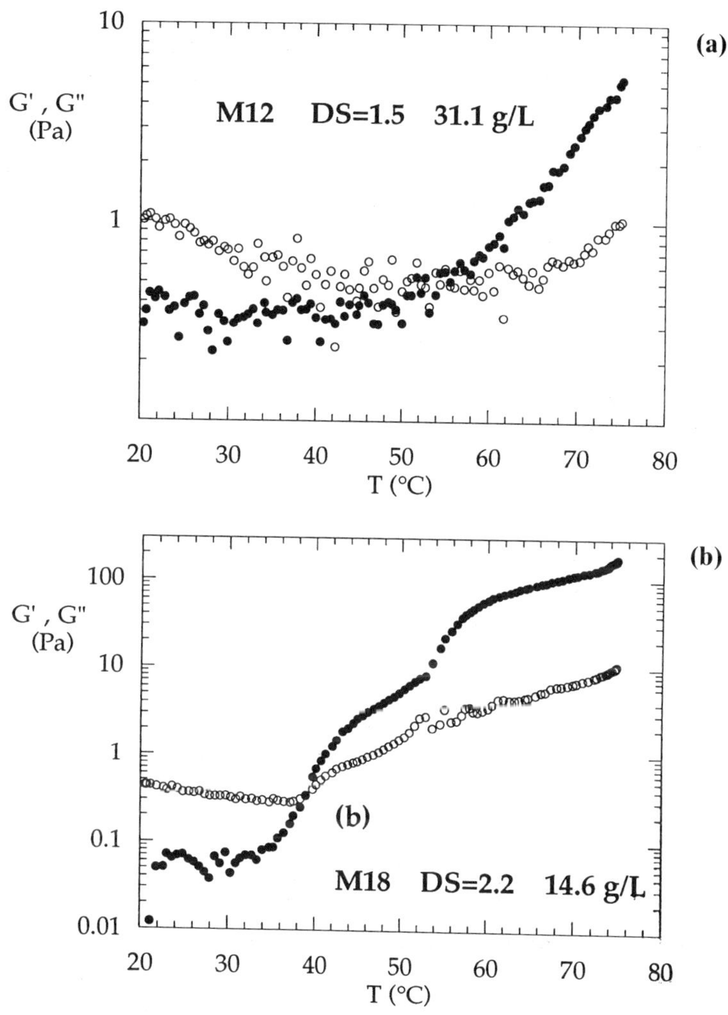

Figure 3. Rheological moduli of solutions of homogemeous methylcellulose in water as a function of temperature (G' and G" moduli at 1 Hz), (a) M12, DS=1.5, c=31.1 g/L; (b) M18, DS=2.2, c=14.6 g/L

Alkylchitosan

Chitosan is a partially deacetylated chitin that is soluble in acid conditions due to the protonation of the amino group on the C-2 position. The net charge of the polymer is controlled by the pH of the solution. But the properties can be modified by alkylation of the polymer on the C-2 position (with different densities of grafting and different lengths of alkyl substituents). Then the balance between electrostatic and hydrophobic interactions can be controlled by the chemical modification, the pH or the temperature.

We have demonstrated that a minimum of six C atoms is necessary on the alkyl chain for observing a hydrophobic effect (gelation, increase in viscosity...) (23). The role of hydrophobic interactions on the viscosity of solutions appears even in dilute solutions (24), and at larger concentrations the roles of the polymer concentration, of the degree of alkylation and of the length of the alkyl chains are clearly demonstrated (25). An example is given in figure 4 for different alkylated derivatives. The increase in the viscosity clearly indicates interaction up to a gel-like behaviour as we have shown from the dynamic rheological measurements (23). Moreover when hydrophobic interactions occur, a non-Newtonian behaviour is observed compared with original chitosan. The hydrophobic nature of these interactions is confirmed from fluorescence spectroscopy measurements (figure 5). The series of data point out the existence of a critical concentration over which the viscosity increases rapidly and the fluorescence indicates that hydrophobic domains are formed where the pyrene is located; this effect is specific of the presence of the alkyl chains. The presence of intermolecular interactions leads to the non-Newtonian behaviour of the alkylchitosan solution compared with the chitosan solution. The junction points for the 3D-network formation are constituted of alkyl aggregates looking like a micellisation. The hydrophobic nature of these interactions was demonstrated as promoted by increase of the temperature (26) or addition of external salt (27). Especially in the presence of external salt, it is well known that for the polyelectrolytes the viscosity decreases due to the screening of electrostatic repulsions; it is the situation of the protonated chitosan. In the presence of external salt, due to the decrease of the electrostatic repulsions and the enhancement of the hydrophobic interactions, the gel-like properties are reinforced (23).

The degree of charge may be decreased by controlled additions of NaOH. The roles of the fraction of net charge α and of the temperature were investigated (figure 6). This allows one to modulate the electrostatic – hydrophobic balance. When the charge is decreased at low temperature, the viscosity increases when $\alpha < 0.5$ and the hydrophobic attractions dominate; the viscosity passes through a maximum and then decreases to phase separation when $\alpha \sim$ 0.1-0.2 for CC12 polymer. When the temperature increases, the hydrophobic interactions are reinforced, the viscosity is less affected and phase separation occurs more and more rapidly; a more tightly interacting system is forced. From these data, one deduces the phase diagram for one of our polymer (CC12) at a given concentration (0.9 g/L) (figure 7). The curve is largely dependent on the polymer concentration and on the substitution.

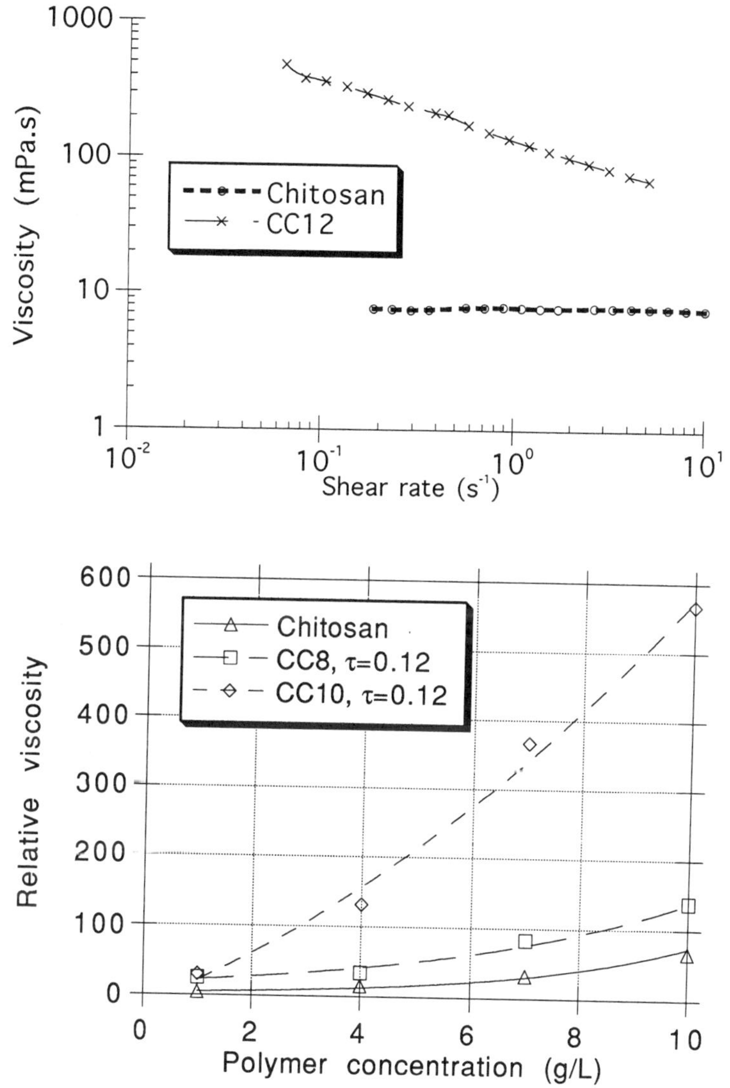

Figure 4. Influence of the alkylation on the rheological behaviour of solutions of hydrophobic derivatives of chitosan (see Table Ib for sample details)

(a) influence of the shear rate (solvent AcOH 0.3M, c=1.5 g/L, T= 20°C)

(b) influence of the polymer concentration (solvent AcOH 0.2M)

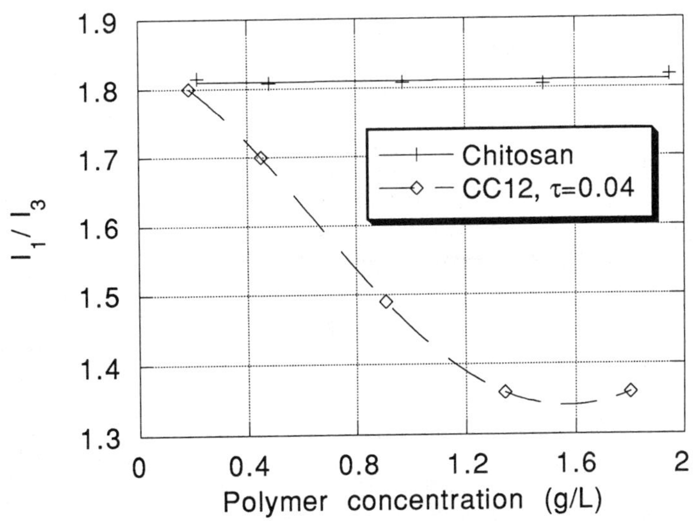

Figure 5. Influence of the polymer concentration on the fluorescence of solutions of chitosan derivatives (chlorhydrate form in water, T=5°C, [pyrene]=10^{-7}M). (Reproduced with permission from reference 25. Copyright 1997 Wiley-VCH Verlag.)

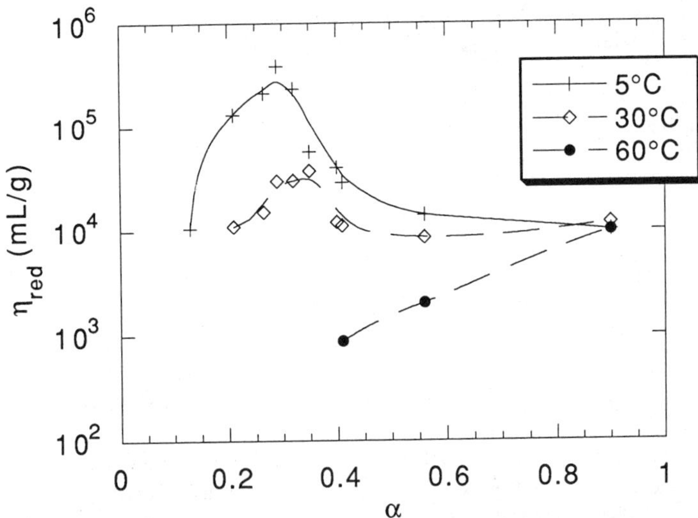

Figure 6. Influence of the charge density and the temperature on reduced viscosity of CC12 solutions (chlorhydrate form in water, τ=0.04, c=0.9 g/L). (Reproduced with permission from reference 25. Copyright 1997 Wiley-VCH Verlag.)

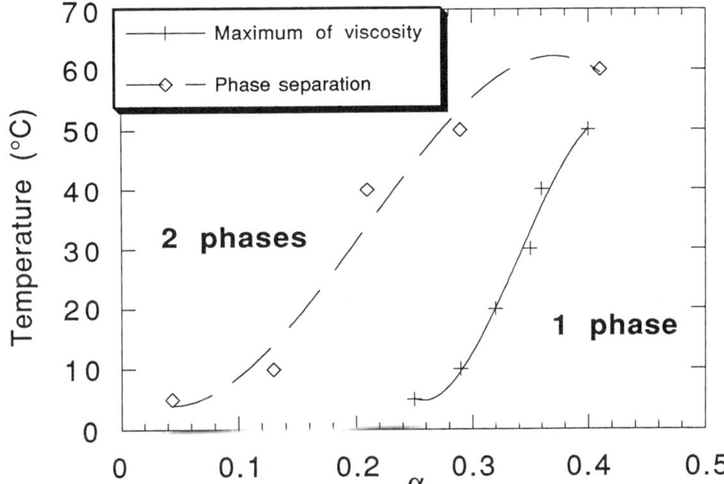

Figure 7. Phase diagram of CC12 alkylchitosan (chlorhydrate form in water, τ=0.04, c=0.9 g/L). (Reproduced with permission from reference 25. Copyright 1997 Wiley-VCH Verlag.)

Conclusion

We are able to prepare amphiphilic polysaccharide derivatives in different conditions with controlled chemical structure allowing us to monitor the hydrophilic-hydrophobic balance. It may be done by adjusting the hydrophilic properties from the density charge (by modifying the pH for example) or the hydrophobic character from the length, nature and the number of the hydrophobic groups. Moreover we may have block-like copolymers (such as methylcellulose) or grafted polymers (such as chitosan derivatives). We have studied the physicochemical properties of such polymers and demonstrated the nature of the interactions and the structural characteristics of the junction zones. As a consequence we are able to adjust the chemical structure of these derivatives to reach the desired properties.

As examples we have demonstrated amphilic polymers that act as tensioactive agents (28) or emulsion stabilizers (29).

References

1. Fringant, C., Desbrieres, J., Rinaudo, M *Polymer* **1996**, *37*, 2663
2. Tanaka, R.; Meadows, J.; Phillips, G.O.; Williams, P.A. *Carbohydr. Polym.* **1990**, *12*, 443
3. Landoll, L.M. *J. Polym. Sci. Polym. Chem.* **1982**, *20*,443
4. San, A.C. *Polym. Mater. Sci. Eng.* **1987**, *57*, 497
5. Schulz, D.N.; Kaladas, J.J.; Marner, J.J.; Block, J.; Pace, S.J.; Schulz, W.W. *Polymer* **1987**, *28*, 2110
6. Hirrien, M.; Desbrieres, J.; Rinaudo, M. *Carbohydr. Polym.* **1996**, *31*, 243
7. Hakomori, S. *J. Biochem. (Tokyo)* **1964**, *55*, 205
8. Desbrieres, J.; Hirrien, M.; Rinaudo, M. *Carbohydr. Polym.* **1998**, *37*, 145
9. Yalpani, M.; Hall, L.D. *Macromolecules* **1984**, *17*, 272
10. Arisz, P.; Kauw, H.H.J.J.; Boon, J.J. *Carbohydr. Res.* **1995**, *127*, 1
11. Heymann, E. *Trans. Faraday Soc.* **1935**, *31*, 846
12. Kuhn, W.; Moser, P.; Majer, H. *Helv. Chim. Acta* **1961**, *44*, 770
13. Sarkar, N. *J. Appl. Polym. Sci.* **1979**, *24*, 1073
14. Haque, A.; Morris, E.R. *Carbohydr. Polym.* **1993**, *22*, 161
15. Vigouret, M.; Rinaudo, M; Desbrieres, J. *J. Chim. Phys.* **1996**, *93*, 858
16. Desbrieres, J.; Hirrien, M.; Rinaudo, M. *Cellulose derivatives. Modification, Characterization and Nanostructures*; ACS Publ., T.J. Heinze, W.G. Glasser Edts, 1998, Chapter 24, p. 332
17. Chevillard, C.; Axelos, M.A.V. *Colloid Polym. Sci.* **1997**, *275*, 537
18. Hirrien, M. Thesis (Grenoble, France), 1996

19. Hirrien, M.; Chevillard, C.; Desbrieres, J.; Axelos, M.A.V.; Rinaudo, M. *Polymer* **1998**, *39*, 6251
20. Chevillard, C. Thesis (Strasbourg, France), 1997
21. Kato, K.; Yokoyama, M.; Takahashi, A. *Colloid and Polymer Sci.* **1978**, *256*, 15
22. Nishinari, K.; Hofmann, K.E.; Moritaka, H.; Kohyama, K.; Nishinari, N. *Macromol. Chem. Phys.* **1997**, *198*, 1217
23. Desbrieres, J.; Martinez, C.; Rinaudo, M. *Int. J. Biol. Macromol.* **1996**, *19*, 21
24. Desbrieres, J.; Rinaudo, M. *Advances in Chitin Sciences, vol II*, J. Andre Publ., Lyon (France), A. Domard, G.A.F. Roberts, K.M. Varum Edts, 1998, p. 339
25. Desbrieres, J.; Rinaudo, M.; Chtcheglova, L. *Macromol. Symp.* **1997**, *113*, 135
26. Ben Aim, A. *Hydrophobic Interactions*, Plenum Press, New York, 1980
27. Barley, F.E.; Callard, R.W. *J. Appl. Polym. Sci.* **1959**, *1*, 56
28. Desbrieres, J.; Rinaudo, M.; Babak, V.; Vikhoreva, G. *Polym. Bull.* **1997**, *39*, 209
29. Desbrieres, J., Rinaudo, M., Klein, J.M., Mahler, B. Patent FR 9408314, EU 95420177.8

Chapter 5

Dissolution and Gelation of κ-Carrageenan

Srividya Ramakrishnan[1] and Robert K. Prud'homme

Department of Chemical Engineering, Princeton University, Princeton, NJ 08544

Abstract

The rheology and conformational helix transition of κ-carrageenan in aqueous solutions of glycerol is studied. Carrageenan forms a hydrated gel in water but a weakly flocculated network of partially hydrated particles in glycerol. The melting transitions of carrageenan in glycerol/water solutions show a single peak in the complex moduli at low and high glycerol concentrations, and two peaks at approximately equal concentrations of glycerol and water. These are thought to be due to the breakage of hydrogen bonds (low temperature peak) and ion dissociation (high temperature peak). The helix-coil transition is shifted to higher temperatures in solutions with higher glycerol content. Comparison of the rheology and OR curves reveals that, though the gelation and conformational transition temperatures correspond for samples in water and also in glycerol, there is more hysteresis between the heating and cooling curves at high water concentrations. This suggests that gelation occurs through helix aggregation in water but the lower solubility of carrageenan results in a more random network of helices in glycerol.

Introduction

Carrageenans are linear sulphated polysaccharides extracted from marine red algae. The backbone is based on a repeating disaccharide sequence as shown in Figure 1 [1, 2]. The polymer is negatively charged and forms three dimensional gels through specific interactions with metal ions [3, 4, 5, 6]. The biocompatability of these gels makes them valuable in a number of applications. For example, they are used in dairy industries and household products such as toothpastes, lotions and water-based paints [7]. In food applications, carrageenans are mainly used to gel, thicken and stabilize [1]. They are also being investigated

[1]Current address: Unilever Research, 45 River Road, Edgewater, NJ 07020.

κ–carrageenan

OSO_3^- HOH₂C O H₂C O O O OH OH

4-*O*-Sulfato-β-D-galactopyranosyl unit 3,6-Anhydro-α-D-galactopyranosyl unit

ι–carrageenan

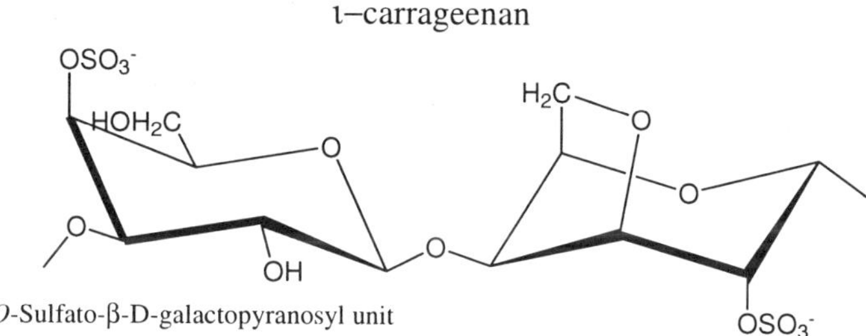

4-*O*-Sulfato-β-D-galactopyranosyl unit

3,6-Anhydro-2-*O*-sulfato-α-D-galactopyranosyl unit

λ–carrageenan

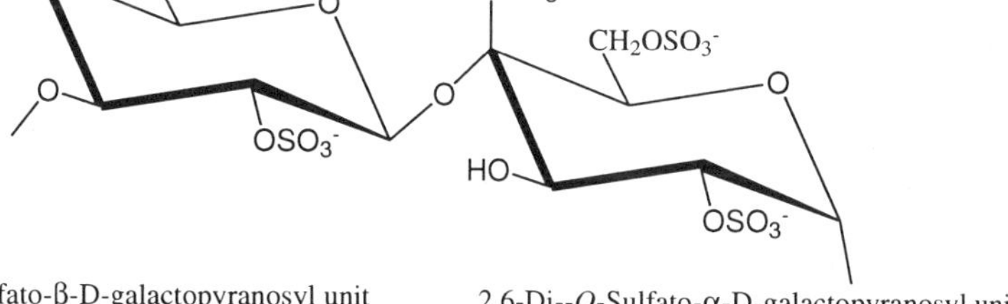

2-*O*-Sulfato-β-D-galactopyranosyl unit 2,6-Di--*O*-Sulfato-α-D-galactopyranosyl unit

Figure 1: *Repeating disaccharide structures of κ, ι, and λ-carrageenan*

for pharmaceutical and environmental applications due to their ability to immobilize microorganisms [8, 9, 10].

It is generally accepted that κ-carrageenan gelation is a two-step mechanism [3, 11]. At high temperatures, the polymer is present as random coils. Upon cooling, it undergoes a conformational transition, forming double helices [12, 13]. In the presence of certain cations which promote gelation, these double helices form aggregates, which result in gelation. Rinaudo et. al [14] found that both the melting temperature, and the degree of order, increased with increasing concentration of gelling cations. A number of studies [4, 5, 6] have shown potassium ions to be significantly more effective in inducing gelation than lithium, sodium or calcium ions. The reason for this specificity is not clear. Long range Coulomb interactions do not explain the preference for monovalent cations over divalent ones. Nilsson and Piculell [15] proposed that the cation specificity may be due to site-specific binding of the ions to the κ-carrageenan helix but the nature of the binding site remained unclear.

Our goal is to study the gelation of κ-carrageenan in mixed solvents. Solvents such as glycerol that lower the activity of water are used in processed foods to prevent loss of water through evaporation. Hence, it is important to gain a fundamental understanding of the behavior of carrageenan in these systems. We study the mechanical properties through rheology, and follow the conformational transition with optical rotation (OR).

Experiments

Three κ-carrageenan samples with different ion contents were supplied by FMC Corporation. One sample contained potassium and calcium ions (5.8% K^+, 0.6% Na^+ and 2.6% Ca^{2+}- referred to as K^+-Ca^{2+}-carrageenan), one contained mainly potassium ions (8.25% K^+ - referred to as K^+-carrageenan) and the last one contained mainly sodium ions (0.2% K^+ and 5% Na^+ - referred to as Na^+-carrageenan).

Steady shear and dynamic oscillatory measurements were performed with a Rheometrics Fluids Spectrometer II (RFSII) using parallel plate geometry (50mm diameter). The melting transitions and gelation of the above samples were studied with dynamic oscillatory measurements during temperature ramps between 0°C and 80°C using the Peltier plate with parallel plate geometry (40mm diameter) on the Dynamic Stress Rheometer (DSR).

The conformational transition was followed by measuring the optical rotation with a Rudolph Research polarimeter (AUTOPOL IV). Carrageenan helices are circularly birefringent, and hence retard the right-handed and left-handed components of circularly polarized light by different amounts, resulting in a rotation of the polarization ellipse. As the random coil state is only minimally optically active, the formation of helices upon cooling is accompanied by a significant increase in the OR signal. The path length of the OR cell was 100 mm and the wavelength of light was 589 nm. A computer controlled circulating bath was used maintain a ramp rate of 1°C/minute.

Results and Discussion

Rheology : Initial Hydration

Phase contrast optical microscopy of samples of K^+- and K^+-Ca^{2+}-carrageenan powders hydrating in water at ambient temperature reveal a highly swollen structure with the volume fraction of particles close to one. Without heating beyond the order/disorder melting transition of the carrageenan helix, the discrete nature of the swollen particles is maintained. The degree of swelling decreases with addition of glycerol. Carrageenan is still slightly swollen in solutions of 50% water, 50% glycerol but the texture is more granular and the sample is less clear. At high glycerol concentrations, the swelling is insignificant. Na^+-carrageenan with added KCl, however, swells to a greater extent than the above samples at high glycerol concentrations.

Rheological measurements were done on samples of 1% K^+-Ca^{2+}-carrageenan in 100:0, 90:10, 50:50 and 0:100 ratios of water/glycerol solutions at ambient temperature. The data obtained from the steady shear measurements, strain sweeps and frequency sweeps are shown in Figures 2 and 3. The steady shear viscosity measurements in Figure 2 show that all four samples are shear thinning, even at very low shear rates. The 1% K^+-carrageenan in water shows less shear thinning at low shear rates as compared to the corresponding sample containing calcium. The initial viscosity is also lower. G' vs. strain showed an initial plateau followed by a decrease in G' with increasing strain for the samples in water and 50% glycerol. G' drops immediately with increasing strain for carrageenan in 100% glycerol. The values of the dynamic moduli decrease as the concentration of glycerol increases.

These results, along with information on the melting transition in the next section, indicate that the polymer is not fully dissolved, but consists of partially hydrated flocculated particles with attractive interactions [16]. This stickiness is probably due to potassium and calcium ions acting as salt bridges between the negatively charged particles. This is supported by the fact that the sample without calcium ions shows less shear thinning and has a lower initial viscosity. The observed decrease in viscosity with increasing glycerol concentration is an effect of the decrease in volume fraction of the particles [17]. Studies on microgel dispersions of ethyl acrylate and methacrylic acid have shown that they undergo similar swelling as the solvent quality is improved [18, 19]. The strain sweep measurements show that the flocculated network breaks down at higher strains. These results confirm that during the hydration at ambient temperature, carrageenan forms a hydrated gel in water but a weakly flocculated network of partially hydrated particles in glycerol/water mixtures.

Rheology : Temperature Cycling

Samples of 1% K^+-Ca^{2+}-carrageenan in solutions of varying water/glycerol ratios were heated to 85°C, and then cooled back to room temperature. All the samples formed stiff gels. Dynamic oscillatory mea-

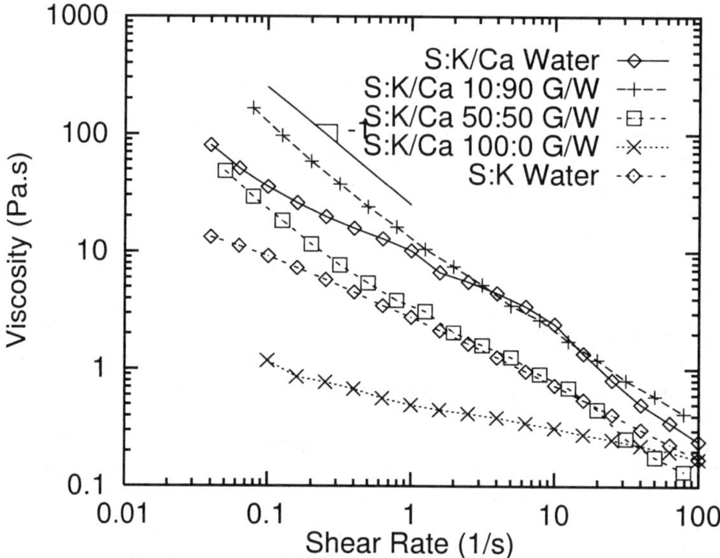

Figure 2: *Steady shear measurements on samples of 1% K^+-Ca^{2+}-carrageenan in water and in 10%, 50% and 100% glycerol, and 1% K^+-carrageenan in water at 25° C. (Reproduced with permission from reference [20])*

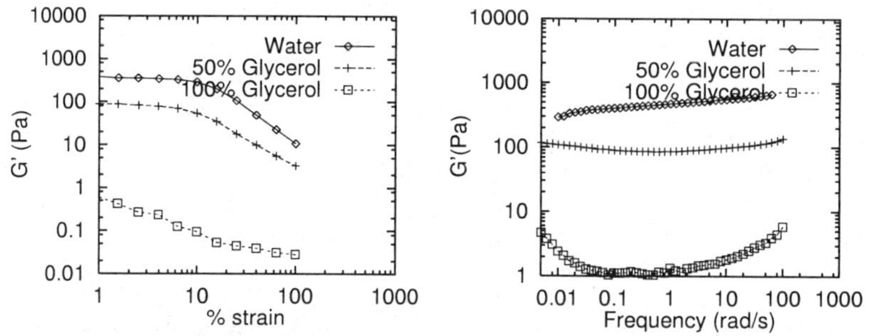

Figure 3: *Strain sweep at a frequency of 1rad/s, and frequency sweep at a strain of 2% on samples of 1% K^+-Ca^{2+}-carrageenan in water, and in 50% and 100% glycerol at 25° C. (Reproduced with permission from reference [20])*

surements at room temperature were carried out on these gels. The frequency sweep experiments (Figure 5) show that the storage modulus remained constant for all samples over the frequency range (0.01 to 100 rad/s) - the signature of a gel. The gel in water was slightly weaker (G' $\approx$ 3000 Pa) than those in mixed solutions of water and glycerol (G' $\approx$ 6000 Pa). The loss modulus decreased rapidly with increasing frequency after an initial plateau for higher glycerol concentrations. Also, in each case, G'' was about ten times less than G'. Subtraction of the solvent contribution to the energy loss from the loss modulus did not significantly change the G'' values. The strain sweep experiments revealed identical behavior for all samples. A master-curve was obtained by plotting G'/G'_0 vs. strain (Figure 4), where G'_0 is the elastic modulus at 0.1% strain and 1 rad/s frequency. This indicates that the type of gel network structure formed after heating and cooling is essentially the same in all four cases, and is independent of the solvent. A comparison of the strain sweep data before melting and cooling the samples (Figure 3) and after (Figure 4) reveals that the melted and cooled samples, especially for high glycerol concentrations, are insensitive to strain as compared to the initial particulate samples.

The results mentioned above indicate that carrageenan goes through a melting transition at some temperature which depends on the solvent quality. We investigated the transition by imposing a temperature ramp while measuring the dynamic moduli. The results are shown in Figure 6 for 1% K^+-Ca^{2+}-carrageenan in various concentrations of water and glycerol (ranging from 0-100% glycerol in water). For carrageenan in water, the moduli decreased with increasing temperature after an initial plateau. At room temperature, the polymer is present as highly swollen but still discrete particles in water. The initial decrease in G' with increase in temperature is due to dissolution of the particles by destruction of the hydrogen bonded carrageenan and weakening of the ion-mediated bonding forces. Carrageenan particles in 30% glycerol have a lower volume fraction than in water and hence, the elastic modulus is nearly zero at 25°C. However, as the temperature is increased, the polymer starts swelling and the moduli increase. This is followed by dissolution, resulting in a peak at 47°C. The transition observed here is similar to DSC melting transitions for polymer systems. Carrageenan in 50% glycerol also has very low dynamic moduli at room temperature. However, in this case, two peaks in the moduli are observed while heating - a main peak at 54°C and a second smaller peak at around 74°C - indicating two transitions during hydration. A possible explanation for this is the following. The first peak that appears on heating occurs at the same temperature as the one observed for 30% glycerol. The helices form random coils which then fuse to form a network. Once they are dissolved, the hydrogen bonds break and the moduli decrease again. The second peak may be due to calcium ion dissociation. The cations form salt bridges with the negatively charged sulphate groups on the polymer chain. Ion pair formation has also been observed in polypeptides [21]. The Born or solvation energy of an ion, i.e., the total free energy of charging the ion in a medium of dielectric constant ε, is given

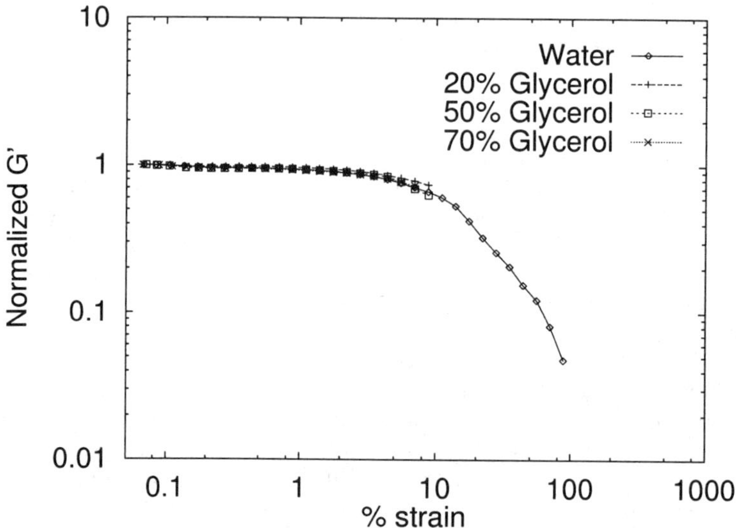

Figure 4: *Normalized (G'/G'_0) strain sweep curves for 1% K^+-Ca^{2+}-carrageenan in water and in 20%, 50% and 70% glycerol (heated) at 25° C.*

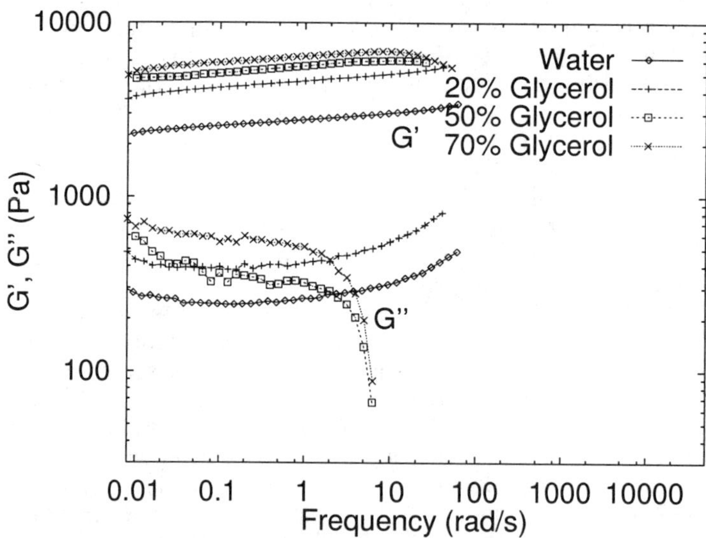

Figure 5: *Frequency sweep plots of G' and G'' for heated samples of 1% K^+-Ca^{2+}-carrageenan in water and 20%, 50% and 100% glycerol at 25° C. The latter rapidly falls at higher frequencies for 50% and 70% glycerol, indicating negligible viscous dissipation from the polymer.*

by [22]

$$\mu^i = \frac{(ze)^2}{8\pi\varepsilon_0\varepsilon a} \tag{1}$$

where z is the valence of the ion, a is the radius, and ε_0 is the permittivity of vacuum. The change in free energy on transferring an ion from a medium of dielectric constant ε_1 to one of dielectric constant ε_2 is equal to

$$\Delta\mu^i = \frac{(ze)^2}{8\pi\varepsilon_0 a}\left(\frac{1}{\varepsilon_2} - \frac{1}{\varepsilon_1}\right) \tag{2}$$

The free energy is positive, i.e, it is energetically unfavorable, for transferring an ion from a medium of high dielectric constant to one of low dielectric constant ($\varepsilon_1 > \varepsilon_2$). The free energy change in separating two oppositely charged ions from contact in a solvent of dielectric constant ε is given by

$$\Delta\mu^i \approx \frac{(ze)^2}{4\pi\varepsilon_0\varepsilon(a_+ + a_-)} \tag{3}$$

where a_+ and a_- are the ionic radii of the cation and anion. The energy required to dissociate the ions increases as the dielectric constant decreases. Since glycerol has a lower dielectric constant ($\varepsilon = 47$) than water ($\varepsilon = 80$), the overall energetics, which include that of solvation and ion dissociation, favor the formation of ion pairs as the concentration of glycerol is increased.

Carrageenan in pure glycerol shows only a single high temperature peak while melting. The disappearance of the second peak is due to the calcium dissociation transition being shifted to temperatures higher than 90°C. As the ability of carrageenan to hydrogen bond with glycerol is lower than with water, the denaturing of the helix is also shifted to higher temperatures at higher glycerol concentrations.

Optical Rotation

Samples of 1% Na^+-carrageenan in solutions of varying glycerol/water content were melted at 80°C. The optical rotation was measured while allowing these to cool to 0°C at a rate of 1°C. The optical rotation was also measured for samples of Na^+-carrageenan to which 0.01M, 0.03M and 0.05M KCl were added. Figure 7 shows that increasing the quantity of glycerol shifts the helix formation transition to higher temperatures. This may be due to two reasons. The lower dielectric constant of glycerol, as compared to water, does not favor ions going into solution. Also, the decreased ability of carrageenan to hydrogen bond with glycerol increases the tendency for intra-molecular hydrogen bonding. Figure 8 shows that increasing the ion concentration also increases the onset temperature of the transition, even in low water activity solvents.

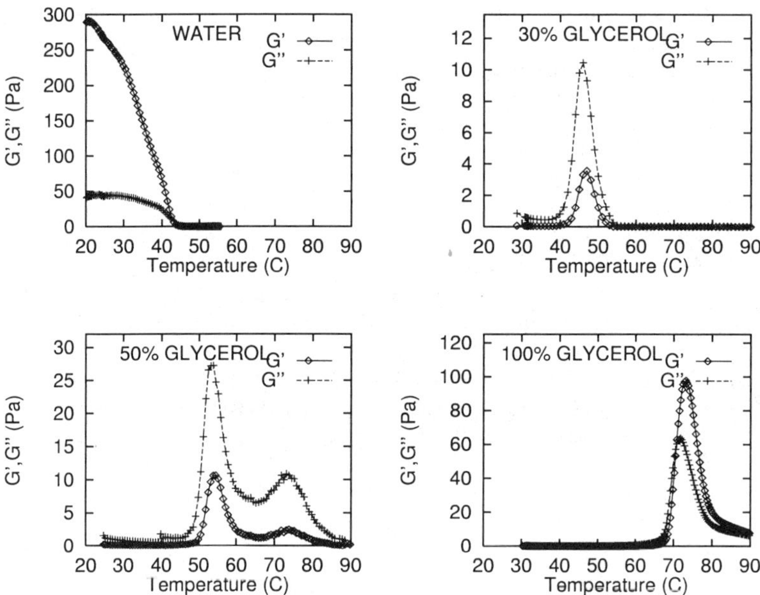

Figure 6: *Temperature ramps at a heating rate of 1 C/min, stress of 0.5 Pa and frequency of 1 rad/s on samples of 1 weight% K^{+}-Ca^{2+}-carrageenan in water, 30%, 50% and 100% glycerol. Note the vertical axes are different for each set of data. (Reproduced with permission from reference [20])*

Comparison of Rheology and OR

Samples of 1% Na^+-carrageenan with 0.01M KCl in solutions of water, 50% glycerol, and 100% glycerol were melted at 80°C. These were then cooled to 0°C and re-heated to 80°C at a rate of 1°C/min using the Peltier instrument on the DSR. During the temperature ramps, the complex moduli were measured at a stress of 0.5 Pa and frequency of 1 rad/s. The optical rotation was measured with the polarimeter at a wavelength of 589 nm. The data are shown in Figure 9. The onset of the conformational transition temperature corresponds to the onset of gelation (indicated by an increase in G') for all the samples. This supports the idea that gelation occurs simultaneously with helix formation. However, the samples at high water concentrations show significant hysteresis between the heating and cooling curves (Figure 9, top right). As the concentration of glycerol is increased, the hysteresis decreases, and for carrageenan in 100% glycerol, the heating and cooling curves are essentially superimposable (Figure 9, bottom right). The hysteresis in water is due to the formation of cooperatively aggregated domains. These crystalline regions are at a lower energy with a concomitant higher melting temperature. Similar behavior is seen for crystallizable synthetic polymers [23]. This co-operative aggregation of double helices causes a sharp increase in the elastic modulus. Aggregates also rotate light in a direction opposite to that of the helices, causing the OR signal to first increase upon cooling and then to decrease. This can be observed for carrageenan in 50% glycerol. The absence of hysteresis in glycerol arises because carrageenan being less soluble in glycerol, comes out of the solution as soon as the double helices are formed. Hence, the polymer forms a more random network of double helices than the ordered network of aggregates. This is also supported by the fact that the OR signal does not decrease at low temperatures for samples in 100% glycerol.This behavior corresponds to increasing the quenching rates for crystallizable synthetic polymers. The random structure results in more junction points between the helices and hence, a more elastic network. G' is around 1400 Pa in glycerol whereas it is just 500 Pa in water. The energy required to break this network while heating is just that required to convert individual double helices to random coils. This explains the lack of hysteresis at high glycerol concentrations.

Effect of ions on gelation

The effect of potassium and calcium ions on the gelation of carrageenan was investigated by comparing the rheological behavior of K^+-carrageenan K^+-Ca^{2+}-carrageenan, and Na^+-carrageenan with 0.01M KCl. The polymer was dispersed in solutions of water, 50% glycerol and 100% glycerol. Figure 10 shows the melting and gelation of the three carrageenans in the above solvents.

For samples in water, maximum swelling is observed in the case of K^+-Ca^{2+}-carrageenan at room temperature (Figure 10, left top). However, in 50% glycerol, Na^+-carrageenan with KCl swells to a greater extent. While K^+-Ca^{2+}-carrageenan in 50% glycerol clearly displays

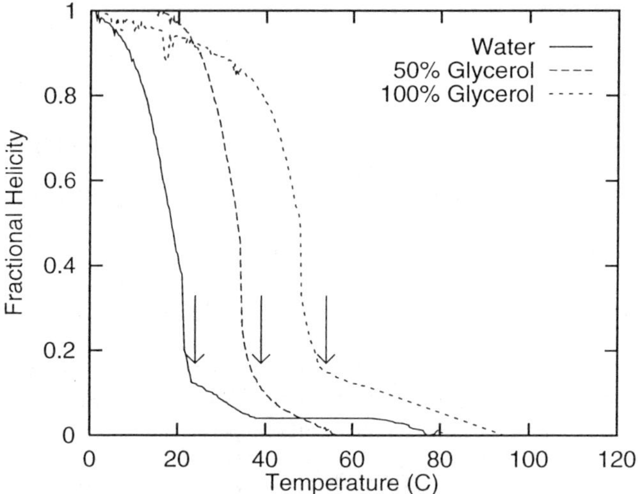

Figure 7: Normalized OR cooling curves of 1% Na^+-carrageenan containing 0.01M KCl in water, 50% and 100% glycerol. The temperature was decreased at 1° C/minute. The fractional helicity was calculated by dividing the optical rotation by the maximum value (taken to be the optical rotation for completely helical network). The arrows indicate the "onset" temperature - the temperature at which the transition begins.

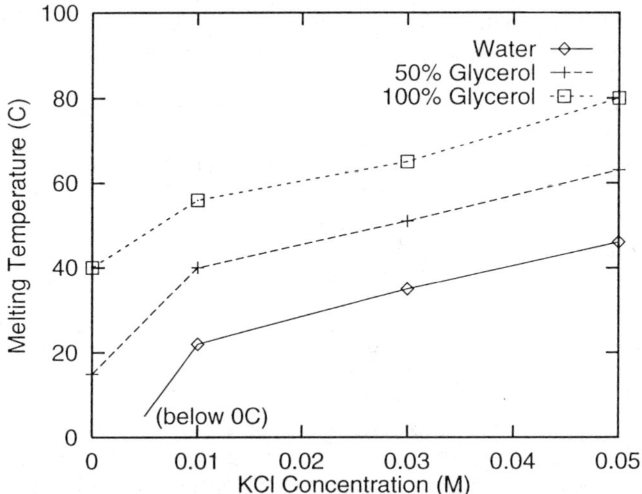

Figure 8: Data from OR curves of 1% Na^+-carrageenan containing no KCl, 0.01M, 0.03M and 0.05M KCl in water, 50% and 100% glycerol. The onset temperature of the conformational transition is plotted as a function of potassium ion concentration. Increase in concentrations of both, glycerol and ions, shifts the transition to higher temperatures.

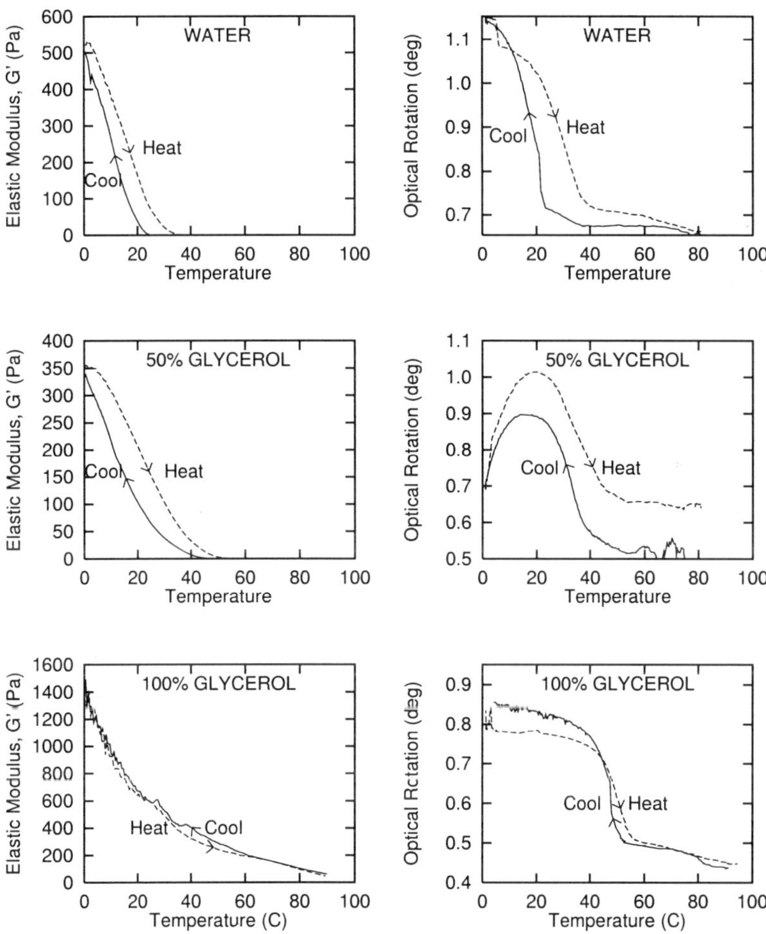

Figure 9: *Rheology (stress of 0.5 Pa at 1 rad/s) and OR data (589 nm wavelength, 100 mm cell) during temperature ramps at 1° C/min on samples of 1 weight% Na^+-carrageenan with 0.01M KCl in water, 50% and 100% glycerol. The samples were initially melted at 80° C, then cooled to 0° C and re-heated to 80° C. (Reproduced with permission from reference [20])*

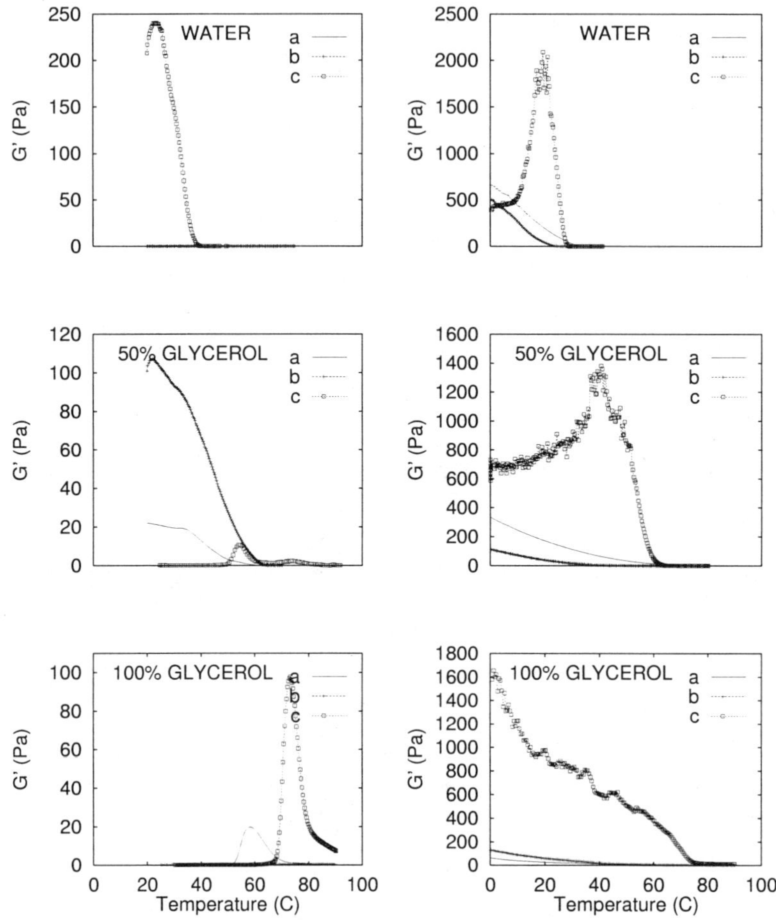

Figure 10: *Comparison of the melting and cooling curves for (a) K^+-carrageenan, (b) Na^+-carrageenan with 0.01M KCl, and (c) K^+-Ca^{2+}-carrageenan. The storage modulus for K^+-carrageenan, and Na^+-carrageenan with 0.01M KCl dispersed in water (top left figure) is negligible.*

two peaks in the moduli during melting (at 54°C and 73°C), only a slight hump around 35°C is observed for the other two samples. This hump corresponds to the breakage of hydrogen bonds. A single peak in the moduli, indicating ion dissociation, is seen for both K^+-Ca^{2+}-carrageenan and K^+-carrageenan in 100% glycerol (Figure 10, left bottom). The transition for the sample without calcium once again occurs at a lower temperature. These results support the conclusion that lowering the ionic strength shifts the transition to lower temperatures. The transition corresponding to ion dissociation is more pronounced in the cases where the counterions are potassium or calcium. These ions bind strongly to the polymer even in the random coil state, especially when other counterions are not present. When KCl is added to Na^+-carrageenan, the potassium ions are dispersed more uniformly in the solvent and do not bind as strongly to the polymer.

The cooling curves show that the storage modulus is an order of magnitude higher for the sample containing calcium. The decrease in G' observed at lower temperatures is due to the shrinking of the gel, resulting in de-bonding from the tool surface. Calcium ions bind to the sulphate groups on the polymer resulting in more junction points, and hence a more elastic gel.

Summary

In this study, we investigated the effect of glycerol on the rheological behavior and conformational transition of κ-carrageenan. While carrageenan forms a swollen gel with water, it forms only a weakly flocculated network of partially hydrated particles with glycerol. Upon melting and re-cooling, all samples form solid gels with similar structure, irrespective of the glycerol content in the solvent. The melting transitions show a single peak in the dynamic moduli for low and high glycerol concentrations, and two peaks in between. We attribute the lower temperature peak to the breakage of hydrogen bonds in the helix, and the higher temperature one to ion dissociation. The transition from fluid to gel with respect to temperature becomes more gradual, and the hysteresis between the heating and cooling curves for G' and the optical rotation decreases as the concentration of glycerol is increased. The lower solubility of carrageenan in glycerol is responsible for this. The helices aggregate to form a gel of oriented helical domains in water whereas they come out of solution and form a more random network in glycerol. This also explains the higher elastic modulus of carrageenan-glycerol gels. Addition of glycerol shifts the conformational transition and gelation to higher temperatures. This is due to its lower dielectric constant and reduced ability to form hydrogen bonds with carrageenan. Higher concentrations of ions increase the onset temperature even in low water activity solvents. The presence of calcium ions results in significantly more elastic gels.

Acknowledgments

This work has been supported by FMC Food Ingredients Division, New Jersey.

References

[1] G.H. Therkelson. Carrageenan. In R.L.Whistler and J.N.BeMiller, editors, *Industrial Gums: Polysaccharides and Their Derivatives*, chapter 7, page 145. Academic Press, Inc., San Diego, California, third edition, 1993.

[2] E.R. Morris, D.A. Rees, and G. Robinson. Cation-specific aggregation of carrageenan helices: Domain model of polymer gel structure. *J. Chem. Soc., Perkin II*, 138:793, 1978.

[3] E.R. Morris, D.A. Rees, and E.J. Welsh. Cation specific aggregation of carrageenan helices - domain model of polymer gel structure. *J. Mol. Biol.*, 138(2):349, 1980.

[4] T.H.M. Snoeren and T.A.J. Payens. On the sol-gel transition of kappa-carrageenan. *Biochim. Biophys. Acta*, 437:264, 1976.

[5] D. Oakenfull and A. Scott. The role of cations in the gelation of kappa-carrageenan. In G.O. Phillips, D.J. Wedlock, and P.A. Williams, editors, *Gums and Stabilizers in the Food Industry 5*. IRL Press, Oxford, 1990.

[6] A.-M Hermansson, E. Eriksson, and E. Jordansson. Effects of potassium, sodium and calcium on the microstructure and rheological behaviour of kappa-carrageenan gels. *Carbohydr. Polym.*, 16:297, 1991.

[7] K.B. Guiseley, N.F. Stanley, and P.A. Whitehouse. Carrageenan. In R.L.Davidson, editor, *Handbook of Water-Soluble Gums and Resins*. Mc-Graw Hill, New York, 1980.

[8] M.B. Cassidy, H. Lee, and J.T. Trevors. Environmental applications of immobilized microbial cells : A Review. *J. of Ind. Microbiol.*, 16:79, 1996.

[9] I.B. Holcberg and P. Margalith. Alcoholic fermentation by immobilized yeast at high sugar concentrations. *Eur. J. of Appl. Microbiol. Biotechnol.*, 13:133, 1981.

[10] Y. Nishida, T. Sato, T. Tosa, and I. Chibata. Immobilization of escherichia-coli-cells having aspartase activity with carrageenan and locust bean gum. *Enzyme and Microbial Technology*, 1(2):95, 1979.

[11] G. Robinson, E.R. Morris, and D.A. Rees. Role of double helices in carrageenan gelation: the domain model. *J.Chem. Soc. Chem. Comm.*, 4:152, 1980.

[12] N.S. Anderson, J.W. Campbell, M.M. Harding, D.A. Rees, and J.W.B. Samuel. X-ray diffraction studies of polysaccharide sulphates: Double helix models for κ- and ι-carrageenans. *J. Mol. Biol.*, 45:85, 1969.

[13] R.A. Jones, E.J. Staples, and A. Penman. A study of the helix-coil transition of iota-carrageenan segments by light scattering and membrane osmometry. *J. Chem. Soc., Perkin II*, 12:1608, 1973.

[14] M. Rinaudo, A. Karimian, and M. Milas. Polyelectrolyte behaviour of carrageenans in aqueous solutions. *Biopolymers*, 18:1673, 1979.

[15] S. Nilsson and L. Piculell. Helix-coil transitions of ionic polysaccharides analyzed within the Poisson-Boltzmann cell model. 4. Effects of site-specific counterion binding. *Macromolecules*, 24:3804, 1991.

[16] Mae Chen. *Colloidal silica gels : Transition and elasticity.* PhD thesis, Princeton University, 1992.

[17] J.C. van der Werff and C.G. de Kruif. Hard-sphere colloidal dispersions: The scaling of rheological properties with particle size, volume fraction, and shear rate. *J. Rheol*, 33:421, 1989.

[18] B.E. Rodriguez, M.S. Wolfe, and M. Fryd. Nonuniform swelling of alkali swellable gels. *Macromolecules*, 27:6642, 1994.

[19] M.S. Wolfe. Dispersion and solution rheology control with swellable microgels. *Progress in Organic Coatings*, 20:487, 1992.

[20] S. Ramakrishnan and R.K. Prud'homme. Effect of solvent quality and ions on the rheology and gelation of κ-carrageenan. *J. Rheol.*, 44(4).885, 2000.

[21] C.R. Cantor and P.R. Schimmel. *Biophysical Chemistry: The Behavior of Biological Macromolecules*, chapter 5, page 289. W.H.Freeman and Company, San Francisco, 1980.

[22] Jacob Israelachvili. *Intermolecular & Surface Forces.* Academic Press, San Diego, CA 92101, 2 edition, 1992.

[23] W.S. Lambert and P.J. Phillips. Structural and melting studies of crosslinked linear polyethylenes. *Polymer*, 31:2077, 1990.

Chapter 6

Biopolymer Additives for the Reduction of Soil Erosion Losses during Irrigation

William J. Orts[1], Robert E. Sojka[2], Gregory M. Glenn[1], and Richard A. Gross[3]

[1]Western Regional Research Center, Agricultural Research Service, U.S. Department of Agriculture, 800 Buchanan Street, Albany, CA 94710
[2]Soil and Water Management Research Unit, Agricultural Research Service, U.S. Department of Agriculture, 3793 N. 3600 East, Kimberly, ID 83341
[3]Department of Chemistry/Chemical Engineering, Polytechnic University, 6 Metrotech Center, Brooklyn, NY 11201

High molecular weight, synthetic polyacrylamides (PAM) are relatively large, water soluble polymers that are used increasingly by farmers to prevent erosion and increase infiltration during irrigation. A lab-scale erosion test was conducted to screen biopolymer solutions for a similar efficacy in reducing shear-induced erosion. In lab-scale mini-furrow tests, chitosan, starch xanthate, cellulose xanthate, and acid-hydrolyzed cellulose microfibrils, at concentrations of 20, 80, 80, and 120 ppm respectively, reduced suspended solids in the runoff water from test soil. None of these biopolymers, however, exhibited the >90% runoff sediment reduction shown by PAM at concentrations as low as 5 ppm. Preliminary field tests results showed that chitosan solutions were only marginally effective in reducing runoff from a 137m long furrow. There were indications that results were dependent on the length of the furrow. Erosion of some clay-rich soils from Northern California was reduced up to 85% by increasing the concentration of exchangeable calcium to >2.5mMole, with or without the addition of polymer additives.

Soil erosion threatens agricultural productivity and water through the loss of valuable "top-soil" and runoff of agricultural chemicals. U.S. row-crop farmers lose an average of 6.4 tons of soil per acre per year, despite being world leaders in practicing soil conservation (*1*). The EPA describes soil erosion as our greatest threat to water quality (*1*), a concern that is increased by evidence that chemical toxins readily pass from our waterways into the air (*2*).

Soil loss is especially significant during furrow irrigation because of the shear of flowing irrigation water, and because arid soils often are low in the natural polysaccharides that stabilize soil structure. Farmers could minimize erosion-induced soil losses by stringent use of settling ponds, sodded furrows, straw-bedded furrows, minibasins, soil stabilizers, and buried runoff pipes; however, these techniques are often too cumbersome or costly for adequate adoption.

Recently, an easy and effective tool has been added to these soil conservation measures -- adding small quantities of polyacrylamide copolymers, PAM, to the in-flowing water (*1,3-8*). Lentz *et al.* (*3*) added 5-20 ppm of high molecular weight, anionic PAM to irrigation water in the first several hours of irrigation, and reduced soil losses of a highly erodible soil up to 97%. This represented an ideological breakthrough in the use of soil conditioners. Rather than treat the "entire" 10 to 20 cm depth tilled surface layer of a field with hundreds of kilograms of polymer per acre, one only modifies the soil in contact with irrigation water. In irrigated agriculture the water itself is the PAM delivery means. For furrow irrigation, this translates to improved soil structure in the ~1-5 mm thick layer at the soil/water interface. Consequently, relatively small quantities of polymer are used (approximately 1 lb per acre per irrigation). Seasonal application totals range from about 2 to 8 lbs of polymer per acre per year.

The benefits of PAM have clearly been seen by farmers. Over a million furrow-irrigated acres were treated in 1998 (*9*). Although the method was introduced in Idaho, it has proven effective in an increasing number of diverse locations throughout the western US and in several countries overseas (*9,10*).

The successful use of PAM in irrigation water raises the question of whether there are other polymers that may have similar benefits. PAM has been used for decades as a flocculating agent -- for various food processing applications, to dewater sewage sludges, to remove heavy metals during potable water treatment, and to process industrial wastewater. PAM is widely recognized as a safe, cost-effective flocculating agent; yet, it is by no means the only flocculating agent available.

There are several reasons to explore other polymer additives. Environmental concerns have been raised about widespread PAM use in open agricultural environments – concerns that have been countered convincingly by experts in the field. PAM is a synthetic polymer that was designed to resist degradation, and is very stable (*11*). Environmental degradation rates are less than 10% per year via deamination, shear-induced chain scission and photo-sensitive chain scission (*11-13*). With such stability, there is some concern for potential accumulation. This concern is unfounded if PAM is used at the low concentrations recommended by the USDA (10 ppm in irrigation water flowing down the furrow in the first several hours). Applied under such guidelines, PAM does not leave the field in the runoff nor accumulate appreciably with time (*10*).

The other environmental concern is the fact that acrylamide, the monomer used to synthesize PAM, is a neurotoxin. Even low levels of monomer impurity in the product must be avoided. This issue has been suitably addressed by suppliers who provide PAM almost devoid of monomer (<0.05%). The EPA recently reviewed the use of PAM with USDA and polyacrylamide industry scientists, and concluded that, at the concentrations recommended for use during furrow irrigation, PAM levels are acceptable, with minimal accumulation (9). Concern that the monomer may occur as a degradation product is countered by studies showing that the most likely route to degradation is early removal of the amine group from the polymer backbone (*11-13*).

There is scientific and anecdotal evidence (*14*) that PAM efficacy varies with certain soil properties, including sodicity, texture, bulk density, and surface charge-related properties. It would be beneficial to have a wide array of polymers with potentially different soil-stabilizing mechanisms, applicable to different soil types. The market price of PAM, ranging from about $2.50 - $5.00/lb, is high relative to many commodity polymers, such as polyethylene, polypropylene, and polystyrene. Treatment for one year can cost up to $40 per acre. Any reduction in price would benefit farmers. Finally, PAM is a synthetic polymer using a non-renewable monomer source from oil refining. It would be advantageous to derive an effective soil amendment from a renewable feedstock, especially from an agricultural waste stream. Not only do natural polymers generally degrade via relatively benign routes, they are often perceived to be safer by the public.

The general objective of this study was to screen a series of biopolymers to quantify their effectiveness as erosion retardants in irrigation water. Natural biopolymers, with potentially similar functional attributes as PAM, were selected specifically because of their environmental advantages (real or perceived). Biopolymer additives were screened using a lab-scale mini-furrow procedure to determine their potential efficacy in reducing erosion losses. The more promising formulations were tested in full field tests.

Materials and Methods

Polyacrylamides (PAM)

Polyacrylamides from a variety of sources were used to establish parameters in the lab-furrow test and as a comparison of the biopolymer samples. PAM samples were generally random copolymers of acrylamide and acrylic acid, varying in molecular weight, charge type, charge amount, and relative solubility in water. A series of samples were provided by Cytec Industries of Stamford, CT, with the trademark, Magnifloc, and product codes 835A, 836A, 837A, 846A, 905N, 442C, 492C, and 494C. Cytec also provided several relatively short-chain samples specifically for this project. Samples were also provided by Allied Colloids, Suffolk, VA, and trademarked Percol (product codes 338, E24, and LT25). Additional PAM samples ranging in molecular weight from 40,000 to 15 million were purchased from Aldrich Chemical. Molecular weight and charge data were generally provided by the manufacturers.

Cellulose and starch xanthates

Cellulose xanthate was produced following the procedure of Menefee and Hautala (*15*) using Whatman filter paper #4 as a source of pure cotton. Starch xanthates were produced according to Maher (*16*) with the following starch sources: Midsol 50 wheat starch (Midwest Grain Products, Inc., Atchison, KS), Hylon VII high amylose corn starch (National Starch and Chemical Corp., Bridgewater, NJ), and potato starch (Avebe, Princeton, NJ). The cellulose or starch was swollen by soaking in ~20% (w/w) aqueous NaOH. After decanting off excess NaOH, molar volumes of carbon disulfide were added with mixing, forming an orange sticky mass. After several hours of stirring, xanthates were diluted to a 2% solution (based on polysaccharide content), which presumably stopped the reaction and facilitated further dilution. The degree of substitution, ds, was estimated using a standard combustion method for sulfur content (*17*). The shelf life of xanthate is generally limited to several days, so samples were tested within 24 hours after production.

Charged cellulose microfibrils

Acid hydrolyzed cellulose microfibril suspensions were formed from cotton fiber (Whatman filter paper #4) following a procedure outlined by Revol *et al.* (*18*). Cellulose fibers were milled in a Wiley Mill to pass through a 40 mesh screen, added at 8% concentrations to ~60% sulfuric acid at 60°C, and stirred for 30 minutes. The reaction was stopped by adding excess water, and the samples were centrifuged and washed repeatedly (at least three times) until clean of salts and acid. According to previous reports (*18*), hydrolysis with sulfuric acid will create sulfate esters at the surface of the microfibrils. The procedure used here corresponds to a sulfonation level of 10%, i.e. a charge distribution of ~10% over the surface anhydroglucose units (or roughly 0.2 sulfate groups per nm^2).

Chitosan

Chitosan samples were provided by Vanson, Inc., Redmond, WA. In general, chitosan stock solutions were put into solution at high concentration by adding weak acid, lowering the pH to ~6.0. Before use, solutions were diluted and the pH was adjusted back to 7.0. This process was not required for a proprietary "water-soluble" chitosan provided by Vanson.

Lab-Scale Mini-furrows

Lab-scale furrows were created based on principles outlined by Stott *et al.* (*19*). Except where noted, the soil tested in the lab-furrow was a Zacharias gravelly clay loam soil obtained from Patterson, CA, a northern California farming community 90 miles south of Sacramento. This soil was chosen because it is typical of California's Northern Central Valley and because anionic PAM has been particularly effective in controlling its erosion (*11*). Soils were dried, sieved and re-moisturized to 18% (w/w) water contents and then formed into miniature furrows roughly 1/100th the size of a full furrow. That is, 1500g of moist soil was packed flat into a 2.5 x 2.5cm well cut into a 1m long bar. A furrow with dimensions 0.63 x 0.63cm was pressed lengthwise down the center of the packed soil, to create a mini-furrow. Test solutions were pumped down the furrow, set at an angle of 5°, at flow rates of 7mL/min. Water was collected at the lower end of the furrow and tested for solids contents. Sediment concentration was determined by measuring turbidity using a Shimadzu UV-Vis

UV1601 spectrophotometer, and calibrating turbidity with those from a set of pre-weighed soil dispersions. The relative suspended solids was determined for at least 5 replicate samples at each condition.

Multiple parameters affected the results of the furrow tests, including soil type, water flow rate, furrow slope, furrow length, and water purity. For example, if the flow rate was too high, not even PAM was effective in clarifying runoff. After some trial and error, conditions were selected so that results were reflective of the field tests results of Lentz *et al.* (*5*). Figure 1 compares the results of the furrow test with those from the field, showing the effect of PAM's molecular weight, MW, on its efficacy. For the lab-furrow test, >95% solids are removed (relative to control) and there is no molecular weight dependence above MW = 200,000. A similar trend is seen for field tests results, although the sediment levels are higher and the molecular weight dependence is apparent up to MW = 6million. Thus, the furrow test appears more "sensitive" than field test results, i.e. a positive result in the lab-furrow does not necessarily imply full effectiveness in the field. Such conditions were chosen to reduce the risk of missing polymers with potential (i.e. it was better to have false positive results).

Field Tests

Field furrow studies were conducted at the USDA-ARS Northwest Irrigation and Soils Research Laboratory at Kimberly, ID. The soil was a highly erodible Portneuf silt loam (coarse-silty, mixed, mesic, Durixerollic Calciorthid). The furrow had a slope of 1.5% and a length of 137m. Irrigation water was from the Snake River, with average electrical conductivity (EC0) of 0.5 dSm^{-1} and sodium adsorption ratio (SAR) of 0.6 (*20*). Irrigation water was applied from individually regulated valves on gated pipe to conventionally prepared furrows between rows of silage corn. During the initial advance stage of irrigation, the water inflow rate was set at approximately 6 gal/min and polymer samples were metered into the flow at the top of each furrow via calibrated peristaltic injection pumps. Details of the irrigation and runoff monitoring procedure were described previously (*3,5*). Sediment content was measured at the end of furrows by collecting 1L of sample and measuring sediment using the Imhoff cone technique, and turbidity measurements (*21,22*).

Results and Discussion

Cellulose Xanthate, Starch Xanthate, and Cellulose Microfibrils

Cellulose and starch xanthates have been reported as soil stabilizers (*15, 23-25*) and in the removal of heavy metals from wastewater (*16,26*). Xanthates are promising because, like PAM, they carry a charge, they can dissolve or disperse readily in water, and they are available with large molecular weights. For example,

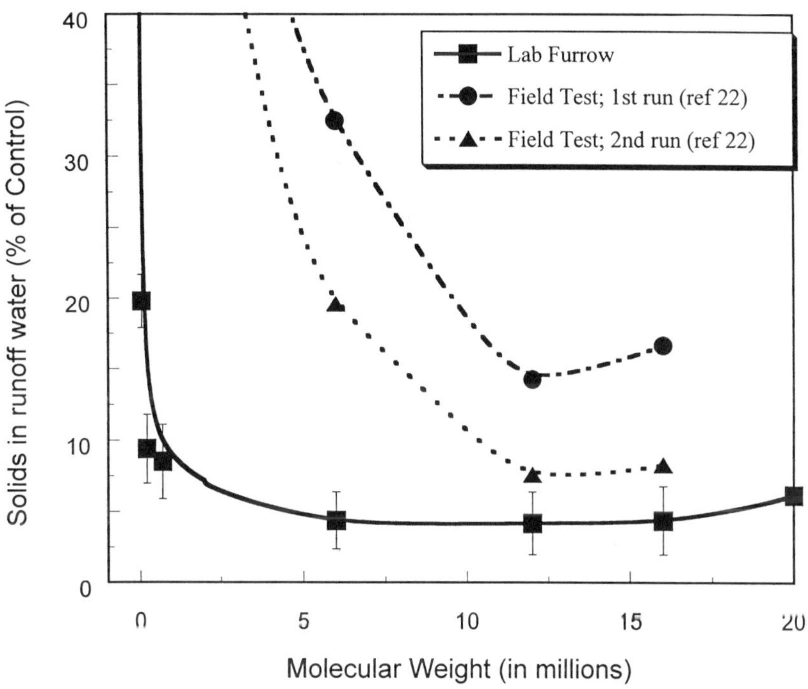

Figure 1. Lab-scale furrow results compared with field test results highlighting the effect of molecular weight on the effectiveness of polyacrylamide, PAM, in controlling erosion-induced soil losses. Curves are drawn between data points in order to guide the eye.

the molecular weight of the amylopectin component of a starch xanthate derivative is in the tens of millions, much greater than that of most PAMs.

Results of lab-scale furrow tests for cellulose xanthate samples are shown in Figure 2. Cellulose xanthate (degree of substitution, ds = 1.7) reduces soil runoff by more than 80% when it is applied at 80 ppm or greater. In comparison, PAM under the same conditions effectively reduces runoff at 10 and even 5 ppm to less than 3% of the control (or 97% sediment removal). Thus, PAM is significantly more effective than cellulose xanthate at concentrations only 1/8th those of the xanthates. Similarly high concentrations are required for starch xanthates produced from wheat, corn and potato starch (Table 1). Interestingly, there is no significant difference between xanthates from wheat, corn or potato, provided the ds is relatively high (ds > 0.38).

Table 1. Soil sediment content in the runoff from lab mini-furrows comparing the efficacy of PAM with several polysaccharide derivatives.

Additive[1]	*Soil Conc. in Runoff (mg/L)*[2]	*Soil conc. in runoff (% of control)*
Control (Tap Water)	51.5 ± 4.2	100
PAM (Cytec 836A); 10ppm	0.99 ± 0.54	1.9
Cellulose xanthate; 80ppm; (ds= 1.7)	9.8 ± 2.2	19.3
Wheat starch; 80ppm; (ds= 0.0)	50.6 ± 6.7	98.6
Wheat starch xanthate; 80ppm; (ds= 0.38)	12.9 ± 5.8	25.0
Wheat starch xanthate; 80ppm; (ds= 0.54)	9.4 ± 4.9	19.1
Potato starch xanthate; 80ppm; (ds= 0.47)	6.7 ± 3.1	12.9
High amylose corn starch (Hylon VII) xanthate; 80ppm; (ds= 0.45)	7.5 ± 1.8	14.5
Cellulose microfibrils; 120ppm	11.0 ± 4.8	21.3

1. The soil used for this study had a pH of 7.5, exchangeable Ca of 7%, and ~5% organics. Calcium nitrate was added to the water, at a concentration of 10 ppm, to ensure ionic bridging.
2. The standard deviations are reported based on 5 replicates per sample.

It is clear from Table 1 that charged polysaccharide derivatives such as cellulose and starch xanthate have some potential to reduce soil runoff during irrigation. Whether xanthates could compete with PAM in the market place, however, is not so clear. The main commercial use of xanthate is as an intermediate in the Viscose production of rayon fiber. Xanthates would generally cost less than PAM ($2.50 -

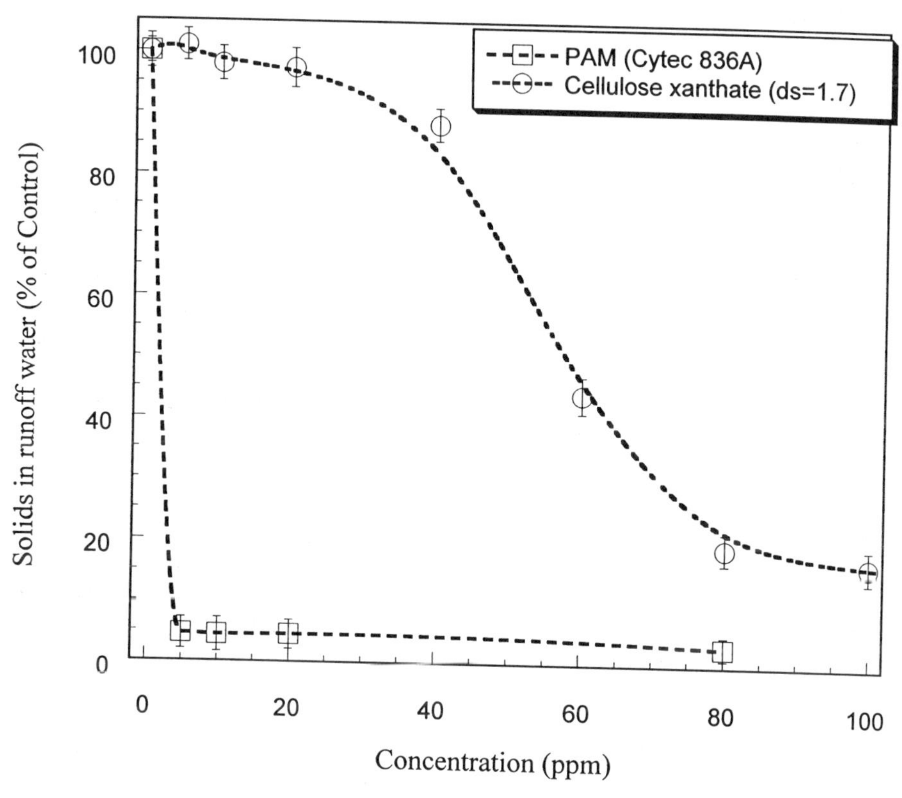

Figure 2. The effect of polymer concentration on the sediment content of runoff water from lab furrow tests. The PAM had a MW = 16 million and charge density of 18% via randomly distributed acrylic acid groups.

5.00 per lb). However, several major drawbacks of xanthate must be overcome. Their relative instability means that they have a shelf life of days or weeks. Meadows (*23,24*) developed strategies to extend the shelf life by removing the water, by storing at cooler temperatures, by storing in vacuum-sealed packages, and by adding dehydrating agents, such as CaO. More critical is the environmental impact of xanthate production. The viscose process for producing rayon fibers via xanthates is becoming increasingly obsolete because of the sulfur-based waste products generated during large-scale production. Interestingly, the sulfur content of xanthates may give it an for sulfur deficient soils or highly calcareous soils. Sulfur could improve both the pH balance and micronutrient availability.

The most applicable conclusion from the data in Table 1 is that charged starch and cellulose derivatives have potential to reduce irrigation induced losses. With this in mind, acid-hydrolyzed cellulose microfibrils were tested using the lab-furrow procedure. Cellulose microfibrils are obtained during acid hydrolysis of pure cellulose, and are the basic crystalline unit of a cellulose fiber. The microfibrils studied here, which were derived from cotton, are rod-like crystallites with a length of 0.1-0.3 microns and a diameter of 5 nm (*18,27*). During acid hydrolysis, microfibrils gain a charge on their outer surface and become stable in water suspensions. Thus, they appear to possess the major attributes required for creating stable soil aggregates, i.e. a large size, an affinity to soil via a surface charge, and easy dispersion in water.

As outlined in Table 1, aqueous suspensions of cellulose microfibrils reduce the irrigation-induced erosion in lab-scale furrow experiments. Concentrations of at least 100 ppm were required to exhibit any significant reduction in runoff sediment, with a concentration of 120ppm resulting in 78% reduction. In contrast, PAM (Cytec Magnifloc 836A) removed 98% of solids at a concentration of 10 ppm.

Despite the fact that relatively high concentrations of cellulose were required, the charged microfibrils are still promising for several reasons. As with starch xanthate, the charge distribution of the microfibrils has not necessarily been optimized. A wide range of charge density, charge type, and microfibril size can be obtained by optimizing reaction conditions and varying the source of cellulose. For example, sugar beet microfibrils are significantly larger than the fibers derived from cotton. In addition, it may be an advantage to add relatively high concentrations of cellulose back to arid soils to improve soil structure, without undue environmental concern. Cellulose microfibrils are attractive because they are readily available from various waste agricultural sources, such as wheat and rice straws, sugar beet fiber, and cotton wastes (including recycled cotton and old clothing). Within the next several years, EPA laws will make it illegal for farmers in several Northwest states, including California, to burn their rice straw between seasons. Having value-added uses for such straws would be advantageous.

Chitosan: Lab-scale and Field Tests

If production costs were no consideration, chitosan would have clear potential as an alternative to PAM. Chitosan has already established itself as a biodegradable flocculating agent for removing heavy metals from plant water (*28*), for reducing suspended biological matter in municipal waste (*29*), and for clarifying swimming pool water in an "environmentally friendly" manner (*30*). The major drawback of chitosan is its market cost of at least $7/lb. This is as much as twice the price of PAM. Chemically, chitosan is similar to cellulose, with the hydroxyl in the 2-position replaced with a primary amino group. It has a net positive charge at neutral or acidic pH values and is available with reasonably high molecular weights.

Lab mini-furrow results outlined in Table 2 show that highly deacetylated chitosan at 20ppm is as effective as PAM in reducing erosion-induced soil losses. With such favorable lab tests results, chitosan was further tested in a series of field tests at the USDA Northwest Irrigation and Soil Research Lab, Kimberly ID. In the field tests, chitosan did not reduce erosion induced soil losses relative to the control (see Table 2). The sediment concentration in the runoff water from chitosan treated furrows was, at best, an order of magnitude higher than that of PAM (although results for the chitosan furrows were highly variable). The only results significantly different from controls were for PAM (Cytec Magnifloc 836A), whereby sediment losses were reduced by 99% relative to controls.

Such poor comparative results, however, do not mean that chitosan had no effect on the irrigation. Observations of the furrows treated with chitosan revealed remarkable results in the first ~20 meters of the furrow. Chitosan acted as such an effective flocculating agent that it removed fine sediments, and even algae from the irrigation water. In the first 20 meters, the chitosan-treated furrows became green in color due, presumably, to algae build-up. In contrast, the control furrows or PAM-treated furrows did not gain a greenish hue during irrigation. Apparently, the flocculation aspect of chitosan works at least on a par with that of PAM. One explanation for the poor results for samples collected at the end of the furrow is that the chitosan binds so readily with sediment that it flocculates out of solution near the top of the furrow. It is not available to reduce sedimentation losses in the lower end of the furrow. Such an explanation implies that the effectiveness of chitosan is dependent on furrow length.

Another explanation is that chitosan, lacking the size of the high molecular weight PAM, does not form a large stable soil/polymer network at the soil surface. Why does molecular weight effect the efficacy of PAM or PAM substitutes? The most common mechanisms invoked for polymeric flocculents is that of bridging. For polymer electrolytes, such as PAM, there is a combination of ionic and polymer bridging. Ionic bridging provides the attractive force between two like-charged entities; for example between two anionic monomers on the PAM chain or between PAM and a negatively charged soil particle. Polymer bridging occurs for large molecular weight polymers if segments of a single polymer chain can adsorb onto more than one particle. For such bridging, the non-adsorbed polymer chain segments must be long enough to exceed the minimum distance of close approach between particles, i.e. the sum of the thicknesses of the electrostatic double layers of the two approaching particles. At very high molecular weights, PAM would have a fully extended end-to-end distance of several microns, and would be able to form stable flocs with multiple soil particles. Ionic bridges further stabilize large flocs by providing interaction between multiple polymer chains, and their associated bound particulates. Longer molecules increase the size of potential polymer/soil flocs.

Forming stable flocs may not be enough, though. Sojka (*10*) points out that the efficacy of PAM depends on two aspects of PAM's interaction with soil. PAM not only forms stable flocs with the soil suspended in water, but also interacts with the soil at the water/soil interface to create a network of PAM/soil aggregates. This network forms a thin protective coating at the furrow surface that prevents shear forces from disrupting the surface structure and improves water infiltration efficiency. The very large molecular weights of PAM allow it to form a larger network, and thus be effective at very low concentrations (5-20 ppm in the irrigation water).

Table 2. A comparison of polyacrylamide, PAM, (MW ~ 16 million, 18% anionic) and chitosan solution in controlling irrigation-induced sediment loss in a lab-scale[1] and a field furrow test.

	Control (Tap Water)	*PAM (10ppm)*	*Chitosan (10ppm)*	*Chitosan (20ppm)*
Lab Mini-Furrow Results[2]				
Solids in runoff (mg/l)	48.1[a]	3.4[b]	5.5[b]	4.2[b]
Solids in runoff (% of control)	100	7.1	11.4	8.7
Field Test Results[2]				
Sediment in runoff (kg/ha)	38745[a]	268[b]	18981[a]	24566[a]
Solids in runoff (% of control)	100	0.7	49.0	63.4

1. No calcium was added to the irrigation water.
2. Results within a row with a different superscript are statistically different from each other.

Calcium Effects

One of the interesting results of this study was the fact that improvements in soil runoff in the lab furrow test were achieved independent of the presence of PAM. To explore this result further, we utilized a soil from Davis, CA that did not respond well to PAM. As shown in Figure 3, calcium alone significantly reduced suspended solids in the runoff from this soil, although PAM and calcium still had a greater effect than

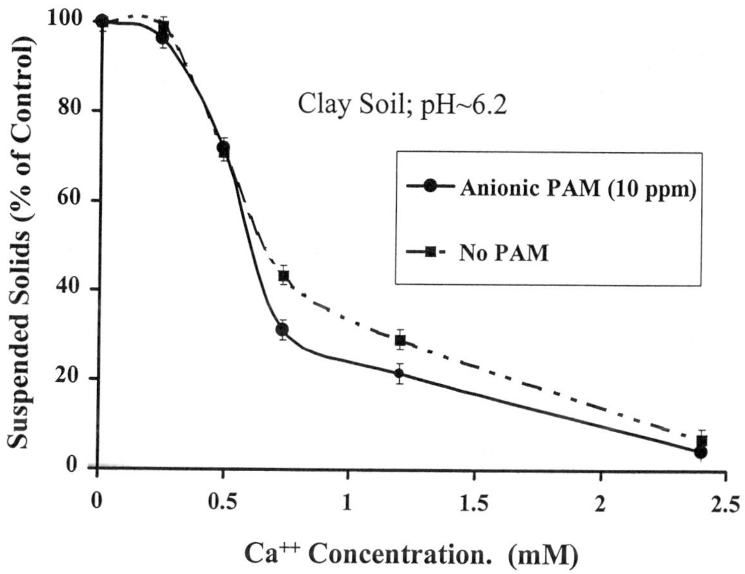

Figure 3. The effect of added calcium on suspended solids in soil runoff from a lab-scale furrow test for water with and without PAM (MW~16 million, 18% anionic).

calcium alone. A similar result was also reported by Wallace and Wallace (*7*), who noted that gypsum or other ions providing higher electrical conductances (*31*) also reduce erosion losses in clay-rich soils. The results in Figure 3 confirm their observations in a lab environment. Note, though, that this result is not necessarily universal. Many sources of irrigation water have a much higher electrical conductance than the tap water used here, with significantly higher levels of exchangeable calcium. Also, the benefit from added calcium is short term. Calcium must be added continuously to the irrigation water. In contrast, PAM can be added for a short period during an initial irrigation series, and provide a lasting effect for weeks without additional doses.

Calcium likely works by precipitating small tenuous floccules from turbid water. Flocs are held together by individual ionic bridges between clay particles. Calcium does not necessarily provide permanent structure to the soil. Floccules can easily be dispersed as they are bathed with water containing less calcium, thereby losing their ionic bridge during cation exchange.

With the addition of polymers, calcium plays an additional role in stabilizing flocs by providing ionic bridges between anionic polymers as well as between polymers and charged particles. As discussed above, the calcium double cation would interact with two separate anionic groups from potentially separate polymer chains, and create a larger, more stable polymer/soil aggregate.

Conclusions

A lab-scale furrow test was utilized to test different polymer additives in irrigation water for their efficacy in reducing soil losses during irrigation. Starch xanthate, cellulose xanthate, and acid hydrolyzed cellulose microfibrils each appear promising, with the ability to reduce soil runoff significantly. The effective concentrations of these derivatives, though, were 10-16 times higher than for PAM, without even matching PAM's full efficacy. These polymers would not cost as much as the "high-grade" PAM employed here, so it is not clear that their high concentrations would automatically exclude them on an economic basis, especially if they are optimized for this application.

The results for chitosan solutions showed that, although lab tests are useful for screening potentially useful polymers for reducing erosion-induced soil losses, field tests, with longer furrows and potentially higher shear forces are required to fully assess polymer properties. In the lab mini-furrow test, chitosan solutions were very effective at controlling erosion-induced soil losses at concentrations approaching those used for PAM. However, in field tests, the effectiveness of chitosan was highly variable, with no clear improvement over controls. This could be because chitosan does not have the molecular size of PAM, preventing it from forming a large stable soil/polymer network at the furrow surface. Another possibility is that chitosan binds "too readily" with sediment and flocculates out of solution near the top of the furrow; an implication that its effectiveness depends on the length of the furrow.

Interestingly, the addition of calcium nitrate to irrigation water, with or without polymer additive, can reduce the amount of soil runoff for the clay-rich soils, and relatively pure water, encountered in Northern California. The addition of exchangeable calcium, however, would not have the long-term effect on soil structure exhibited by PAM.

Although PAM's efficacy was not reached in these tests, there is promise that biopolymers can play a role in reducing erosion-induced sediment loss, especially if their properties are optimized for their specific application. This study explored a very specific application, furrow irrigation. However, the performance of the biopolymers may already be in a range of efficacy to be tried in various other related uses. These include erosion reduction at construction sites and highways, "tackifying" straw beds (so they stay in place), and hydroseeding. In such applications, the costs of material are not the limiting restriction. The environmental advantages of degradable polymers derived from renewable sources may give these biopolymers a market edge.

Literature Cited

1. Becker, H. *Agricultural Res.* **1997**, April, 4.
2. McConnell, L.L.; Bidleman, T.F; Cotham, W.E.; Walla, M.D. *Environ. Pollut.* **1998**, 101, 1-9.
3. Lentz, R.D.; Shainberg, I.; Sojka, R.E.; Carter, D.L. *Soil Sci. Soc. of Am. J.* **1992**, *56*, 1926.
4. Lentz, R.D.; Sojka, R.E. *Soil Sci.*, **1994,** *158*, 274.
5. Lentz, R.D.; Sojka, R.E. 1996. IN Proceedings: Managing Irrigation-Induced Erosion and Infiltration with Polyacrylamide. Editors R.E. Sojka, and R.D. Lentz, Twin Falls ID, May **1996**.
6. Levy, G.J.; Ben-Hur, M.; Agassi, M. *Irrig. Sci.***1991**, *12*, 55.
7. Wallace, A.; G.A. Wallace. IN Proceedings: Managing Irrigation-Induced Erosion and Infiltration with Polyacrylamide. Editors R.E. Sojka, and R.D. Lentz, Twin Falls ID, May **1996**.
8. Mitchell, A.R. Soil Sci. 1986, 141, 353-358.
9. Sojka, R.E.: Lentz, R.D.; Westermann, D.T. *Soil Sci. Soc. Am. J.* **1998,** 62:1672-1680.; R.E. Sojka, 1998, USDA-ARS Soil and Water Management Research Unit, Kimberley ID, personal communication.
10. Sojka, R.E.; Lentz, R.D.; Ross, C.W.; Trout, T.J.; Bjorneberg, D.L.; Aase, J.K. *J. Soil and Water Cons.* **1998**. 53(4) 325-331.
11. Kay-Shoemake, J.L.; M.E. Watwood. IN Proceedings: Managing Irrigation-Induced Erosion and Infiltration with Polyacrylamide. Editors R.E. Sojka, and R.D. Lentz, Twin Falls ID, May **1996**.
12. Kawai, F. *Appl. Microbiol. Biotech.* **1993**, *39*,382.
13. Barvenik, F.W.; Sojka, R.E.; Lentz, R.D.; Andrawes, F.F.; Messner, L.S. IN Proceedings: Managing Irrigation-Induced Erosion and Infiltration with Polyacrylamide. Editors R.E. Sojka, and R.D. Lentz, Twin Falls ID, May **1996**.
14. McElhiney, M.A., USDA Natural Resources Conservation Service, Modesto, CA personal communication, 1998.
15. Menefee, E.; Hautala, E. *Nature* **1978**, *275*, 530.
16. Maher, G.G. **1981**. US Patent 4,253,970.

17. Scroggins, L.H. *J. AOAC* **1973**, 56, 892.
18. Revol, J.-F.; Bradford, H.; Giasson, J.; Marchessault, R.H.; Gray, D.G. *Int. J. Biol. Macromol.* **1992**, *14*, 170.
19. Stott, D.E.; Trimnell, D.; G.F. Fanta. IN Proceedings: Managing Irrigation-Induced Erosion and Infiltration with Polyacrylamide. Editors R.E. Sojka, and R.D. Lentz, Twin Falls ID, May **1996**.
20. Carter, D.L.; Robbins, C.W.; Bondurant, J.A. *ARS-W-4* **1973**, Western Region USDA-ARS Reports, Albany, CA.
21. Sojka, R.E.; Carter, D.L.; Brown, M.J. *Soil Sci. Soc. Am. J.* **1992**, 56, 884.
22. Trout, T.J.; Sojka, R.E.; Lentz, R.D. *Trans. ASAE,* **1995**, *38*,761.
23. Meadows, G.W. **1956**. US Patent 2,761,247.
24. Meadows, G.W. **1959**. US Patent 2,884,334.
25. Swanson, C.L.; Wing; R.E.; Doane, W.M. **1975**. US Patent 3,947,354.
26. Coltrinari, E. **1994**. US Patent 5,320,759.
27. Marchessault, R.H.; Morehead, F.F.; Walter, N.M *Nature* **1959**, *184*, 632.
28. Deans, J.R. **1993**. US Patent 5,336,415.
29. Murcott, S.E.; Harleman, D.R.F. **1994**. US Patent 5,543,056.
30. Nichols, E., Vanson Inc., Redmond, WA, company literature and personal communication, 1997.
31. Shainberg, I.; Letey, J. *Hilgardia* 1984, *52*, 57.

Acknowledgements

The authors would like to thank Youngla Nam and Jim Foerster and for their superior technical support, and Michael McElhiney, USDA Natural Resources Conservationist, Modest, CA, for helpful discussions and for supplying characterized soil samples. This work was partially supported by the Washington and Idaho Wheat Commissions under ARS agreement 58-5325-7-850.

Poly(amino acids): Natural and Synthetic

Chapter 7

Films and Coatings from Commodity Agroproteins

Nicholas Parris[1], Leland C. Dickey[1], Peggy M. Tomasula[1], David R. Coffin[1], and Peter J. Vergano[2]

[1]Eastern Regional Research Center, Agricultural Research Service, U.S. Department of Agriculture, 600 East Mermaid Lane, Wyndmoor, PA 19038
[2]Department of Packaging Science, 225 Poole Agricultural Center, Clemson University, Clemson, SC 29634–8370

Edible films and coatings from agricultural proteins have been investigated as an alternative to synthetic wrapping material for food protection and preservation. Milk protein films are hydrophilic, which limits their utility for long-term storage. Films prepared from casein precipitated from skim milk with CO_2 were found to be more hydrophobic and exhibited better water vapor barrier properties than calcium caseinate films. Unplasticized corn zein films, cross-linked with 20% polymeric dialdehyde starch, exhibited the best water vapor barrier properties. Zein proteins isolated from dry milled corn were less denatured and could be recovered more economically than commercial zein preparations from corn gluten. Paper coated with zein isolate (80-85% protein and 15-20% oil) exhibited good grease resistance and water vapor barrier properties.

Agriculturally derived alternatives to the polyolefin packaging materials currently used by the food industry provide opportunities to strengthen the agricultural economy and reduce importation of petroleum and petroleum products. Interest in the preparation of biopolymer films and coatings from agricultural proteins has recently been renewed. Such films alone or in combination with polysaccharides, lipids etc. have the potential to control mass transfer of gases and thus extend food

Mention of brand or firm name does not constitute an endorsement by the U.S. Department of Agriculture above others of a similar nature not mentioned.

shelf life. Comprehensive review articles have been published on the use of edible films and coatings for food preservation (*1,2,3,4*). Polymeric films are inexpensive to use when applied to food as wrapping. Films made from soluble agricultural proteins, although more expensive may be applied to food by dipping or spraying, thus eliminating packaging waste and manufacturing costs. Edible films, however, are sensitive to relative humidity and are less likely to replace synthetic packaging materials for prolonged storage of food (*2*).

In addition, films prepared from water soluble proteins such as gelatin, casein, serum albumin, and egg albumin are poor water vapor barriers. Films and coatings of soy and grain (wheat and corn) proteins are better water vapor barriers. In this study we have reported isolation techniques and compositional factors affecting the physical properties of films and coatings made from milk and corn protein. The research was directed to the preparation of biodegradable biopolymer films and coatings from renewable resources which could potentially replace synthetic materials.

Milk Protein Films

(Nonfat Dry Milk) NDM

The principal proteins in cow's milk are caseins (~30g/L) which exhibit amphipathic character and assemble to form micelles, and whey proteins (~4.5g/L) which are globular and easily denatured. NDM powder prepared from skim milk has become increasingly important as a food ingredient because of its ease of handling, storage, and nutritional value. To meet some NDM product specifications on protein denaturation, skim milk is heated before spray-drying. For example, high pre heat treatment (85°C for 30 min) of NDM used as an additive in baking is essential to improve the extensibility and water absorption of dough (*5*). A whey-casein complex has been shown chromatographically to form in NDM preheated to 74 and 85°C for 30 min and is responsible, in part, for the functional properties of the product (*6*). Maynes and Krotchta (*7*) found that formation of NDM films was impeded due to the high concentration of lactose (~50g/L) in the reconstituted milk which crystallized during film drying. To overcome this problem, lactose was enzymatically hydrolyzed, complexed with potassium sorbate or removed by ultrafiltration (UF). They obtained homogeneous films from NDM after the lactose had been hydrolyzed with β-D-galactoside galactohydrolase and then heated at 100°C for 30 min. The monomers, galactose and glucose, are more soluble and do not crystallize during film formation. Solutions of NDM containing 10% potassium sorbate and ultrafiltered milk with 25% glycerol formed films after similar heat treatment. The presence of potassium sorbate interferes with the crystalline environment of lactose and thus inhibits crystal growth. UF-total milk protein (TMP) films had the lowest permeability followed by potassium sorbate complexed-NDM and lastly, NDM with lactose hydrolyzed to monomers (Table I). The relative humidity (RH), at the inner surface of the film which was corrected for stagnant air effects, was between 65 and

Table I. Water Vapor Permeability (WVP) of NDM Edible Films

Film[a]	*Thickness (mm)*	*RH (%)*[b]	*Permeability (g-mm/m²-h-kPa)*
NDM, hydrolyzed lactose	0.12	72	3.76
NDM: K Sorbate (9:1)	0.11	74	3.30
UF-TMP[c]:Glycerol (4:1)	0.07	65	2.93

[a]Heated at 100°C for 30 min.
[b]Relative humidity (RH) at the inner surface of the film.
[c]TMP = Total milk protein (ultra filtered rehydrated NDM).
SOURCE: Adapted from ref (*7*).

74%. These values were considerably lower than those obtained for synthetic films (compare Table I and Table IV).

Casein

Casein can be isolated from milk through the action of lactic acid bacteria during cheese making or by acidification of skim milk by an acid such as HCl or H_2SO_4. Acid precipitation of casein has also been carried out by dissolution of CO_2 in skim milk (*8*). The use of CO_2 as a precipitant is attractive because it eliminates the formation of ionic byproducts from mineral acids. Typically casein was precipitated by injecting CO_2 into skim milk at ~5000 kPa and 38°C. The precipitate was washed with distilled water to remove whey proteins, lactose and unbound minerals. The calcium content of casein precipitated in this way was higher than that precipitated using mineral acids. Films from CO_2-casein were found to be more hydrophobic and only 7% of the material dissolved in water compared to 90% of the calcium caseinate films (Table 2). Increased solubility was observed in both cases for films containing glycerol. The increased solubility for the films containing GLY appears to be due to the presence of the plasticizer, because both film have the same protein content. The solubility of plasticized CO_2-casein films is comparable to that of plasticized zein films made from corn zein, a prolamine, which is soluble in aqueous alcohol mixtures. Water vapor permeability (WVP) values for CO_2-casein films are lower than those for the calcium caseinate films (Table 3). The lower RH for the calcium caseinate films, at comparable thickness to the CO_2-casein films, is attributed to greater absorption of water by the protein resulting in swelling of the film. Both types of films showed an increase in WVP values with increasing film thickness which is characteristic of hydrophilic films due to swelling. Such differences in the properties of the two types of films are indicative of structural dissimilarities.

Table II. Water Solubility of CO_2-Casein Films

Protein film	*Material Dissolved in Water (g/100ml)*	*Reference*
CO_2-casein	7.1	Reference (*20*)
CO_2-casein-30% GLY	16.8	Reference (*20*)
Calcium caseinate	90.0	Reference (*20*)
Calcium-caseinate-30% GLY	100.0	Reference (*20*)
Zein-30% GLY	14.2	Reference (*14*)

Table III. WVP of Casein Edible Films

Film[a]	*Thickness (mm)*	*RH (%)*	*Film Swelling*	*Permeability*[b] *(g-mm/m²-h-kPa)*
CO_2 casein	0.112	85.8	No	2.22a
	0.163	87.7	No	2.58b
	0.184	87.9	No	3.21c
	0.277	89.7	No	3.80d
Ca caseinate	0.171	86.5	Yes	3.18c
	0.222	85.5	Yes	4.45e

[a]30% Glycerol.
[b]Means in the same column with no letter in common are significant at $P < 0.05$ using the Bonferroni LSD multiple-comparison method.
SOURCE: Adapted from ref (*20*).

Whey Protein

Whey proteins can be isolated after precipitation of casein from skim-milk. Whey films have been prepared by heating an 8-12% solution of whey protein isolate (WPI) to temperatures from 75 to 100°C (*9*). The effects of heat treatment, protein concentration, and pH on plasticized films were examined. The best film formation occurred at neutral pH for a 10% solution heated at 90°C for 30 min. Heat was essential to form intermolecular covalent bonds and intact films (*9*). The pure whey protein films, however, were too brittle and required plasticizer, which significantly increased their WVP values (Table IV). At comparable concentrations and relative humidity conditions, sorbitol plasticized whey protein films had lower WVP values than glycerol plasticized films. Despite many attempts by researchers to make hydrophilic films more hydrophobic, the WVP values for synthetic films are still orders of magnitude lower. Comparisons of oxygen permeability, however, showed that hydrophilic milk protein films are less permeable than synthetic packaging materials (*4*). Covalent crosslinking of whey protein films with transglutaminase results in very strong films. One should be aware that the effectiveness of this crosslinked coating depends on the acidity and the proteolytic activity of the surface to be coated (*10*).

Table IV. WVP of Whey Protein Edible Films

Film	*Thickness (mm)*	*RH (%)*	*Permeability (g-mm/m²-h-kPa)*
Hydrophilic:[a]			
WPI, 50% glycerol[b]	0.12	59	6.44
WPI, 37.5% glycerol	0.12	65	4.99
WPI, 50% sorbitol	0.14	75	3.53
WPI, 37.5% sorbitol	0.13	79	2.58
Synthetic:[c]			
polyethylene	0.0078	99.8	0.0022
poly(vinylidene chloride)	0.0084	99.8	0.0016

SOURCE: Adapted from ref[a] (*9*) and ref[c] (*21*).
[b](w/w)

Corn Protein Films

Zein

Isolation and film-forming properties of corn zein was investigated in order to identify a more inexpensive and more hydrophobic film or coating. Classification of corn proteins into albumins, globulins, prolamines and glutelins traditionally has been based on solubility (*11*). Newer techniques of protein identification have shown that zein, a mixture of prolamine rich proteins, comprises approximately 50% of the protein in the corn kernel (*12*). The most interesting property of zein for use as an industrial protein is its ability to form a tough, glossy, and grease-proof coating. It can be cured with various aldehydes to form an essentially insoluble product. Zein has been commercially isolated from corn gluten meal (60% protein) at 60°C using 88% isopropanol containing 0.25 wt% NaOH (*13*). Less than half of the protein in the meal, however, is recovered at low temperatures, after the supernate is decanted. The principal proteins in commercial zein are the α-zeins which are a complex group of closely related prolamines with molecular weights of 19 and 22 kDa. Their molecular weights can readily be estimated by sodium dodecyl sulfate capillary electrophoresis (SDS-CE) by comparing their migration time to a calibration plot of molecular weight standard proteins. Zein proteins gel, or polymerize, upon storage in aqueous alcohol solutions at concentrations greater than 20% protein as shown in the capillary electropherogram of zein (Figure 1a). These proteins associate through disulfide linkages, since only the monomer plus a small amount of dimer was present after reduction with 2-mercaptoethanol (Figure 1b).

Commercial Zein Films

Zein films have been prepared by heating 10% solutions of zein and plasticizer at 60°C for 10 min. The solutions were then dried in a vacuum oven adjusted to 10 inches mercury, at approximately 50°C (*14*). Cross-linked films were prepared by adding the cross-linking agent to the mixture which was then heated at 70°C for 30min in a sealed container and dried as above. Zein films containing no plasticizer had the lowest WVP values (Table 5). Of these, films cross-linked with 20% polymeric dialdehyde starch (PDS) exhibited the best water barrier properties. Films prepared in acetone had lower WVP values than those prepared in ethanol. Incorporation of plasticizer into zein films significantly reduced water vapor barrier properties resulting in an almost doubling in WVP values (Table 5). In addition, WVP values, and the corresponding variation between zein films with the same plasticizer composition, decreased with increasing poly(propylene glycol) (PPG) concentrations (Figure 2). This variation could be attributed to preferential separation of GLY over PPG from the film. Films containing a GLY:PPG ratio of 1:3 exhibited elongation values almost fifty times greater than GLY plasticized film. Incorporation of cross-linking agents into zein films resulted in approximately a 2-3-fold increase in tensile strength values (*14*).

Table V. Water Vapor Permeability of Zein Films

Film Type	*Casting Solvent*	*Plasticizer*	*Permeability (g-mm/ m²-h- kPa)*
zein	ethanol	none	0.620c
zein	acetone	none	0.577c
zein	ethanol	GLY:PPG (30%)[a]	1.060ab
zein/PDS (20%)	ethanol	none	0.533c
zein/PDS (20%)	ethanol	GLY:PPG (30%)[a]	1.150a

[a]The ratio of mixed plasticizers was 1:3.
SOURCE: Adapted from ref (*14*).

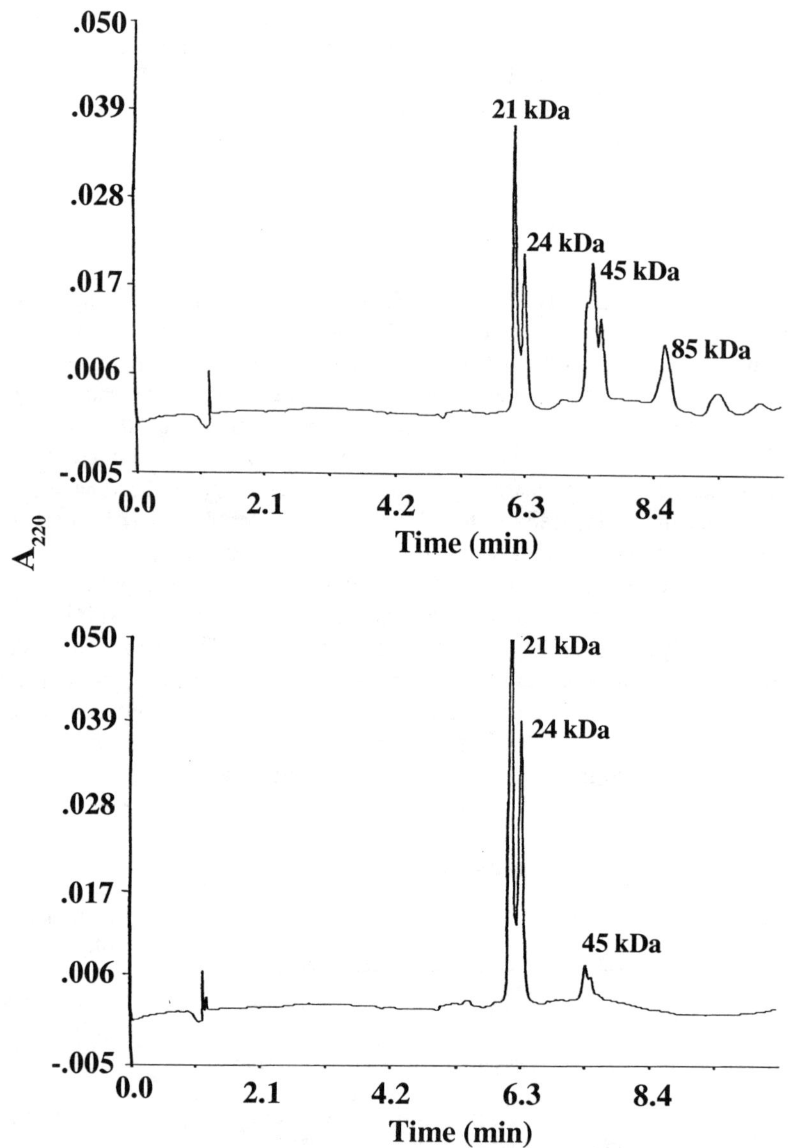

Figure 1. Sodium dodecyl sulfate capillary electrophoresis of zein: unreduced (a) and reduced (b).

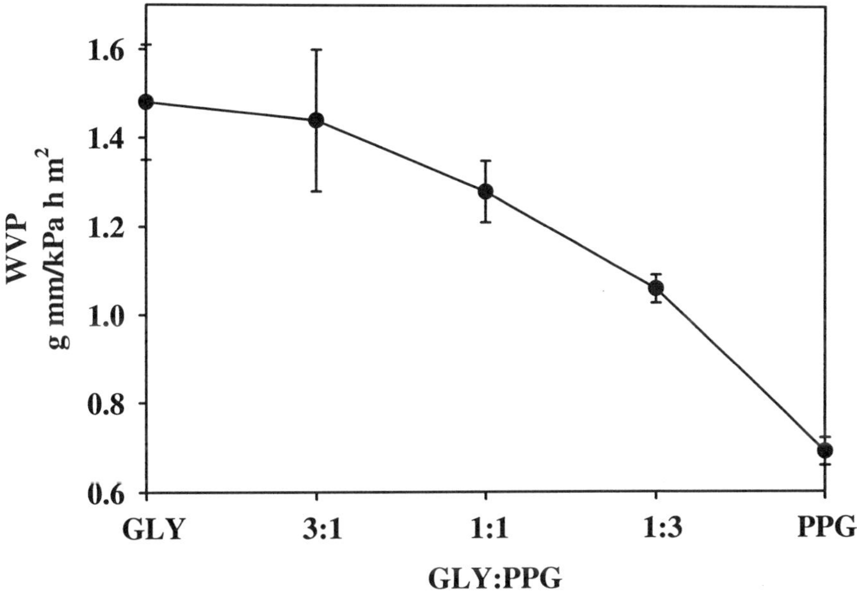

Figure 2. Effect of PPG concentration on zein film water vapor permeability (WVP).
SOURCE: Adapted from ref (14)

Zein Isolate

New markets for zein are severely limited because of its cost. Presently commercial zein sells for about $10/lb which is too much to compete with markets presently held by other raw materials such as edible shellac at $2-3/lb. Dickey et al., (*15*) estimated the cost of ethanol extraction of zein/oil mixtures from ground dent corn to be significantly cheaper than commercial purified zein. Dent maize, milled to a median size of 2mm was extracted with 70% ethanol as described in Figure 3. The extract was separated from the corn by centrifugation and diluted to 40% ethanol by vacuum evaporation. The precipitated product was scraped from the bottom and walls of the evaporation vessel and dried in a lyophilizer. The typical analysis for the dried product was 80-85% protein; 15-20% oil; <0.2% starch and it costs approximately $1/lb. Comparison of HPLC chromatographic profiles of alcohol-extractable corn protein in commercial zein and the zein isolate indicated that the isolates contained a greater number of resolved peaks (Figure 4). The broader, poorly resolved peaks displayed in the profile for commercial zein (Figure 4a) could be attributed to deamidation reactions which occur during steeping of the corn prior to gluten separation (*13*). Early eluting peaks (10-17 min) in Figures 4b and 4c represent β- and γ -zeins. They are not found in commercial zein but were quantitatively extracted from the endosperm of corn kernels (*16*). These subunits reportedly contribute to improved antioxidant activity of zein against docosahexaenoic acid ethyl ester (*17*).

Zein Isolate Films

WVP for glycerol plasticized films, containing various amounts of zein protein isolated from dry milled corn, varied from 0.91 to 1.21 (Table 6). Except for zein isolate 3, WVP values for the isolates and commercial zein were the same. As mentioned above, zein films were much less water soluble than milk protein films. For the zein isolates, the amount of material dissolved in water was between 14.2 and 15.7%, which was similar to glycerol plasticized CO_2-casein films (Table 2).

Table VI. Water Vapor Permeability and Solubility of Zein Isolate Films

Film[a]	*Thickness (mm)*	*Protein (%)*	*WVP (g-mm/kPa-h-m²)*	*Material Dissolved in Water (g/100ml)*
zein isolate 2	0.080	77	1.21a	15.7b
zein isolate 3	0.084	88	0.91b	14.2c
zein isolate 30	0.086	86	1.18a	14.4c
commercial zein	0.093	91	1.21a	16.9a

[a]contains 30% glycerol.

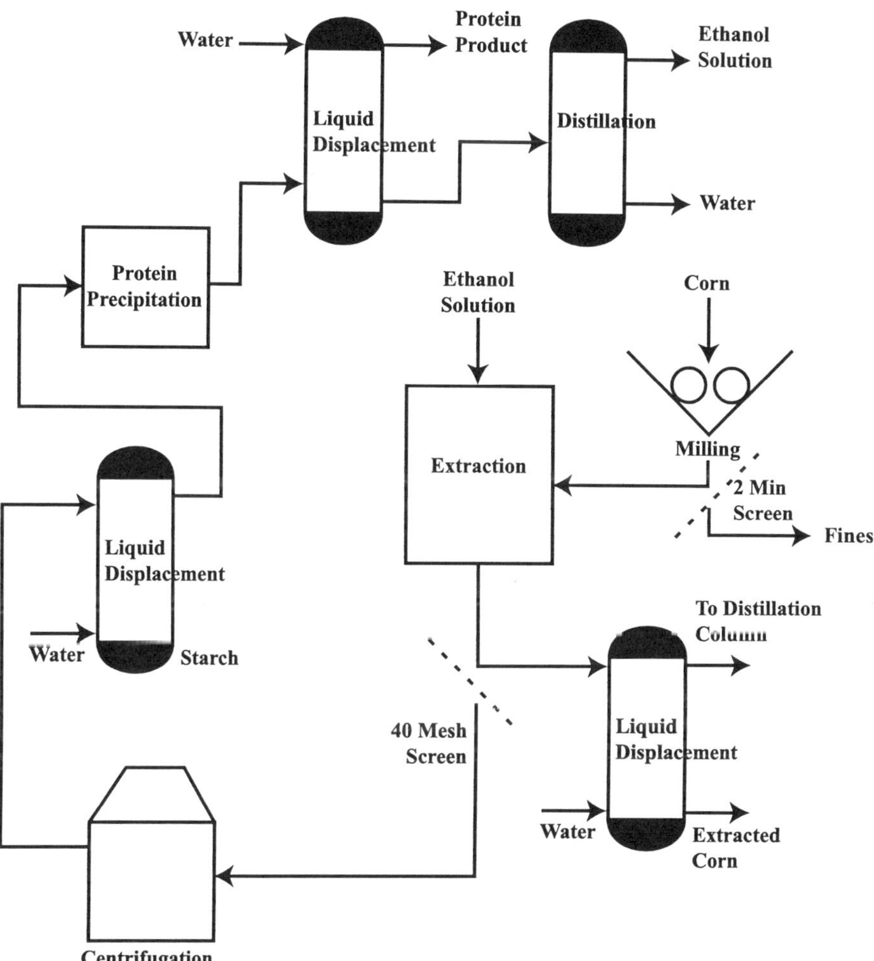

Figure 3. Schematic diagram of zein extraction.

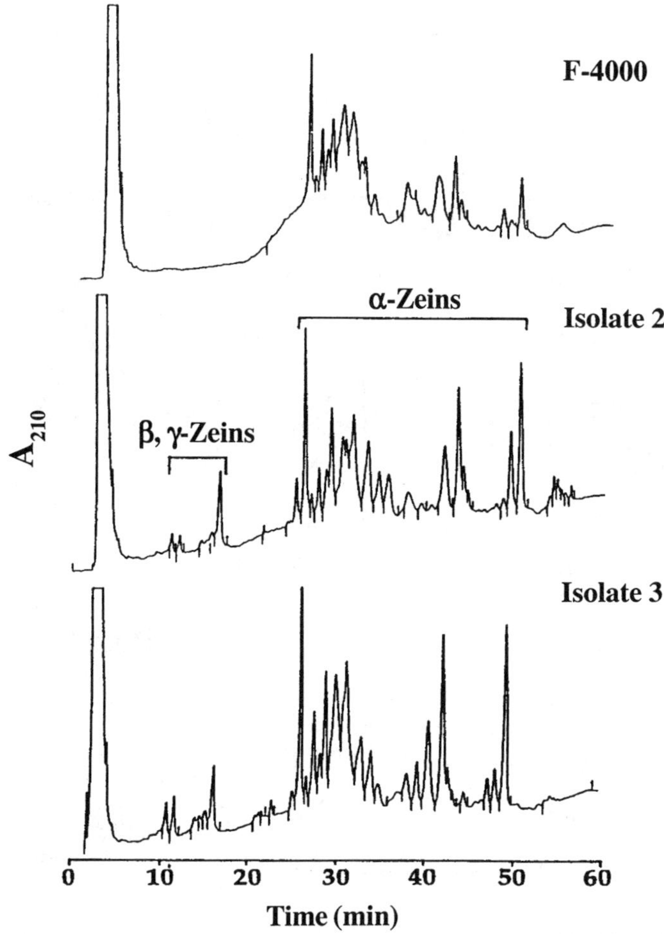

Figure 4. RP-HPLC separation of alcohol soluble zein proteins (a) commercial zein (Freeman F-4000). (b) zein isolate 2. (c) zein isolate 3.

Zein Isolate Coatings

Trezza and Vergano, (*18*) reported the grease-proofing properties of paper coated with commercial zein as a function of coating level, plasticizer addition and time of exposure. We found that grease permeation for zein applied to brown Kraft paper was dependent on the amount zein applied. As shown in Figure 5, the strong dependence of area stained on the amount of zein coating on the paper appeared to stop below 50 mg/16 in^2. Suggesting above that monolayer coverage, at that loading level of coating, the increase in grease resistance with increased zein was lower. Grease permeability measurements were determined for Kraft paper with approximately this level of coating for various time periods (Table 7). The average reduction in the average area stained for paper coated with five different zein isolate samples was greater than 50% relative to the uncoated paper. In another study of recyclable zein-wax coated paper, we demonstrated that the zein layer of the bilayer coating contributes grease-proofing and the paraffin wax layer water resistance (*19*). Since the isolates contained between 15-20% corn oil, coating Kraft paper with zein isolate at levels which produced good grease resistance may also exhibit water barrier properties. For the five isolates listed in Table 8, there was a 14% decrease in water vapor transmission relative to the uncoated paper.

In conclusion, hydrophilic milk protein films can be made more hydrophobic through protein crosslinking, type of plasticizer, or technique used to isolate protein. They have better oxygen barrier properties than synthetic films. Zein protein films are more hydrophobic than milk protein films but are expensive, which limits their commercial applications. The more economical zein isolated from dry milled corn is capable of forming films and coating. Grease proofing and water barrier properties indicate that the isolate is a suitable coating for packaging materials used for the storage and shipping of perishable foods. This material is biodegradable.

Table VII. Grease Permeability of Zein Isolate[a]

Isolate[b,c]	*Coating(mg/16 in^2)*	*Time (hr)*	*% Area Stained/hr*
blank	0	1,2,3	36
123B	52	1,3,4	14
124-2	53	1,2,4	15
125-2	40	1,3,4	17
125-4	57	1,2,4	18
127-3	68	1,3,4	17

[a]TAPPI T507, (22).
[b]80% protein; 15% oil; 0% starch.
[c]n=3.

Table VIII. Water Vapor Transmission Rate of Zein Isolate[a]

Isolate[b,c]	*Coating (mg/16 in²)*	*WVTR (g-m²/day)*
blank	0	1524
123B	73	1279
124-2	46	1350
125-2	43	1403
125-4	32	1385
127-3	88	1165

[a]ASTM E96, ref (*23*); 50% RH; 25°C.
[b]80% protein; 15% oil; 0% starch.
[c]n=3.

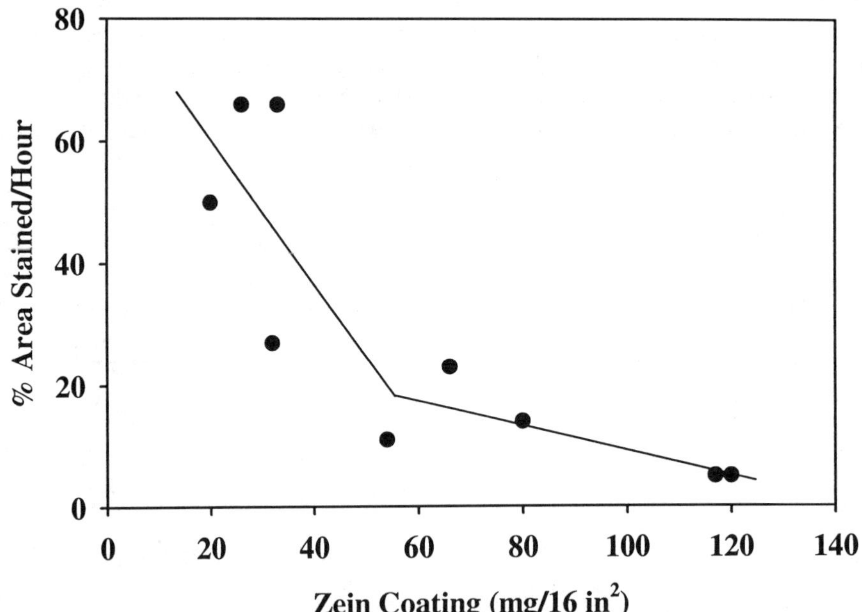

Figure 5. Effect of the amount of zein coating on area stained.
Source: Adapted from ref (19).

References

1. Guilbert, S. In *Food Packaging and Preservation;* Mathlouthi, M., Ed.; Elsevier Science Publishing: New York, 1986; pp 371-394.
2. Kester, J. J.; Fennema, O. R. *Food Technol.* **1986,** *40*(12), 47-59.
3. Gennadios, A.; Weller, C. L. *Food Technol.* **1990**, *44*, 63-69.
4. McHugh, T. H.; Krochta, J. M. *Food Technol.* **1994,** *48*(1), 97-103.
5. Guy, E. J. In *By-Products from Milk;* Webb, B. H.; Whittier, E. O. Eds.; AVI Publishing Co.: Westport, CT, 1970; pp 197.
6. Parris, N.; White, A. E.; Farrell, H. M. *J. Agric. Food Chem.* **1990**, *38*(3), 824-829.
7. Maynes, J. R.; Krotchta, J. M. *J. Food Sci.* **1994**, *59*(4), 909-911.
8. Tomasula, P. M.; Craig, J. C.; Boswell, R. T.; Cook, R. D.; Kurantz, M. J.; Maxwell, M. *J. Dairy Sci.* **1995,** *78*, 506-514.
9. McHugh, T. H.; Krochta, J. M. *J. Agric. Food Chem.* **1994**, *42*(4), 841-845.
10. Mahmoud, R.; Savello, P. A. *J. Dairy Sci.* **1993,** *76*, 29-35.
11. Osborne, T. B. *The Vegetable Proteins;* 2nd ed. Longmans, Green and Co.: London, 1924; pp 154.
12. Hamaker, B. R.; Mohamed, A. A.; Haben, J. E.; Huang, C. P.; Larkins, B. A. *Cereal Chem.* **1995,** *72*, 583-588.
13. Reiner, R. A.; Wall, J. S.; Inglett, G. E. In *Industrial Uses of Cereals;* Pomeranz, Y., Ed.; American Association of Cereal Chemists, Inc.: St. Paul, MN, 1973; pp 285-302.
14. Parris, N.; Coffin, D. R. *J. Agric. Food Chem.* **1997,** *45*(5), 1596-1599.
15. Dickey, L. C.; McAloon, A; Craig, J. C.; Parris, N. *Can. J. Chem. Eng.* **1998,** (in press).
16. Dombrink-Kurtzman, M. A. *J. Cereal Sci.* **1994**, *195*, 57-64.
17. Matsumura, Y.; Andonova, P. P.; Hayashi, Y.; Murakami, H.; Mori, T. *Cereal Chem.* **1994,** *71*, 428-433.
18. Trezza , T. A.; Vergano, P. *J. Food Sci.* **1994,** *59*, 912-915.
19. Parris, N.; Vergano, P. J.; Dickey, L. C.; Cooke, P. H.; Craig, J. C. *J. Agric Food Chem.* **1998,** 46 (10), 4056-4059.
20. Tomasula, P. M.; Parris, N.; Yee, W.; Coffin, D. *J. Agric. Food Chem.* **1998**, 46 (11) 4470-4474.
21. Parris, N.; Coffin D. R.; Joubran, R. F.; Pessen, H. *J. Agric. Food Chem.* **1995,** *43*(6), 1432-1435.
22. TAPPI. *TAPPI Test Methods.* Technical Association of the Pulp and Paper Industry, Atlanta GA.
23. ASTM. Standard test method for water vapor transmission of materials. In ASTM *Book of Standards;* American Society for Testing and Materials: Philadelphia, PA, 1980; pp E96-80.

Chapter 8

Novel Materials from Agroproteins: Current and Potential Applications of Soy Protein Polymers

Xiuzhi S. Sun

Department of Grain Science and Industry, Kansas State University, Manhattan, KS 66506

Advanced technology in petroleum polymers has brought incredible benefits to human beings. However, environmental pollution from petroleum-based disposable plastic wastes has become a severe global problem. Protein polymers have recently been considered as alternatives to petroleum polymers for disposable items, such as fast food utensils, shopping/trash bags, and biodegradable composites. Protein polymers also have been considered as environmentally-friendly adhesives in plywood, particle boards, and packaging and labeling. The objective of this paper is to give an overview of current and related research and potential applications of soybean protein polymers. Molecular structure and related properties (thermal and mechanical properties and water absorption) of the soy proteins, which could help to better understand and utilize the soy protein polymers, also are discussed.

Introduction

The environmental impact of nondegradable plastic wastes is growing worldwide. About 22 billion pounds of plastic waste were discarded in 1992, and this figure is expected to reach 34 billion pounds by 2002 (*1*). Both government and private agencies have recommended recycling these plastics, but only 3%, was recycled in 1996. Alternative disposal methods also are insufficient. Incineration can generate toxic air pollution, and satisfactory landfill sites are limited. Also, petroleum resources are finite and becoming limited, and must be imported by most countries, so the cost of petroleum-based plastics is increasing steadily.

On the other hand, abundant proteins are available from renewable resources and agricultural processing by-products, such as soybean proteins from oil processing and gluten proteins from corn or wheat starch production. For example, soybeans contain about 40% protein, and the U.S. produces about 52% of the total world soybean crops. Also, sufficient proteins are available from dairy and animal processing, such as whey proteins and animal blood proteins, which in themselves often cause environmental problems. Utilizing these protein by-products for biodegradable resins would help to ease the environmental problems.

Research related to using proteins as alternatives to petroleum polymers is under way worldwide. Protein polymers have been found to have great potential for many unique applications in industrial, medical, and military areas. Proteins have been considered as environmentally-friendly adhesives, which could be used for plywood, particle boards, and in packaging and labeling. Biodegradable composites that incorporate protein resins with natural fiber have several unique applications in both industrial and military uses. Proteins have been considered for edible coatings and medical capsules. They also could be manufactured into many articles and films through extrusion and injection molding.

The purpose of this paper is to give an overview of current and related research works and application potentials of soybean protein polymers. Soy protein molecular structure and related properties (thermal and mechanical properties and water absorptions), which could help to better understand and utilize the soy protein polymers, also are discussed.

Soy Protein Structures

Soy proteins are complex macromolecules that contain certain amino acid monomers, which determine their properties. A number of side chains and residuals are connected to these monomers which significantly affect the physical and chemical properties of the proteins. They can react or interact with chemical additives and plasticizers during processing, resulting in different properties. Also, protein structure often is modified easily by physical, chemical, or enzymatic methods to obtain desirable properties and different thermal behaviors.

Soy proteins contain many subunits. Major components (up to 80% of the total proteins) include glycinins (11S-rich globulins, 52%) and conglycinins (7S-rich globulins, 35%). The 11S fraction has a molecular weight of about 300,000 to 600,000 D and is made up of 12 main subunits (six acidic and six basic subunits). The molecular weights of the basic and acidic subunits are about 37,000 D and 20,000 D, respectively. Basic subunits have higher hydrophobic amino acid contents than acidic subunits. The 11S molecules contain two hexagonal rings that stack on top of each other and enclose the acidic and basic subunits (*2*). The subunits are held together by hydrophobic interactions and disulfide bonds, and the two layers are held together by electrostatic and or hydrogen-bonding forces (*2, 3*). The 11S contains about 23.5% hydrophobic amino acids and about 46.7% hydrophilic amino acids (*4, 5*).

The 7S fraction has a molecular weight about 180,000 to 200,000 D and is the combination of three main subunits (α, α′, and β) with molecular weights of about 57,000 to 83,000 D (*6*). These three major subunits unfold in phosphate buffer containing urea and then reassociate and reconstitute after the removal of the urea. The pH values and ionic strength of a buffer significantly affect the association-dissociation of the 7S globulins (*7*). The amino acid compositions of the major subunits of 11S and 7S and selected physical characteristics of soy proteins are listed in Table 1.

The covalent disulfide (SS) bonds play an important role in the stabilization of the tertiary structure of many proteins. The formation of SS bonds has been implicated in the formation of SS polymers, heat denaturation, and inter-subunit association of the proteins (*8, 9*). Most SS bonds are buried inside the protein molecules and are brought outward during denaturation (*9*). More SS bonds are formed with the 11S fraction than with the 7S upon heating, and proteins with more SS bonds absorb less water than those with less SS bonds (*10*).

Soy Protein Modification

The properties of proteins can be modified using physical, chemical, and enzymatic methods. Physical modification methods mainly include heat (*11*) and pressure (*12*) treatments. The heat provides the protein with sufficient kinetic energy to break hydrophobic bond which disassociate the subunits (*11*). The disassociation and unfolding expose hydrophobic groups, previously buried within the contact area between subunits or inside the folded molecules (*13, 14*). In solution, soy protein disassociates and coagulates at high pressure opening large hydrophobic regions resulting in greater viscosity than at atmospheric pressure (*12*).

Chemical modification methods can cause alteration of these protein properties. Chemical treatments can cause reactions between functional groups, resulting in either addition of a new functional group or removal of a component from the protein. Such treatment include acetylation, succinylation, phosphorylation, limited hydrolysis, and specific amide bond hydrolysis. Acetylation is the reaction between protein amino or hydroxyl groups and the carboxyl group of an acylating agent. Acetylation reactions increase the surface hydrophobicity of a protein (*15*). Succinylation treatment converts the cationic amino groups in the protein to anionic residues, increasing the net negative charge and resulting in an increase in hydrophobicity at specific succinylating conditions (*16*). Succinylation treatment also causes increased viscosity (*17*). Phosphorylation is another effective method to increase negative charges (*18*) and improve gel forming ability and cross-linking (*15*). Chemical hydrolysis is one of the most popular methods for protein modifications by acid-based agents. Peptide bonds on either side of aspartic acid may be cleaved at a higher rate than other peptide bonds during mild acid hydrolysis (*19*). Hydrophobicity of a protein greatly increases at a specific mild acid hydrolysis condition (*20, 21*). Proteins also can be modified simply by using various acids, salts, and alkali to unfold the protein main chains and increase adhesive strength and hydrophobicity (*22*).

Table 1. Amino acid composition and physical characteristics of soy proteins subunits

	11S	Acidic	Basic	7S	α′	α	β
Try	0.75	Nd[a]	ND	0.30	ND	ND	ND
Ile	4.24	3.95	4.72	6.40	4.88	5.76	6.23
Tyr	2.81	1.70	2.76	3.60	2.34	2.35	2.55
Phe	3.85	3.13	4.17	7.40	5.10	5.12	6.52
Pro	6.85	7.51	5.76	4.30	6.58	7.04	5.10
Leu	7.05	5.36	10.15	10.30	7.43	8.74	10.48
Val	4.83	4.00	7.61	5.10	4.88	4.45	5.38
Lys	4.44	4.96	3.59	7.00	7.21	6.18	5.38
Met	0.98	0.94	0.89	0.30	0.42	0.43	0.00
Cys	1.44	0.93	0.69	0.00	0.00	0.00	0.00
Ala	5.16	3.52	7.59	3.70	4.25	4.48	5.38
Arg	5.81	6.49	5.99	8.80	7.21	8.74	7.08
Thr	3.91	3.64	4.09	2.80	2.33	2.13	2.54
Gly	7.50	7.99	7.23	2.90	4.88	4.48	4.53
Ser	6.66	6.01	7.15	6.80	6.58	6.61	7.08
His	1.89	2.25	1.96	1.70	3.61	1.28	1.98
ASx	11.88	12.42	12.72	14.10	11.04	11.73	13.03
GLx	19.97	25.23	12.93	20.50	21.23	20.47	16.71
Hϕ[a]	953	827	1044	1092	1048	988	1091
pI[b]		4.6-5.4	8-8.5		5.2	4.9	5.7-6
CHGS[c]	0.20			0.47			
Carbohy (%)[d]	0	0	0	5	5	5	3.5

[a]Hϕ = Hydrophobicity, cal per residue

[b]pI = Isoelectric pH

[c]CHGS= Proportion of charged groups at pH 6

[d]Carbohy = Carbohydrate

Source: Reproduced with permission from reference 59, Copyright 1985 Academic.

Thermal Behavior

A polymer goes through a number of phases during thermal processing, from glassy to rubbery and to liquid flow as temperature increases, which is called thermal transition. After the liquid flow, thermoset polymers become harder and harder due to cross-link reactions or curing. Different polymer structures have different thermal and flow-curing behavior and result in different rheological properties at each thermal state. These characteristics often are used as processing windows to select correct processing approaches and parameters. The interactions between polymer structure and processing design significantly affect the final product quality.

A few proteins, such as cheese proteins, are thermoplastics. Soy proteins are pressure-sensitive thermoset polymers. Epoxy, for example, is a petroleum-based thermoset polymer. At elevated temperature, epoxy will melt and then will cure at atmospheric pressure. Soy protein polymers will unfold/melt (paste like) and cure under both elevated temperature and pressure. At atmospheric pressure, the soy proteins with less than 20% plasticizers still remain powder-like. However, soy proteins with a high plasticizer content (i.e., 80%) was cured into films by casting method at atmospheric conditions (*23*).

A differential scanning calorimeter (DSC) thermogram of a 1:1 mixture of 7S and 11S powders at 10% moisture content is shown in Figure 1. The thermal transition temperatures were at 163.4°C for 11S-RG and 137.6°C for 7S. The peak temperature of 11S globulin was about 25.8°C higher than that of 7S globulin, which dould be attributed to differences in chemical structure. Previous studies have reported that the thermal transition temperature of 11S globulin was always higher than that of 7S globulin at any water content, and the peak temperature decreased as moisture content increased (*24, 25*). Soy proteins are not thermally reversible polymers, because proteins are denatured at elevated temperature. A heat-cool-heat cycle DSC curve of the 11S powder with 10% moisture content is given in Figure 2. A thermal transition at 163°C was observed during the first time heating, and no thermal transition was observed during the second heating.

The decomposition temperatures of 7S and 11S and their mixture were all in the ranges from 200 to 500°C, and their weight loss curve patterns were almost identical. Figure 3 shows the weight loss curve of a 7S sample with 10% moisture content. The first weight loss occurred from room temperature to about 100°C, mainly because of the water evaporation. The sample weight remained constant from 100 to 200°C, and then the sample started to lose weight, which was caused mainly by decomposition of the components in the 7S sample. After 500°C, the sample weight loss became slow and remained almost constant again, and the solids left were mainly ashes.

Mechanical Properties

The curing quality of soy protein was affected significantly by curing temperature (*23, 26, 27*). If the temperature was too high, the proteins were brittle and burned, and

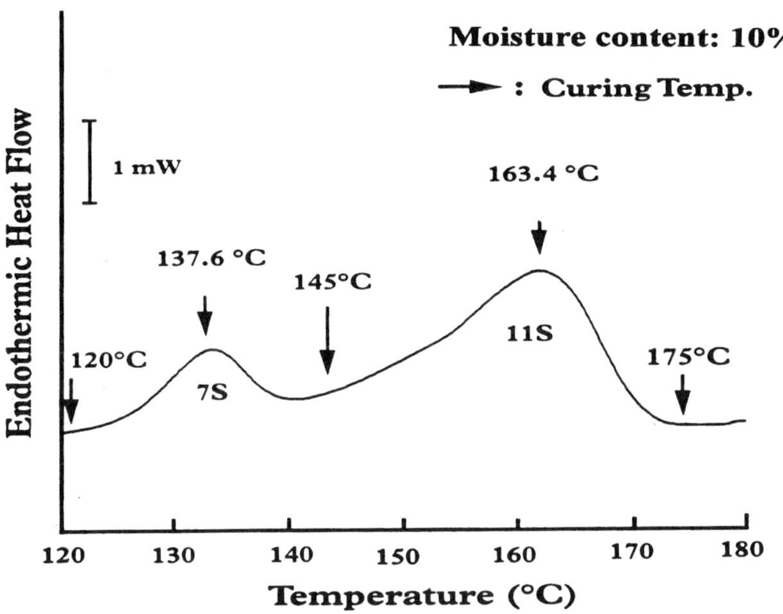

Figure 1. DSC thermogram for the mixture (1:1) of 7S- and 11S- rich soybean globulins with 10% moisture content. Large DSC sample pan was used at a temperature scan rate of 10 °C/min from 30 °C to 180 °C (reproduced with permission from reference 29. American Oil Chemistry Society Press, 1999)

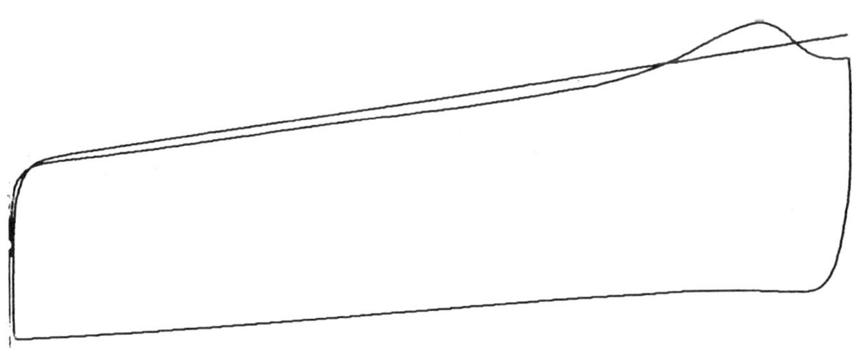

Figure 2. DSC heat-cool-heat cycle thermogram of 11S-rich globulins with 10% moisture content. Large DSC sample pan was used. The sample was first heated from 80 to 180 °C at 10 °C/min, then cooled from 180 to 80 °C at 10 °C/min, and then heated from 80 to 180 °C again at 10 °C/min.

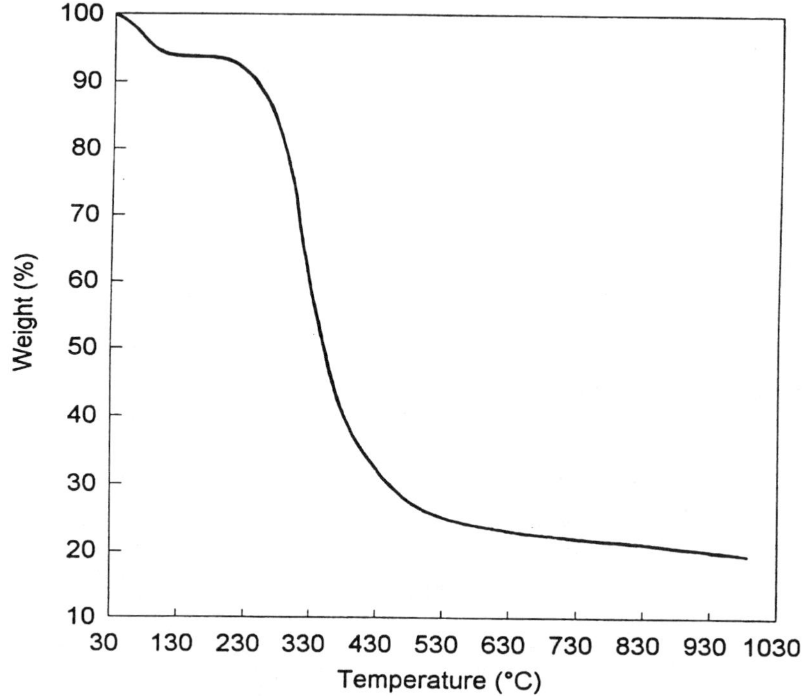

Figure 3. Weight loss curve (TGA) of 7S-rich globulins with 10% moisture content. About 5.8 mg sample was used and scan temperature range was from room temperature to 1000 °C at 20 °C/min increment (Adapted with permission from reference 29. American Oil Chemistry Society Press. 1999)

if the temperature was low, the proteins remained uncured. Soy protein plastics prepared from soy protein powder with 11.7% moisture content had a tensile strength (TS) of 40 MPa at 140°C molding temperature, whereas the TS was 35 MPa at 125°C molding temperature (*26*). Jane et al. (*28*) reported that TS of an injection-molded soy protein-based specimen was 8.73 MPa at 135°C (zone 1) and 145°C (zone 2) molding temperatures but was 5.39 MPa at 135°C (zone 1) and 150°C (zone 2) molding temperatures.

Sun et al. (*29*) studied the relationships among thermal transition and molding temperature and associated curing quality (strength and water absorption). Molding temperatures were chosen based on the DSC thermograms (Fig. 1), before, at, and after the thermal transition peak temperatures. As shown in Figures 4a and 4b, the TS and

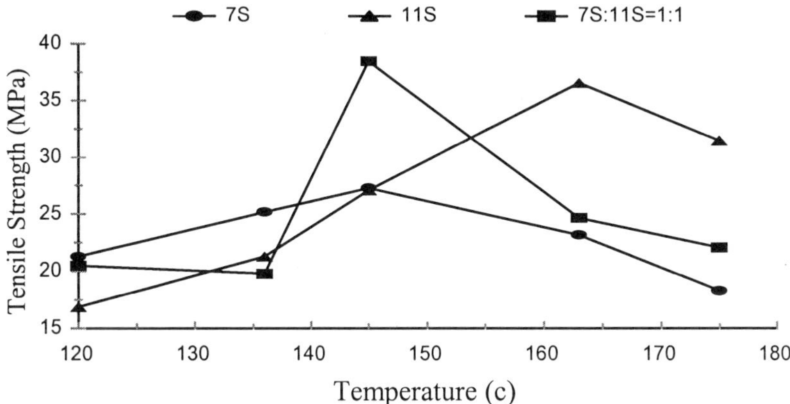

Figure 4A.

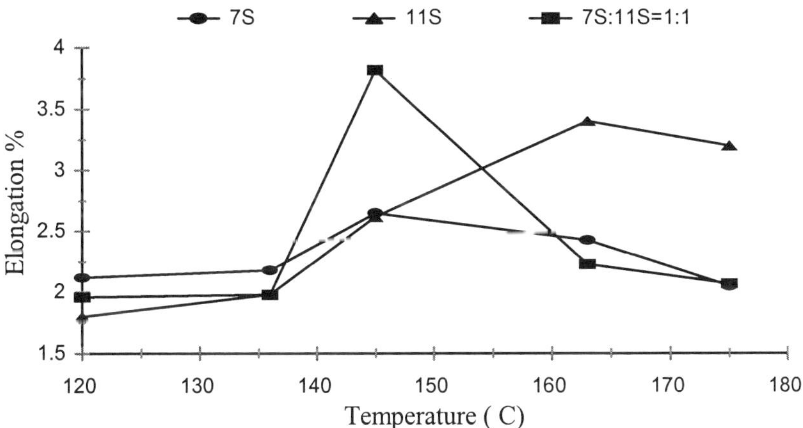

Figure 4B

Figure 4. Effects of molding temperature on tensile strength (A) and elongation (B) of plastics (about 10% moisture content) made from 7S- and 11S-rich soybean globulins and their mixture (1:1). Each value was averaged from three successful measurements with a maximum standard deviations of ±3.5 MPa for tensile strength and ±0.12% for elongation (Adapted with permission from reference 29. American Oil Chemistry Society Press. 1999)

elongation (EL) of the plastics from different protein fractions and the mixture were influenced significantly by the molding temperatures. The trends for changes in TS and EL were similar in regard to the molding temperature used. The TS and EL of the plastic from 11S increased as molding temperature increased up to its DSC thermal transition temperature (about 163°C). However, the TS and EL of the molded plastic from 7S reached their maximum values at 145°C, which was about 7°C higher than its DSC thermal transition (about 138°C). The maximum TS and EL values for 11S were 35MPa and 3.4%, respectively, higher than those for 7S (26MPa and 2.5%). The decreases in TS and EL at higher molding temperatures could have been caused by the thermal degradation of the protein by exposure to a temperature higher than its denaturation temperature. The color of the molded plastic was white powder like at molding temperature (before the thermal transition), brown transparent (at the thermal transition), and dark brown (after the thermal transition).

Sun et al. (29) also found that the plastic from the mixture of 7S and 11S had its maximum TS and EL values at a molding temperature of 145°C. These TS and EL values were about 39MPa and 4%, respectively, which were higher than those from either 7S or 11S alone. This difference suggests that an interaction between 7S and 11S-globulins could take place during the molding process. Evidence for interactions between soybean protein globulins has been observed in gelation behavior by other researchers. Gels from soybean protein isolates or a mixture (1:1) of 7S and 11S globulins were stronger than those from either 7S or 11S globulins alone (*30*). Utsumi et al. (*31*) conducted an electrophoretic analysis of a precipitate prepared by heating a mixture (1:1) of 7S and 11S globulins at 0.5% protein concentration at 80 °C. The precipitate was found to have a molecular weight of over one million, which was made up mainly from the interactions between basic (11S) and β (7S) subunits and the complex of acidic (11S) and αα'(7S) subunits located in the supernant, respectively. This result suggested that the interactions between these subunits were predominantly electrostatic in nature and, to some extent, involved the disulfide bonds. Similar results were reported by Yamagishi (*32*).

The temperature at which the interactions occurred in the dilute protein solution (0.5%) was 80 °C (30). This was about 3 °C above the DSC peak temperature for 7S globulins and about 13 °C below the DSC peak for 11S globulins in the presence of excess water (Fig. 3). This suggests that the interactions between 7S and 11S globulins occurred when the 7S was thermally denatured and 11S was still in its native state. This may explain why the plastic from the mixture showed maximum TS and EL values at 145°C, which was about 7°C above the DSC peak temperature for 7S and about 18 °C below the DSC peak temperature for 11S globulin at 10% moisture content.

The plastics prepared from native soy proteins in the presence of 5 - 30% water content without other plasticizers are strong but brittle (*23, 26*). Several plasticizers, including glyceral, ethylene glycol, and propylene glycol, improved the plastic flexibility (*23, 29, 33*). The flexibility of the plastics increases as the amount of plasticizers increases, and the tradeoff is the strength. The maximum flexibility of a soy protein-based plastic could reach about 200% with a TS of about 8MPa (*23*).

Water Absorption

Soy proteins contain more hydrophilic amino acids and fewer hydrophobic amino acids (*34*). Most hydrophobic amino acids are buried inside the molecular structure but can be brought outward by heating or chemical modifications. The water absorption of a plastic made from native soy proteins could reach about 200% with a few hours of water soaking. Jane and co-workers (*28*) reduced the water absorption of a molded soy protein-based tensile bar to about 23% after 26 hours of soaking using a formaldehyde cross-link agent.

Sun and co-workers (*29*) studied the water absorption of soy protein components as a function of molding temperature (Table 2). Water absorption decreased as molding temperature increased up to 163°C, and then water absorption increased as temperature continued to increase to 175°C. When exposed to higher temperatures, more protein molecules unfold resulting in more entanglement (aggregation) and packing tighter during the molding process, which could inhibit water absorption. If the molding temperature was too high, for example at 175°C, proteins would become thermally degraded. This degradation could destroy the protein structure, resulting in smaller peptides that could absorb water more easily. A similar phenomenon has been observed by several researchers and explained from different angles. The heat provides the protein with sufficient thermal energy to reduce hydrophobic interactions and disassociate the subunits (*11*). The disassociation and unfolding cause exposure of hydrophobic residues previously enclosed within the contact area between subunits or on the interior of the folded molecules (*13, 14*). Soy protein is disassociated and coagulated at high pressure exhibiting large hydrophobic regions and high viscosity, resulting in decreased water absorption (*12*). Paetau et al. (*26*) cited the statement by McCrum et al. (*35*) that higher molding temperatures increased the mobility of the polymer chains, which would enhance the flowability of the protein molecules, resulting in an improvement of molecular alignment and interactions, and a decrease in water absorption. Takagi et al. (*5*) found that the number of SS bonds formed in a protein affected water absorption. The protein with more SS bonds absorbed less water than that with fewer SS bonds, and more SS bonds were formed in high heat treatment (*8, 9*).

Table 2. Water absorptions *(%) of plastics made from 7S- and 11S- rich globulins and their mixture soaked in distilled water at room temperature for 2 hours (Reproduced with permission from reference 29. American Oil Chemistry Society Press. 1999)

Molding temperature °C ±5°C	7S-rich globulins	11S-rich globulins	7S/11S (1:1) mixture
120	335	127	209
138	257	---	133
145	172	57	73
155	102	57	64
163	68	49	54
175	---	58	65

NOTE: Average value of three repeating measurements with maximum standard deviation ±1.5 %

Plastics made from 7S absorbed more water than those made from 11S, and the plastics made from the mixture of 7S and 11S had medium water absorption compared to those made from 7S and 11S alone. Peng et al (*10*) found that more SS bonds formed in 11S than in 7S during heating. Combined with the findings described above (*5, 8, 9*), this explained why that 11S was more hydrophobic than 7S.

Processing and Applications

Soy proteins have been used as plastics and adhesives since the 1910s (*36*) and recently have been promoted (*28, 37, 38*). Soy proteins can be used in three major forms: solution, semisolid, and powder. Current existing equipment and processing techniques for petroleum polymers, such as extrusion, injection, and compressive molding, can be used with modification to process soy proteins in semisolid or powder forms. Protein in solution forms often are used for adhesives, coatings, and casting films.

Extrusion and Molding Articles

Jane and co-workers at Iowa State University have investigated the feasibility of extrusion and injection molding of soy proteins and explored new applications during the past decade. Jane and Wang (*39*) used up to about 65% of soy proteins blended with modified starch by extrusion. The composites acted as thermoplastics during processing. The tensile strength of the compressive molded extrusion pellets ranged from 7 to 44 MPa, and the elongation of the tensile bar ranged from 4.5 to 260% depending on the amount of plasticizers and additives. The blends were water insoluble, but water absorption ranged from 77 to 120% after 24 hours of water soaking. The extruded pellets of the soy protein-based composites could be injection molded into various articles. Huang et al. (*40*) prepared soy protein-based composites containing modified starch or natural fibers using injection molding. The TS of the protein/starch (60/40) reached 4MPa, and was about 11MPa with 30% fiber fillers (protein/fiber=70/30).

The soy protein and starch composite prepared by Jane and co-workers could be used to produce low water containing/cold fast food utensils, golf tees, and many other short-term containers. A lot of labor is required to collect the petroleum-based or wooden golf tees left on golf courses. The lifetime of a golf tee is very short, and golf usually is played in good weather, so water absorption is not a problem. A biodegradable golf tee made from the soy protein and starch blends is a unique application. Jane and co-workers also are exploring the feasibility of recycling soy protein-based molding articles for further applications as animal feed ingredients.

Mungara et al. (*41*) prepared foams using soy proteins in the presence of plasticizers and blowing agents by extrusion. The density of the foams ranged from 0.4 to 0.7 g/cc, and the TS ranged from 2 to 4MPa. The density was comparable to medium density (0.1 - 0.5 g/cc) for petroleum-based foams. The TS of the protein foam was lower than that of the petroleum-based foam. This foam could be used as packaging loose-fill peanuts because TS in this application is not critical. The thermal conductivity of the protein foams was very low, so it could be used for short-term insulation applications.

Chang et al. (*42*) prepared rigid foams from defatted soy flour blended with polyurethane. The processing method was simple. All ingredients were mixed using a high speed industrial mixer, the mixture was allowed to rise at ambient conditions for about 1 hour, and then the foam was cured at room temperature for about 24 hours. The density of the foam ranged from 0.03 to 0.045g/cc, the compressive strength ranged from 180 to 300kPa, and the thermal conductivity ranged from 0.019 to 0.028 W/mK. The volume change for testing dimensional stability was within +2.2 to -1.0% at 97% RH and 38°C. This research indicated that soy flour could be used as a filler for rigid foam production to reduce the formulation cost, while maintaining or increasing the desirable end-use properties.

Narayan (*43*) used up to about 60% soy protein hydrophilic fractions blended with selected aliphatic polyesters to produce a fully biodegradable thermoplastic composite. This blend could be blown into film. The elongation at break of the film ranged from 300 to 500%, and the TS ranged from 2000 to 3000 psi, which are compatible to values for nondegradable low-density polyethylene films. This soy protein-based film could be used in applications like lawn and leave bags, carry-out bags, trash bags, and agricultural mulch films. Many other application potentials of this blends, such as fast food utensils and containers for seedlings, are being pursued currently using the injection molding process.

Soy Protein and Fiber Composites

Soy proteins have been found to be a good resins for binding various fibers and even waste newspapers to produce composites with desirable application properties. Haumann (*44*) reported that the composite made from soy proteins and waste newspapers looked like granite but handled like hard wood. This composite could be used for picture frames, furniture, walls, and flooring materials.

Liang et al. (*45*) at Kansas State University investigated incorporating soy proteins onto high strength fibers, such as glass fiber and carbon fiber, to produce an impact-resistant composite using precoating and three-dimensional braiding systems. The fibers coated with soy proteins were buried inside and covered with fibers coated with petroleum resins through the braiding machine. The modified proteins are viscous and ductile, which enables the composite to absorb more impact energy locally in a collision. This composite could have great application potential in the automobile industry.

Sun and co-workers (*46*) at Kansas State University have developed a biodegradable composite using soy proteins and natural fibers. The strength and water resistance or related properties of this composite can be adjusted by modifying the soy proteins and formulations as well as processing conditions. This composite currently is being characterized for many applications. One unique application is for feed packaging. Millions of pounds of nutritional gel supplements are distributed to pastures in steel drums or nondegradable plastic tubs. These containers could be replaced by the biodegradable composite compressively molded into a barrel shape. In addition, the formulation of the composite can be designed so it is edible by animals. This composite also is being modified to have desirable properties for military applications.

Adhesives

A large amount of adhesives has been used annually for both interior and exterior applications of plywood and particle board; in paper manufacture; in bookbinding and textile sizing; for abrasives, gummed tape, and matches; and in many other packaging and labeling applications. Soy protein was used as an adhesive in ancient time before the replacement by petroleum-based adhesives (*22*). Soy protein adhesives lack gluing strength and water resistance, whereas petroleum-based adhesives are strong and water resistant. However, petroleum-based adhesives often contain phenol-formaldehyde as cross-link agents, which cause a severe environmental problem during processing. Developments in biotechnology allow modification of the proteins to have high gluing strength and water resistance. Therefore, soy protein polymers have been reconsidered recently as an alternative to reduce the usage of petroleum polymers and to prevent environmental pollution.

The major principle of protein gluing is that the protein molecules disperse and unfold in solution. The unfolded molecules increase the contact area and adhesion onto other surfaces, and the unfolded molecules entangle with each other during the curing process to retain bonding strength. Therefore, the adhesive performance of soy protein could be improved by using protein modification techniques to increase gluing strength.

Protein modification also could increase the hydrophobic property and increase water resistance.

Alkali has been used commonly in soy protein adhesive preparations (*37, 47*). Kreibich et al. (*48*) investigated the feasibility of combining the alkaline hydrolysis soy protein fractions with phenol-resorcinol-formaldehyde resin to speed up the curing process. They found that the adhesive system formulated with these components formed a gel within a few seconds and further polymerized at room temperature. This adhesive could provide the strength and durability required for the use in structural lumber.

Dunn and Hojilla (*49*) has developed a foam sheet adhesive from modified soy proteins through extrusion to replace the foam adhesives from animal blood that currently are used in plywood. Foam glue technology currently is used to glue wet veneer by laying long ribbons of the foam across the wood surface, laying another piece of veneer on top of the ribbons, and pressing. Soy proteins produce more foam volume and foams with greater stability than animal blood proteins. The foam is low in moisture content and dries fast. Soy concentrates and soy flours currently are being tested for such applications.

Bian and Sun (*22*) prepared formaldehyde-free adhesives from soy proteins that were chemically modified. The adhesive was high in both gluing strength and water resistance. For selected wood samples, such as walnut, cherry, and pine, 0% delamination was found after three cycles of 48 hours of water soaking tests. The gluing strength of the dried specimens after the water soaking tests remained the same and ranged from 30 to 60kg/cm^2. In many cases, the wood samples were broken first. This adhesive could be used for plywood, particle boards, and packaging or labeling.

Coatings and Casting Films

Soy proteins have been used to produce edible films through both casting films and coating technology (*50, 51, 52*). Most protein films are rigid and brittle. Glycerol often has been used as a plasticizer to improve the extensibility, but the strength was reduced (*53, 54*). The addition of soy protein to wheat gluten increases the tensile and bursting strengths of the films (*53*). Storage conditions, such as relative humidity and temperature (*55*) and pH values (*56*), are also major factors affecting mechanical properties of protein-based films.

The TS of films prepared from soy protein and wheat gluten in the presence of glycerin ranged from 1.4 to 10 MPa with 30 to 150% elongation at break depending on formulation and amount of plasticizers (*52, 57*). The films made from soy protein fractions, such as 11S, had higher TS than those made from bulk soy proteins (*52*). Soy protein-based films are good oxygen barriers but poor water vapor barriers (*53*). The films could be utilized in the manufacture of multilayer packaging, where they would function as the oxygen barrier layer (*52*). Those authors also suggested that soy protein coatings on precooked meat products could prevent lipid oxidation of the meat. Incorporation of antioxidants and flavoring agents in soy protein coatings could improve overall quality of food products.

Krinski et al. (*58*) prepared an adhesive binder from modified soy proteins. The soy protein was modified with cationic epoxide and cationic acrylate monomers to increase the binding ability and provide good rheological properties. They suggested that these properties made the binder suitable for paper coating compositions containing pigments.

Future Research and Development

Soy proteins have been found to have great potential in many application areas. Research and development of soy proteins as biodegradable resins should be continued in the future. More knowledge of soy protein structure at the molecular level is needed to better understand the functional properties of soy protein fractions. Soy proteins could be bioconverted or modified to have desirable properties through biotechnology. The thermal and rheological behaviors of soy proteins could be better characterized to facilitate current and future research and development of soy protein-based products. New properties of soy proteins could be identified for new uses and/or unique applications could be identified to utilize their properties.

Acknowledgment

The author greatly appreciates the information offered by Mr. Bill Williams, United Soybean Board.

References

1. Anonymous. *Plastics Eng.* 1994, *50*(2), 34.
2. Catsimpoolas, N.; Ekenstam, C. *Arch. Biochem. Biophys.* 1969, *129*, 490-497.
3. Iyengar, R. B.; Ravestein, P. *Cereal Chem.* 1981, *58*, 325-330.
4. De Bravo, E. N. Ph.D. dissertation, The Pennsylvania State University-University Park, State College, PA, 1987.
5. Takagi, S.; Okamoto, N.; Akashi, M.; Yasumatsu, K. *Nippon Shokuhin Kogyo Gakkai-shi.* 1979, *26*, 139, 1979, cited by Bravo (1987) In "Effect of blending on the structure and functional properties of soybean major storage proteins", Ph.D. Dissertation published by University Microfilms International, A Bell & Howell Information Co. MI, U.S.A.
6. Thanh, V. H.; Shibasaki, K. *J. Agric. Food Chem.* 1978, *26*, 695-698.
7. Koshiyama, I. *Agric. Biol. Chem.* 1969, *33*, 281-284.
8. Wolf, W. J.; Tamura, T. *Cereal Chem.* 1969, *46*, 331-344.
9. Draper, M.; Catsimpoolas, N. *Cereal Chem.* 1978, *55*, 16-23.
10. Peng, I. C.; Quass, D. W.; Dayton, W. R.; Allen, C. E. *Cereal Chem.* 1984, *61*(6), 480-490.

11. Niwae, E.; Wang; T.; Kanoh, S.; Nakayana, T. *Bulletin of the Japanese Society of Scientific Fisheries* 1988, *54*(10), p. 1851.
12. Kajiyama, N.; Isobe, S.; Uemura, K.; Noguchi, A. *Int. J. Food Sci. Tech.* 1995, *30*(2), p. 147.
13. Sorgentini, A. D.; Wagner, R. J. *J. Agri. Food Chem.* 1995, *43*(9), p. 2471.
14. Petruccelli, S.; Anon, M.C. *J. Agri. Food Chem.* 1994, *42*(10), p. 2161.
15. Kim, S. H.; Rhee, J.S. *J. Foods Biochem.* 1989, *13*(3), p. 187.
16. Kim, S. H.; Kinsella, J.E. *Cereal Chem.* 1986, *63*, p. 342.
17. Kim, S. H.; Kinsella, J.E. *J. Food Sci.* 1987, *52*(5), p. 1341.
18. Frederick, F. S. 1992. In *Biochemistry of Food Proteins*; Hudson, B. J., Ed.; Elsevier Science Publishers Ltd.: Oxford, England, 1992; pp 235-270.
19. Han, K.K. R.; Biserte, G. *Inte. J. Biochem.* 1983, *15*, p 875.
20. Matsudomi, N.; Sasaki, T.; Kato, A.; Kobayashi, K. *Agri. and Biol. Chem.* 1985, *49*(5), p 1251.
21. Wagner, J. R.; Feebey, A.D. *J. Food. Chem.* 1995, *43*(8), call no. 583.
22. Bian, K.; Sun, X..S. *Polymer Preprints*; American Chemistry Society Symposium: Boston, MA, August, 1998; *39*(2). pp 72-73.
23. Mo, X.; Sun, X.; Wang, Y. *J. of Applied Polym*er Sci. 1999, 73, pp 2595-2602.
24. Tolstoguzov, V.B. *Food Hydrocolloids*. 1988, *2* (5), p 339-344.
25. Kitabatake, N.; Tahara, M. *Agr. and Biol. Chem.* 1990, *54*(9), pp 2205-2212.
26. Paetau, I.; Chen, C-Z.; Jane, J. *Ind. Eng. Chem. Res.* 1994, *33*(7), pp 1821-1827.
27. Huang, H. M.S. Thesis, Iowa State University, Ames, IA, 1994.
28. Jane, J. L.; Lim, S.; Paetau, I. *Biodegradable Polymers and Packaging*. Technomic Publishing Company: Lancaster, PA, 1993; pp 63-73.
29. Sun, X.; Kim, H.R.; Mo, X.; *J. of American Oil Che*mistry Society, 76 (1), pp 117-123.
30. Babajimopoulos, M.; Damodaran, S.; Rizvi, S.S.H.; Kinsella, J.E. *J. Agr.and Food Chem.* 1983, *31*, pp 1270-1275.
31. Utsumi, S., Domodaran, S., and Kinsella, J. E. *J. Agr. and Food Chem.* *32*(6), pp 1406-1412.
32. Yamagishi, T.; Miyakawa, A.; Noda, N.; Yamauchi, F. *Agr. and Bio. Chem.* *47*(6), pp 1229-1237.
33. Wang, S.; Sue, H.-J.; Jane, J. *Pure Appl. Chem.* 1996, *A33*(5), pp 557-569.
34. Catsimpoolas, N.; Kenney, J. A.; Meyer, E. W.; Szuhab, B.F. *J. Sci. Food Agric.* 1971, *22*, pp 448-450.
35. McCrum, N.G.; Buckley, C.P.; Bucknall, C.B. *Principles of Polymer Engineering*; Oxford University Press: New York, 1988; pp 101-166.
36. Satow F.C.; Sadakichi, S. U.S. Patent 1245975-6, 1917.
37. Lambuth, A.L. In *Handbook of Adhesive Tehcnology*", Pizzi, A.; Mittal, K.L., Eds.; Marvel Dekker, Inc.: New York; 1994.

38. Kalapathy, U.; Hettiarachchy, N.S.; Myers, D.; Hanna, M.A. *JAOCS* 1995, *72*(5), p 507.
39. Jane, J.L.; Wang, S. U.S. Patent 5,523,293, 1996.
40. Huang, H.; Chang, T.C.; Jane, J.L. *Abstracts of Papers*, 6th Bio/Environmentally Degradable Polymers Annual Meeting, San Diego, CA; 1997; poster 4.
41. Mungara, P.; Zhang, S.; Jane, J. *Abstracts of Papers*, 6th Bio/Environmentally Degradable Polymers Annual Meeting, San Diego, CA; 1997; poster 8.
42. Chang, L.C.; Li, Y.; Xue, Y.; Hsieh, F.H. *Abstracts of Papers*, *New Industrial Products Based on Soy Protein*; United Soybean Board: Kansas City, MO, 1998.
43. Narayan, R. *Feedstocks: News about Industrial Products Made from Soy*; United Soybean Board: Kansas City, MO, 1998; Vol. 3(3), pp 2-3.
44. Haumann, B.F. *Inform*. 1993, *4*(12), pp 1324-1333.
45. Liang, F.; Youqi W.; Sun, X.S. 1998. Protein-based-polymers. The Fifth International Conference on Composites Engineering, July 5-11, Las Vegas, NE, U.S.A.
46. Sun, X. Courtesy information. Kansas State University, Manhattan, KS, 1998.
47. Hettiarachchy, N.S.; Kalapathy, U.; Myers, D.J. *JAOCS*. 1995, *72*(12), pp 1461-1464.
48. Kreibich, R.E.; Hemingway, R.W.; Steynberg, J.P.; Rials, T.G. *Abstracts of Papers*, *New Industrial Products Based on Soy Proteins*; United Soybean Board: Kansas City, MO, 1998, pp 1.
49. Dunn, L.B.; Hojilla, M.P. *Abstracts of Papers*, *New Industrial Products Based on Soy Proteins*; United Soybean Board: Kansas City, MO, 1998, pp 1.
50. Krochta, J.M.; Baldwin, E.A.; Nisperos-Carriedo, M. *Edible Coatings and Films to Improve Food Quality*; Technomic Publishing Company, Inc.: Lancaster, PA, 1994, pp 5-500.
51. Gennadios, A.; McHugh, T.H.; Weller, C.L.; Krochta, J.M. 1994. In *Edible Coatings and Films to Improve Food Quality*", Krochta, J.M.; Baldwin, E.A.; Nisperos-Carriedo, M., Eds.; Technomic Publishing Company, Inc.: Lancaster, PA, 1994, pp 201-275.
52. Kunte, L.A.; Gennadios, A.; Cuppett, S.;L.; Hanna, M.A.; Weeler, C.L. *Cereal Chem*. 1997, *74*(2), pp 115-118.
53. Gennadios, A.; Weller, C.L.; Testin, R.F. *Trans. ASAE* 1993b, *36*(2), pp 465-470.
54. Park, H. J.; Bunn, J.M.; Weller, C.L.; Vergano, P.J.; Testin, R.F. *Trans. ASAE* 1994, *37*(4), pp 1281-1285.
55. Gennadios, A.; Park, H.J.; Weller, C.L. *Trans. ASAE* 1993a, *36*(6), pp 1867-1872.
56. Gennadios, A.; Brandenburg, A.H.; Weller, C.L.; Testin, R.F. *Journal of Agriculture and Food Chemistry* 1993c, *41*, pp 1835-1839.
57. Ghorpade, V.M.; Gennadois, A.; Hanna, M.A.; Weller, C.L. *Cereal Chem*. 1995, *72*(6), pp 559-563.
58. Krinski, T.L.; Coco, C.E.; Steinmetz, A.L. U.S. Patent 4,689,381, 1987.
59. Kinsella, J.E.; Damodaran, S.; German, B. 1985. In *New Protein Foods*; Altschul, A.M.; Wilcke, H.L., Eds.; Academic Press, Inc.: New York, 1985; Vol. 5, pp 108-172.

Chapter 9

Novel Materials from Agroproteins: Adhesives from Modified Soy Protein Polymers

Xiuzhi Sun and Ke Bian

Department of Grain Science and Industry, Kansas State University, Manhattan, KS 66506

Soy protein polymers recently have been considered as alternatives to petroleum polymers to ease environmental pollution. The use of soy proteins as adhesives for plywood has been limited because of their low water resistance. The objective of this research was to enhance the water resistance of adhesives made from urea modified soy proteins. Alkali-modified and heat-treated soy proteins were used for comparison. Two ASTM standard methods for gluing strength and water resistance of wood were tested. The urea-modified soy proteins had stronger gluing strength for plywood than the nonmodified soy proteins. Urea-modified proteins had higher water resistance than alkali-modified and heat-treated proteins. After three cycles of 48 hours of water soaking followed by 48 hours of drying, 0% delamination was found for both walnut and pine specimens glued with the urea-modified soy protein adhesives.

INTRODUCTION

Large amounts of adhesives are used annually for both interior and exterior applications of plywood, particle board, paper manufacture, bookbinding, textile sizing, abrasives, gummed tape, matches, and many other packaging and labeling applications. Soy protein was used as an adhesive in the 1930s before replacement by petroleum-based adhesives (1), which have provided many advantages. However, most of them contain phenol formaldehyde, a major cross-linking agent that causes unpleasant environmental and even toxicity problems during in both processing and product distribution. Also, petroleum resources are naturally limited and politically controlled. In addition, petroleum-based adhesives are not biodegradable, which causes waste accumulation. A need exists to develop biological polymer-based adhesives and explore their potential applications.

Soy protein polymers recently have been reconsidered as alternatives to reduce the usage of petroleum polymers and to prevent environmental pollution. Soy proteins are complex macromolecules that contain about 18 different amino acid monomers connected through peptide bonds to form the primary structure (polypeptide chain) that dominates their properties. A number of side chains are connected to these monomers and interact with many materials containing inorganic and organic chemicals and cellulosic fibers. These side chains often are modified easily by physical, chemical, or enzymatic methods to obtain desirable properties.

The principle of protein gluing is that the protein molecules disperse and unfold in solution. The unfolded molecules increase the contact area and adhesion onto other surfaces and become entangled with each other during the curing process to retain bonding strength. Selected protein modification techniques could increase the tendency to unfold and, consequently, increase the bonding strength. Protein modification also could move some hydrophobic amino acids, which are buried inside, outwards to increase water resistance.

Alkalis, such as sodium hydroxide, have been the most common chemicals used to increase the gluing strength and water resistance of soy protein-based adhesives. Hettiarachchy et al. (2) prepared soy protein-based adhesives using alkali (NaOH)- and trypsin-modification methods. They found that the adhesive strength and water resistance of the modified soy protein adhesives were enhanced compared to those of the nonmodified soy protein adhesives. The alkali-modified soy protein adhesive was stronger and more water resistant than the trypsin-modified soy protein adhesive. The authors believed that alkali might increase the unfolding of protein molecules, resulting in an increased contact area and exposure of the hydrophobic bonds.

Urea was found to be a useful denaturation chemical to unfold the secondary helical structure of a protein (3). It has oxygen and hydrogen groups that would interact actively with hydroxyl groups of the soy proteins, which might break down the hydrogen bonding in the protein body and, consequently, unfold the protein complex. The objective of this research was to study the adhesion properties of urea-modified soy protein in plywood applications. Adhesion properties of alkali- and heat-modified soy protein were evaluated for comparison.

MATERIALS AND METHODS

Soy protein isolate (SPI), prepared by acid precipitation, was provided by Archer Daniels Midland (Decatur, IL) and contained more than 90% protein (dry basis) (4,5). Urea (Mallinckrodt Chemical Works, St. Louis, MO) and alkali (NaOH, Fisher Scientific, Fair Lawn, NJ) were analytical reagent grades. Nonmodified SPI was used as the control.

Protein Modification

Urea solutions (1, 3, 5, and 8M) were prepared at room temperature. Ten grams of the SPI powder were mixed with each urea-distilled water solution (150 ml) and stirred for about 1 hour to get a uniform, modified, protein dispersion.

The method of Hettiarachchy et al (2) was followed for alkali modification. Thirty grams of the SPI powder were mixed with 400 ml distilled water at room temperature and stirred for about 120 min. The pH value of the mixture then was adjusted to 11 using sodium hydroxide, and the mixture was stirred at 50°C for another 120 min to hydrolyze the SPI.

To compare the effects of alkali and heat modification on adhesive properties, a mixture of SPI powder and water was prepared in the same way but without the sodium hydroxide treatment. All mixtures were freeze-dried and milled into powders.

Wood Specimen Preparation

Four woods ranging from soft to hard were used: pine, maple, poplar, and walnut. The ASTM standard method for adhesives (D-906) with modification was followed to prepare the test specimens (6). The dimensions of each wood sample were 3 mm thick x 20 mm wide x 50 mm long. Each SPI powder was added to distilled water at a ratio of 1:6 (SPI:water) and allowed to disperse at room temperature for about 5 min. The adhesive slurry was brushed onto the wood sample until it was completely wet. The amount of the slurry on the wood sample was controlled so that no slurry flow was observed as the wood sample was placed vertically. The slurry-brushed wood sample was allowed to rest at room temperature for about 5 min and then was placed in top of another sample with the adhesive side down and pressed using a Hot Press (Model 3890 Auto "M", Carver Inc., Wabash, IN) at 20 kg/cm^2 pressure. The specimen was placed in a plastic bag and kept at room conditions until analysis.

Adhesive Quality Measurements

Shear strength of wood specimens (as an indication of glue strength) was tested following the ASTM standard method (D-906) using an Instron testing machine (Model 4466, Canton, MA) with a crosshead speed of 2.4 cm/min (6). The maximum shear strength at breakage was recorded.

Water resistance (for exterior application) of the adhesive was tested using the modified method described by Hettiarachchy et al (2). Specimens were soaked in tap water at room temperature for 48 hours and then dried at room temperature in a fume hood for 48 hours. The soaking and drying were repeated three times, and delamination of the specimen was recorded after each cycle.

Water resistance (for interior application) of the adhesive was tested following the ASTM standard method (D-1183) (7). Two cycles were used. For the first cycle, the specimen first was conditioned in a chamber at 90% relative humidity (RH) and 23°C for 60 hours and then was conditioned at 25% RH and 48°C for 24 hours. For the second cycle, the conditioning parameters were 90%RH and 23°C for 60 hours and 25%RH and 48°C for 24 hours. The shear strength of the specimen was tested after each step of the cycles. Ten duplicates were used for each experiment described above.

Viscosities of the modified and nonmodified soy proteins were measured using a Rapid Viscosity Analyzer (RVA) (Model 3D, Foss North America, Inc., MN) following method #18 recommended by Newport Scientific Methods for testing soups. Twenty eight grams of the adhesive solution with about 16% solids were used. The RVA was run for 30 min with paddle speed of 960 rpm for the first 10 seconds and then 160 rpm for the rest of the test. Solution were held at 95°C for 10 min.

Wood Expansion and Surface Microstructure

Wood samples were soaked in distilled water at room temperature for 48 hours, and the dimensions were measured before and after soaking. Linear expansions in both length and thickness and volume expansion were calculated. The microstructure of the surface of dry wood samples was observed using a scanning electron microscope (SEM) (AutoScan, ETEC Corporation) at an accelerated voltage of 20 KV.

RESULTS and DISCUSSION

Table I shows the effects of modification degree on gluing strength and water resistance. Soy proteins modified with urea solution from 1M to 3M (about 6% to 18%) yielded higher gluing strength than those modified using urea solution more than 5M as well as the nonmodified SPI adhesives, especially after the two cycles of humidity chamber incubation tests. The delamination of the wood specimens glued using the 3M urea-modified adhesive was 0% for all three wood samples after three cycles of 48 hours water soaking, compared to delamination of 90 to 100% for the nonmodified SPI adhesive. As mentioned before, unfolding proteins should increase their surface contact area and result in high gluing strength, and meanwhile, some of the hydrophobic amino acids buried inside should be moved outward to increase water resistance. With up to 8M of urea, the soy protein was about 100% denatured resulting in random coils, which might reduce significantly the cross-linking effects during curing. The partially denatured (about 20 to 50%) soy proteins resulted in partially random coils, which should balance the cross-linking effects and the surface contact area as well as the water resistance.

Table I. Shear strength (kg/cm², SD < ±10) and delamination (%) of urea-modified and nonmodified soy protein adhesives

Sample	non modified	Urea concentration 1M	3M	5M	8M
Strength before RH incubation					
Cherry	41	42	59	37	33
Pine	31	41	42	40	36
Walnut	48	54	46	26	30
Strength after RH incubation					
Cherry	38	42	49	29	25
Pine	21	41	39	31	21
Walnut	25	49	45	33	21
Delamination after three cycles of water soaking					
Cherry	100	0	0	30	100
Pine	90	0	0	0	0
Walnut	100	10	0	20	90

Curing temperature and time were also important factors affecting the gluing strength. Table II shows the results of shear strength of the walnut specimens glued with the 3M urea-modified soy protein adhesive. Longer curing time was needed

Table II. Effects of press temperature and time on shear strength (kg/cm², SD < ±10) of walnut wood glued with 3Murea-modified soy protein adhesive.

Time(min)	Temperature (°C) 80	90	104	130
5	26	43	46	52
10	26	42	55	54
15	35	48	55	52

at lower press temperatures. For example, with fixed press pressure (20kg/cm^2), about 5 minutes were required to reach the highest gluing strength at 130°C , but about 15 minutes were required to obtain similar curing quality at 90°C.

Surface structure of gluing materials is another important factor affecting the gluing strength. The shear strengths of wood samples are given in Table III. Alkali-modified and heat-treated soy protein adhesives also were evaluated for as comparison. The gluing strengths of all modified SPI adhesives were high with

Table III. Shear strength (kg/cm^2, SD < ±10) and delamination (%) of wood specimens glued with modified and nonmodified soy protein adhesives*.

Sample	U-SPI	A-SPI	H-SPI	SPI
Strength before RH incubation				
Walnut	46	60	54	48
Maple	58	65	56	50
Poplar	62	64	49	46
Pine	42	30	28	31
Strength after RH incubation				
Walnut	45	40	39	25
Maple	57	45	41	42
Poplar	53	50	43	41
Pine	39	39	24	21
Delamination after three cycles of water soaking				
Walnut	0	40	60	100
Maple	20	60	80	80
Poplar	40	80	80	100
Pine	0	60	60	90

*U-SPI=urea modification; A-SPI=alkali modification;
H-SPI=heat modification; SPI=nonmomified soy protein isolates

walnut, maple, and poplar specimens and ranged from 46 to 62 kg/cm^2. In most cases, the wood itself was broken first in strength testing. However, the gluing strengths of all adhesives with pine averaged only about 30 kg/cm^2. Kalapathy et al. (8) observed similar results using trypsin-modified soy protein adhesives with five wood species. They found that the gluing strength with pine was much lower than that with walnut, cherry, maple, and poplar samples. Examination of surface microstructure of the pine sample showed a smooth and oriented structure compared to that of other wood samples, which could be the major reason for the low gluing strength. After two cycles of incubation in a humidity chamber, gluing strengths still remained the same for U-SPI but were reduced significantly for A-SPI and H-SPI as well as for the nonmodified SPI.

Water soaking and drying tests showed that the adhesives made from U-SPI had the highest water resistance (Table III). The U-SPI-glued specimens had the lowest delamination rate. About 50% of the A-SPI-glued specimens were delaminated. Although 0% of the U-SPI-glued specimens of walnut and pine wood were delaminated, 20% and 40% of the specimens of maple and poplar were delaminated, respectively. The gluing strength of the U-SPI adhesive with walnut was reduced about 10% after 72 hours of water soaking.

Table IV presents the expansion results of the four woods in water soaking tests. Maple had the largest swelling, and poplar had zero swelling but also had the highest linear expansion, resulting in higher total bulk volume expansion. Maple and poplar also had higher linear expansion than other two wood samples. Woods with higher linear or bulk volume expansion would have higher shrinkage stress during drying. If the shrinkage stress is higher than the adhesive bonding strength, delamination would occur. That could be the major reason that maple and poplar showed higher delamination in the water soaking test.

Table IV. Dimension expansion (%) properties of wood samples after 48 hours of water soaking at room temperature.

Sample	Linear (±1%)	Swelling(±0.1%)	Bulk Volume(±1%)
Walnut	4.7	0.0	14.3
Maple	6.3	0.4	16.0
Poplar	7.6	0.0	16.0
Pine	5.1	0.1	13.9

The viscosity of the nonmodified soy protein decreased at the beginning as temperature increased from room temperature to about 65°C and remained constant for about 4 min (Figure 1). This is because the protein molecular chain became more unfolded at higher temperature, resulting in lower viscosity. As temperature increased up to about 80°C, the protein started to be thermally denatured, the unfolded molecular chains became entangled, and the viscosity increased rapidly. However, the viscosity of the protein started to decrease immediately after reaching

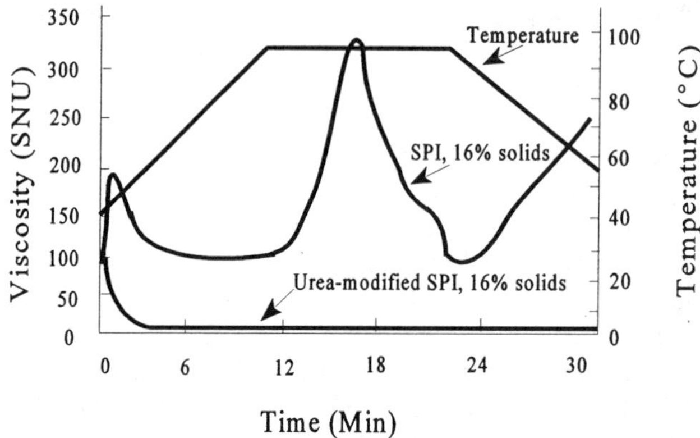

Figure 1. RVA viscosity curves of urea-modified (3M) and nonmodified soy protein slurries at 16% solids content.

its maximum value at the holding temperature, presenting a shear-thinning behavior. Continued spinning of the RVA paddle at 95°C could destroy the entanglement structure, resulting in a low viscosity. During cooling, the viscosity of the protein increased again because of gelling. Compared to the nonmodified proteins, the viscosity of the urea-modified soy proteins was low and thermally stable. This means that the modified protein adhesives had a longer working time and longer shelf life.

REFERENCE

1. Lambuth, A. L. In "*Handbook of Adhesive Technology*", Pizzi, A.; and K. L. Mittal, K. L. Eds.. Marvel Dekker, Inc.: New York, 1994, pp 259-281.
2. Hettiarachchy, N. S., Kalapathy, U., and Myers, D. J. *J. Amer Oil Chem Soc.* 1995, *72*(12), pp 1461-1464.
3. Tanford, C. In "*Advanced in Protein Chemistry*", Anfinsen, C. B.; Anson, M. L.; Edsall, J. T.; and Richards, F. M., Eds., 1968, *23*, Academic Press, Inc., NY, USA.
4. Wolf, W. J., *J. Agric. Food Chem.*, 1970, *18*, pp 969-976.
5. Kinsella, J. E. *J. Amer Oil Chem Soc.*, 1979, *56*, pp 242-258.
6. ASTM Standards for Wood and Adhesives (D-906), Annual Book, pp. 243-246. American Society of Testing and Materials, Philadelphia, PA
7. ASTM Standards for Wood and Adhesives (D-1183), Annual Book, pp. 407-409. American Society of Testing and Materials, Philadelphia, PA
8. Kalapathy, U., Hettiarachchy, N. S., Myers, D., and Hanna, M. A., *J. Amer Oil Chem Soc.* 1995, *72*(5), pp 507-510.

Chapter 10

Environmental Factors That Influence Biodegradation of Thermal Poly(aspartate)

Yin Tang and A. P. Wheeler

Department of Biological Sciences, Clemson University, Clemson, SC 29634–1903

Thermal poly(aspartate) (TPA) has potential for use in various water treatment applications. TPA also can enhance nutrient uptake in crops. Because of its capacity to adsorb to particles, it was assumed the availability of TPA for degradation might be limited. This was confirmed in assays in which TPA mineralization was reduced in the presence of the model adsorbent kaolin. However, virtually no correlation exists between adsorption and mineralization of TPA when seven soil types are compared. Thus, a multitude of factors is likely involved in controlling TPA degradation. Indeed, it was shown that complex microbial communities and relatively high C:N ratios optimize mineralization of TPA. Accordingly, flora and nutrient composition represent additional degradation-controlling factors for TPA that would vary in natural environments, including soil.

Each year, millions of pounds of polyanionic water treatment and related polymers are released into environmental waters and soils, either directly or indirectly through wastewater treatment plants. The most commonly used polymers, such as the polyacrylates, are essentially non-degradable (1). Because of this lack of degradability and because these polymers, being water soluble, are essentially non-recoverable, a search for biodegradable substitutes has been on-going. For example, based on observations made as to the surface reactive properties of some natural proteins, polyanionic polyamides have been suggested as possible alternatives to existing technologies (2). In particular, partly because of its simplicity, poly(aspartate) (PA) was examined in this light.

PA can be made by numerous chemical and biological methods. However, for commercial scale production various thermal polymerization processes have been studied (3, 4). The thermal poly(aspartate) (TPA) used in this study was obtained from lots synthesized directly from aspartic acid as part of normal production by

Donlar Corp. (3). This synthesis involves two steps (Fig. 1). In the first step, polycondensation of dry aspartic acid at elevated temperature results in the production of polysuccinimide. The product remains dry and the only significant byproduct by weight is condensed water. The process can be extremely efficient with a yield of more than 97% of polysuccinimide routinely achieved. For the purposes of this study, the polysuccinimide was washed to remove residual aspartic acid. In the second step, the ring structure of polysuccinimide is hydrolyzed with base to form TPA. Unlike its natural counterpart, the polyamide formed in this reaction contains a racemic mixture of aspartic acid (5). Also, it is actually a copolymer because the hydrolysis can occur on either side of the ring. Thus, the amide bonds are formed from both α- and β-carboxyl groups (6), whereas natural aspartyl amides would contain only the former.

In addition to the bulk syntheses, uniformly labeled ^{14}C-TPA was prepared by a small scale thermal method following the recrystallization of L-[U-^{14}C] aspartic acid with unlabeled L-aspartic acid. During the thermal synthesis, 99.6% of the aspartic acid was converted to polymer as determined by a ninhydrin analysis. Although extensive analysis of the radiolabeled material was not possible, parallel syntheses of non-radiolabeled polymers have been noted to give TPA with characteristics virtually identical to the polymer made by the bulk reaction.

Recently TPA has become available for pilot studies or full market development for use in a wide variety of water treatment and related applications. For example, it has been shown to be effective as a water-treatment additive in off-shore oil production (7) and mining (8) applications and is currently commercially employed in these industries. Also, the polymer is being marketed for agricultural applications. Specifically, TPA has been shown to increase the efficiency of plant nutrient uptake (9), which not only enhances growth and yields, but also lessens the time for maturation of some crops (10). It is anticipated, as a result of enhanced nutrient utilization, that TPA will help to lessen the negative impact of fertilizer runoff.

Understanding the environmental fate of TPA is critical if the polymer is to continue to replace existing water treatment polymer technologics and be directly applied to soils. Further, understanding the fate of the polymer in soil could also help in the preparation of guidelines for field application rates. There are two types of factors that influence the fate of the polymer. The first are the structural properties of the polymer and the second result from the interaction of polymer with the environment and its biota. Although TPA was modeled after natural proteins, it contains structures atypical of such proteins that may affect its biodegradability. For example, the rate of TPA degradation in screening assays appears to depend on its mode of synthesis, which in turn may be correlated to the degree of branching of the polymer (4).

To date, very little is known about the environmental factors that may limit TPA degradation. Accordingly, the effect of several environmentally relevant variables, such as the bioavailability for degradation, and the nutrient condition and the biota composition of degradation media were investigated.

Bioavailability

Availability of substances to microorganisms and their enzymes has been cited previously as a limiting factor in their biodegradation (e.g., 11). Usually this results from the adsorption of the substance to environmental particles. The reason that TPA is effective for many of its intended applications results from the fact that the polymer is highly adsorptive to many kinds of particulates (12), a finding that is further confirmed in this study.

If only TPA that is in soluble form is readily available to microbial degradation, it is conceivable that degradation might be reduced in environments where particulate levels are high. This suggestion may help to explain why in batch activated sludge (BAS) assays, which have much higher solids than the typical screening assay (Sturm test), less than 50% theoretical CO_2 production was observed compared to typically greater than 70% for the Sturm assay (12, Fig. 2). Further, in a preliminary soil degradation (biometry) assay, the level of mineralization was even further decreased from what observed in the BAS assays (Fig. 2). In these soil assays, the particulate level is quite high, as the soil water is only a fraction of the total weight of the sample.

However, any conclusions regarding the importance of adsorption as a limiting factor in degradation would be tenuous as the three assays were performed under different conditions. Thus, the effect of adsorption of TPA on degradation was tested using a miniature mineralization assay designed to control variables such as concentration and source of TPA, inoculum type, medium and type of adsorbent, all of which vary when one compares the assays in Fig. 2 (15). Kaolin was chosen as a model adsorbent. An adsorption isotherm of TPA onto kaolin is shown in Fig. 3 with the dissociation constant (Kd) and capacity (N) calculated from the data set assuming a Langmuir model (16). From these constants, the concentration of kaolin and ^{14}C-TPA was selected for the mineralization assay to insure that most of the polymer would be adsorbed.

The results of a representative miniature assay are given in Fig. 4 and show that mineralization of TPA in 5 days is in fact reduced when it was mostly adsorbed to the kaolin. In parallel experiments, aspartic acid, which is non-adsorptive, was also run. As expected, the mineralization of aspartic acid was high due to uptake and utilization by the bacteria. The fact that aspartic acid mineralization was not affected by the presence of the kaolin suggests that the adsorbent had no effect on the bacteria. Thus, the finding that TPA mineralization was inhibited in the presence of kaolin suggests the adsorption of TPA to particulate matter can reduce its availability to degrading microorganisms.

An exception to the relationship between adsorption and degradation would likely result if TPA could adsorb directly to bacteria. This possibility was explored by performing adsorption assays using various pure or mixed bacterial cultures. From these assays it was determined that the organisms only bind by weight a few percent of the TPA that can bind to sludge particulates (Fig. 5). These results suggest that living bacterial cells are not principal binding sites for TPA in the environment and thus it appears the cells would not gain access to the polymer by such a direct route.

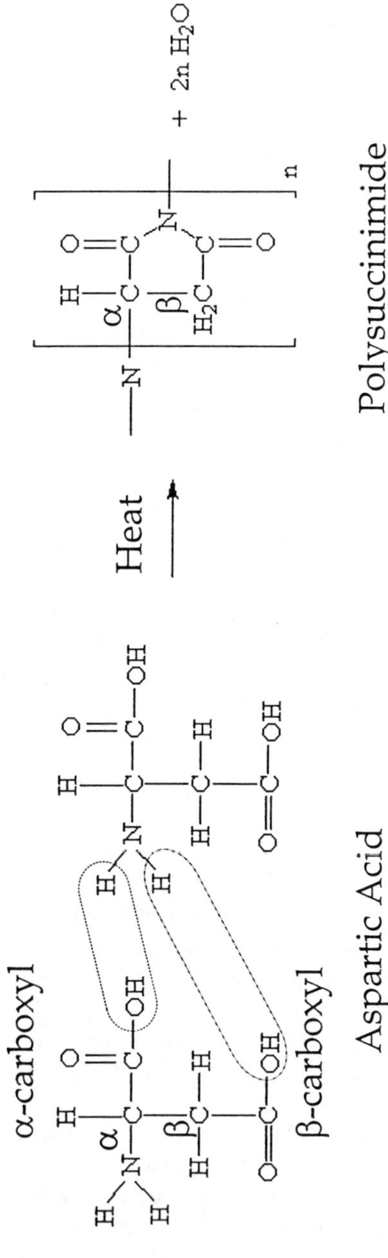
A.
α-carboxyl
β-carboxyl
Aspartic Acid
Heat
Polysuccinimide
+ 2n H2O

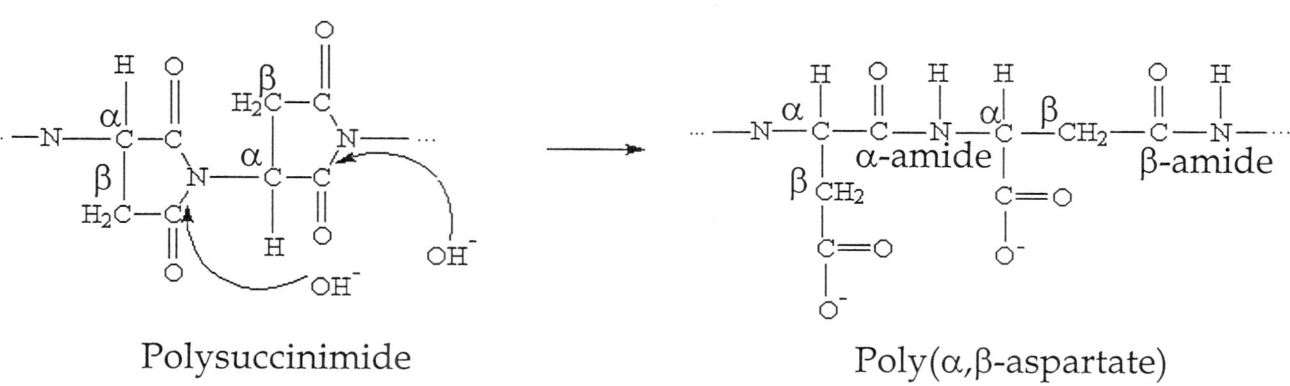

Figure 1. Synthesis of α,β-D,L-poly(aspartate), TPA. A. Thermal condensation of aspartic acid to form polysuccinimide (PS). B. Base hydrolysis of PS to form TPA. The final TPA product has a Mw of 5000 and 70% of the amide bonds are formed from the β-carboxyl groups.

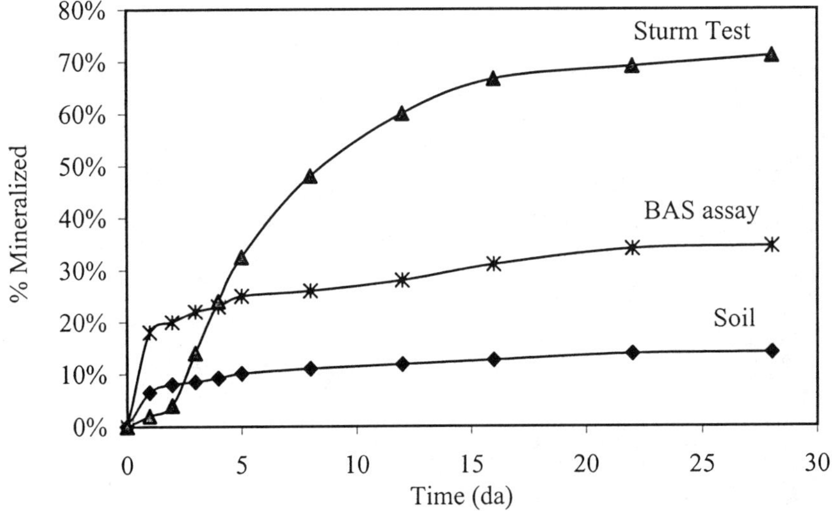

Figure 2. Percent of theoretical CO_2 evolved following incubation of TPA with a dilute sludge supernatant inoculum (Sturm test), whole sludge (BAS assay) or mixed soil (standard soil biometry assay). 20 ppm TPA was used in the former and 1ppm ^{14}C-TPA in the latter two. The particulate load was nominal in the Sturm test, approx. 2g L^{-1} in the BAS assay and the soil was adjusted to a moisture content of 35%. Most of the unmineralized TPA could be accounted for as soluble carbon in the Sturm assay but less than 5% of the C-14 was soluble at the end of the BAS assay. As a reference, aspartic acid at the same concentration of TPA tested was 80-90% mineralized in all the assays. The fact that in excess of 60% of TPA was mineralized in the Sturm test meets an OECD criterion for ready biodegradability (13). Data in part from reference 12 and 14.

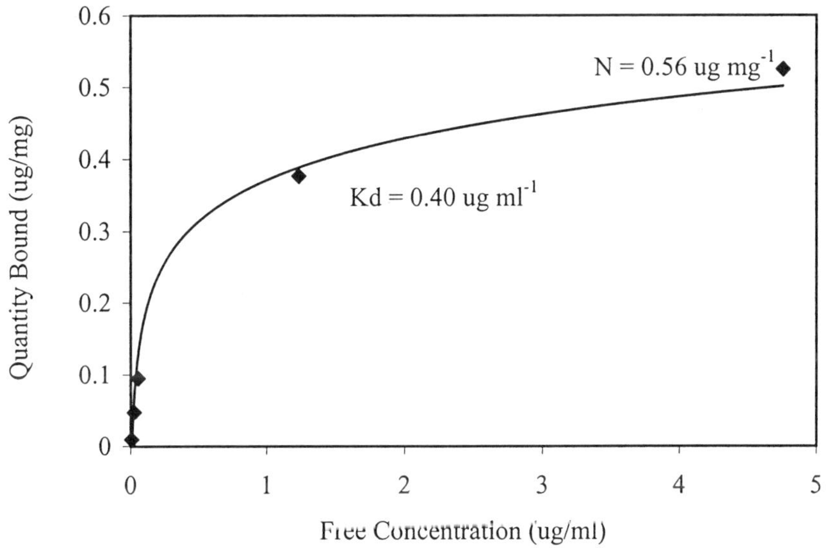

Figure 3. Adsorption isotherm for ^{14}C-TPA onto kaolin. TPA was incubated with acid-washed kaolin for 15 min in Sturm test medium at a pH of approx. 7. The quantity bound is determined as the difference between total counts and those in the supernatant following centrifugation. Points on the curve represent means of data from triplicate assays. The constants are calculated according to reference 16, assuming a Langmuir model.

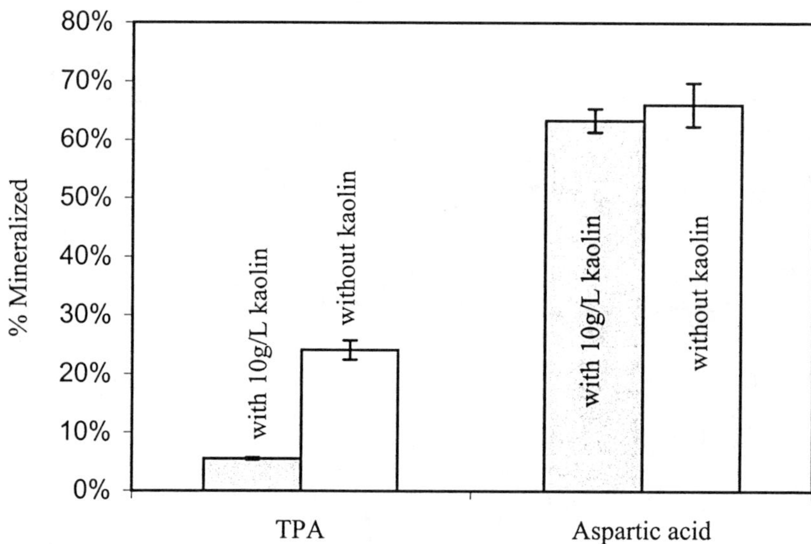

Figure 4. Mineralization of ^{14}C-TPA and ^{14}C-aspartic acid in the presence and absence of kaolin. Assays were performed in 10 ml flasks sealed with a serum stopper. The flasks contained 3 ml of balanced salt medium and 1 ppm TPA inoculated with a mixed culture of bacteria from sludge. Disposable plastic cups containing 1 N NaOH were suspended from the serum stopper to trap $^{14}CO_2$. The entire cup could be cut into scintillation fluid for counting. Under conditions of the experiment, approximately 95% of the TPA was bound to kaolin.

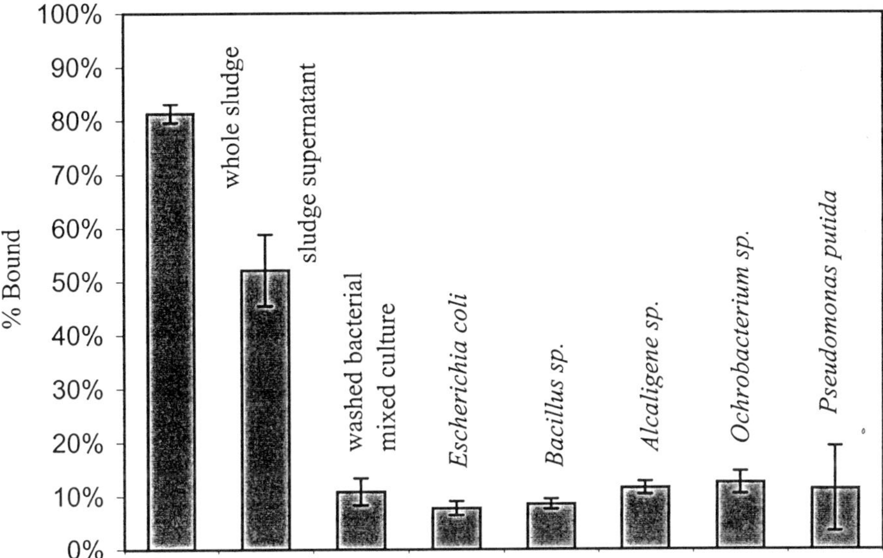

Figure 5. Adsorption of 1ppm ^{14}C-TPA onto sludge and bacterial isolates from sludge. Isolated bacteria were identified using GC-based cellular fatty acid analysis (17) and were further confirmed with API 20E biochemical test strips (BioMérieux Vitek, Inc.). One ml of adsorbent solution along with microliter quantities of TPA were added into 1.5 ml microcentrifuge tubes. Samples of suspension (0.3 ml) were removed after 15 min agitation at room temperature and the total counts were determined by a liquid scintillation method. The residual portion of the suspension was then centrifuged at approx. 13,000 g for 15 min and 0.3 ml of supernatant was removed for counting. The difference between the total counts and the counts in the supernatant was used to calculate the percent of bound TPA. The concentration of sludge solids in the supernatant was approx. 0.5mg/ml while the concentrations of bacteria in the assays ranged from 1.5 to 4.7 mg/ml. Concentrated cultures were used to increase the sensitivity of the assay. Each sample was assayed in triplicate.

Environmental Fate in Soil

A direct application for the effect of adsorption on degradation would result from the case in which TPA is added directly to soil as a fertilizer enhancer. Soils are inherently quite variable in composition and thus it is expected that they would vary in their TPA binding characteristics. In fact, when seven soils taken from areas under test for use of the polymer are compared as to their capacity (N) in short term assays such as described in Fig. 3 for kaolin, they varied by approximately two orders of magnitude. However, in contrast to kaolin, it was determined that the approach to equilibrium for adsorption of polymer to soils was relatively slow and displayed a complex time course in most cases. Accordingly, the quantity bound to the various soils for 100 ppm TPA was determined after 24 h (Table I). This concentration of TPA was chosen as it is in the range one might expect for surface soil exposed to normal application rates of the polymer. As with the short-term assays, the quantity bound was highly variable. The explanation for the slow rates of adsorption may result from the diffusion of TPA into pores and the likelihood that there are classes of sites to which the rate of TPA binding was in fact quite slow.

In separate studies, the mineralization of 100 ppm ^{14}C-TPA in these same soils also was determined using standard soil biometry studies (18, Table I). If adsorption of TPA to particulates is indeed the only major factor that reduces mineralization rates in an environment such as soil, it can be anticipated that TPA degradation in soil with a higher binding capacity will be slower than in soil with a lower binding capacity. However, as evident in Table I, there is minimal correlation between adsorption and mineralization ($r^2 \approx 0.1$). In conclusion, although the adsorption findings suggest that bioavailability can be a factor for TPA degradation in soil, clearly other soil factors are significant as well. The importance of other potential factors is discussed below.

A second way in which adsorption may affect the environmental fate of TPA is in the degree to which the polymer is retained in soil. Residence time of TPA would affect the degree of degradation that could take place in soil and also would be important in determining application rates of the polymer. Accordingly, the mobility of TPA in six of the soils was studied using a continuous flow column system. As expected from the binding data, the profile of elution of TPA varied dramatically from soil type to soil type. The total percent eluted for these soils was determined at the point where less than 0.5% of TPA was recovered per volume of soil water collected and is given in Table I. In contrast to mineralization, the adsorption was negatively correlated to mobility (as % eluted) with $r^2 = 0.98$. In fact, despite the inherent differences in the two protocols, TPA eluted in the mobility assay could be nearly predicted as 100 - % bound in the adsorption assay, suggesting that simple adsorption assays run under the proper conditions may be used to predict mobility in soils. It also can be suggested that adsorption measurements along with biodegradation studies may be able to approximate environmental fate and dose rates for TPA. Obviously, complete carbon balance experiments, including field studies, must be performed before such generalizations can be made.

Table I. Adsorption, biodegradation and mobility of TPA in soils.

Soil Type	*% Adsorb.*[a]	*% Mineral.*[b]	*% Eluted*[c]	*100-%Adsorb.*[d]
Cecil, SC	96.5	4.3	3.8	3.5
Norfolk, VA	20.1	19.3	75.3	79.9
N. Dakota	47.6	13.7	61.7	52.4
Pullman, TX	79.8	32.1	14.8	20.2
Amarillo, TX	66.2	23.8	34.3	33.8
Hort. mix	1.4	2.8	93.5	98.6
California	88.2	19.0	--	11.8

[a] Adsorption was conducted on a distilled water slurry of 100 mg soil as obtained from the field in a total volume of 1.1 ml with 100 ppm ^{14}C-TPA on a soil weight basis. The data is for a 24-h incubation. Percent adsorbed was determined by difference following centrifugation.
[b] Mineralization for 100 ppm ^{14}C- TPA on a dry soil weight basis was conducted in standard soil biometry flasks for 50 g of soil (Horticulture mix, 12.5 g). Soils were adjusted to field water capacity before assays were initiated. Data is for 35 da accumulation.
[c] Mobility of 100 ppm ^{14}C-TPA on a soil weight at field capacity basis in a column system containing 55 g of soil (Horticulture mix, 15 g), as obtained from the field. Flow varied with soil type, but was approx. 1 ml/min. The data represents total eluted at the point when the quantity of TPA eluted per soil water volume was less than 0.5% of the total TPA loaded.
[d] 100-% adsorbed in adsorption assay (footnote "a").

Nutrient Composition

It has been reported that in natural environments the addition of easily utilized compounds can affect the biodegradation of a variety of organic wastes (19, 20). In the present study, the effect of nutrient status in the medium on the degradation (mineralization) of ^{14}C-TPA was investigated using the miniature assay described in Fig. 4 above. Five different media were tested and the percent TPA mineralized and the number of colony forming units (CFU) per ml after a 5-day incubation are shown in Fig. 6.

The results reveal that the highest level of TPA mineralization, over 10% in 5 days, was obtained in a medium in which carbon was supplemented with 0.1% (w/v) glucose but TPA was the only exogenous nitrogen source. The glucose-stimulated mineralization of TPA was reduced in assays to which 0.004% (w/v) ammonia, a ready source of nitrogen was added. The decrease was not due to a decrease in cell growth as indicated by higher cell counts in this medium compared to the TPA/glucose medium. Among all the media tested, nutrient broth, a medium containing a ready source of nutrients to facilitate bacterial growth, gave the lowest level of TPA mineralization. As was the case for the glucose/ammonia assay, the reduced TPA mineralization in nutrient broth could not be attributed to reduced overall bacteria growth, as the number of CFU/ml was the highest for all the assay conditions. It is suggested from these findings that TPA is a more adequate source of metabolic nitrogen than carbon and as the environmental media becomes more complete, a lower level of TPA degradation will result.

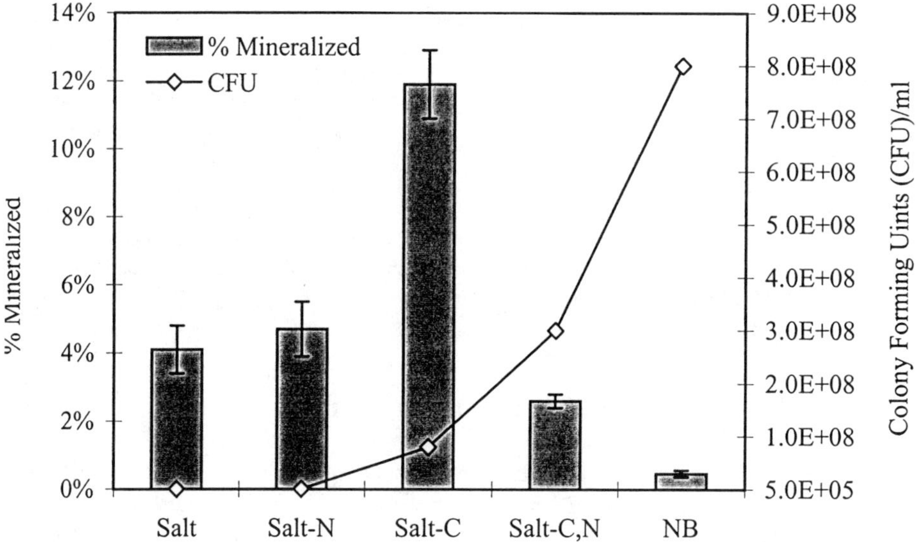

Figure 6. Effect of nutrient composition on mineralization of 1 ppm TPA by sludge supernatant. Data were collected at day 5. Colony forming units (CFU) per ml of culture was determined by plate count. Salts: Sturm salt solution, prepared by following the recipe for the modified Sturm Test (13) with the nitrogen source, $(NH_4)_2SO_4$ omitted; -C, -N: addition of carbon in the form of glucose or nitrogen in the form of ammonia (Salts-N is the same as for the complete Sturm Test medium); NB is nutrient broth (Difco®).

Inoculum Complexity

The inherent competency of the environmental biota represents another factor that may limit the degradation of a substance. For example, Johanides and Hrsak (21) showed that no growth by any pure culture alone occurred in a medium with linear alkylbenzene sulphonate (LAS) as the single nutrient source, whereas the highest rate of LAS degradation was obtained with a four-member community. In the case of TPA, this idea was tested, again using the miniature mineralization assay.

The relative percents of mineralization of ^{14}C-TPA by single bacterial species and by mixed cultures are shown in Fig. 7. The results suggest that the mineralization of TPA increased with the complexity of the inoculum. That is, by day 10 no more than 2% of TPA was mineralized by any single bacteria culture tested while approximately 37% was mineralized by sludge supernatant. This finding, taken together with the assumption that fragmentation of TPA would be the initial step of the polymer degradation, suggests that effective hydrolysis of TPA requires proteolytic activity from a variety of microorganisms. The combination of microorganisms that is optimal for TPA degradation has not yet been determined.

Summary

Results from these laboratory studies suggest that the physical and biological condition of the environment affects TPA utilization. Adsorption of the TPA to kaolin reduced its mineralization. Presumably this was due to a decrease in availability of the polymer to bacteria or their enzymes. However, when various soils are compared there was little correlation between adsorption of TPA and mineralization, suggesting that other environmental factors were equally important. One factor may be related to the community of bacteria available for degradation. Effective TPA mineralization only occurred in the presence of a complex mixture of bacteria. The addition of glucose enhanced TPA mineralization, whereas addition of ammonia decreased it. These findings suggest that TPA is a better source of metabolic nitrogen than carbon for bacteria and that the presence of environmental nitrogen, such as that supplied by fertilizers, would suppress utilization of the polymer.

Acknowledgements

We acknowledge support from Donlar Corporation (Bedford Park, IL) and the South Carolina Sea Grant Consortium, without which these projects would not have been possible. We also thank Donlar for providing the polysuccinimide and the soil samples used in this study and Horace Skipper for technical advise regarding the soils.

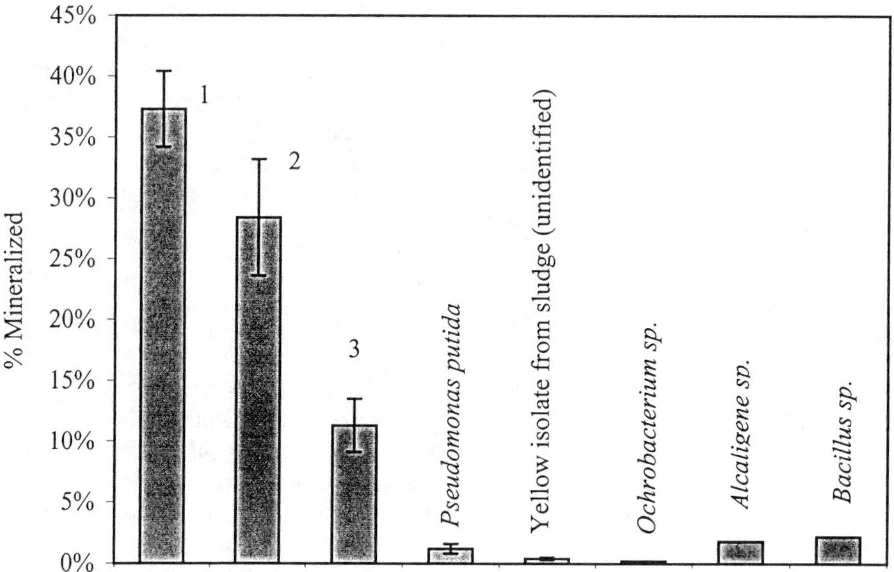

Figure 7. The effect of inoculum composition on 1ppm TPA mineralization after 10-da incubation in glucose supplemented Sturm salt solution. 1: sludge supernatant; 2: washed mixed sludge bacteria; 3: mixed culture of nine sludge bacterial isolates obtained from selection studies in which TPA was the sole carbon and nitrogen source (data not shown). They include the five isolates studied individually plus Acinetobacter sp., Burkholderia sp, Micrococcus sp. and Klebsiella pneumoniae. Sludge bacterial isolates were identified as outlined in Fig. 5. The initial inoculum in each case was 10^5 CFU per ml.

References

1. Paik, Y. H.; Simon, E. S.; Swift G. In *Hydrophilic Polymers – Performance with Environmental Acceptance;* Glass J. E. Ed.; ACS Advances in Chemistry Series 248, ACS: Washington, D. C., 1996, pp 79-98.
2. Sikes, C. S.; Wheeler, A. P. *Chemtech*, **1988**, 18, 620-626.
3. Low, K. C.; Wheeler, A. P.; Koskan, L. P. In *Hydrophilic Polymers – Performance with Environmental Acceptance;* Glass J. E. Ed.; ACS Advances in Chemistry Series 248, ACS: Washington, D. C., 1996, pp 99-111.
4. Freeman, M. B.; Paik, Y. H.; Swift, G.; Wilczynski, R.; Wolk, S. K.; Yocom, K. M. In *Hydrogels and Biodegradable Polymers for Bioapplications*. Ottenbrite R.; Huang S.; Park K.; Eds.; ACS Symposium Series 626; ACS: Washington, D. C., 1996, pp 118-136.
5. Kokufuta, E.; Shinnichiro, S.; Harada, K. *Bulletin of Chemical Society of Japan,* **1978**, 51, 1555-1556.
6. Pivcova, H.; Saudek, V.; Drobnik, J.; Vlasik, J. *Polymer*, **1982**, 23, 1237-1241.
7. Wilson, D.; Hepburn, B. J. *Corrosion 97*, **1997**, Paper No. 172, 172/1-172/22.
8. Hater, W. *Corrosion 98*, **1998**, Paper No. 213, 213/1-213/16.
9. Below, F. E.; Wang, X. T. *Agronomy Abstracts.* 1996, p 107.
10. Murphy, L. S.; Sanders, J. L., Presented at: National Meeting, Div. of Fertilizer and Soil Chemistry, ACS, Las Vegas, NV, Sept. 7-11, 1997.
11. Shimp, R. J.; Young, R. L. *Ecotoxicity and Environmental Safety*. **1988**, 15, 31-45.
12. Alford, D. D.; Wheeler, A. P.; Pettigrew, C. A. *Journal of Environmental Polymer Degradation*, **1994**, 2, 225-236.
13. *OECD Guidelines for Testing of Chemicals*, Organization for Economic Cooperation and Development (OECD), Paris, **1981**, Section III.
14. Wheeler A. P.; Alford, D. D.; Koskan, L. P. Presented at: Polyamino Acids, the Emergence of Life and Industrial Applications, ACS symposium, 1994, San Diego.
15. Tang, Y. MS Thesis, Clemson University, Clemson, South Carolina, 1997.
16. Wheeler, A. P.; Low, K. C.; Sikes, C. S. *Biomineralization - Chemical and Biochemical Perspectives.* Sikes C. S.; Wheeler A. P., Eds; ACS Symposium Series 44, ACS, Washington, D. C., 1991, pp 72-84.
17. Miller, L.; Berger, T. *Hewlett-Packard GC Application Note*, Microbial ID, Inc., 1985, pp. 228.
18. Xiong, K.; Skipper, H. P.; Kim, J. H.; Wheeler, A. P. *Agronomy Abstracts*. **1998**, in press.
19. Kunc, F.; Rybarova, J.; Lasik, J. *Folia Microbiology* **1984**, 29, 148-155.
20. Zaidi, B. R.; Mehta, N. K. *Biotechnology Letters* **1996**, 18, No. 5, 565-570.
21. Johanides, V.; Hrsak, D. *Abstract and Papers, 5th International Fermentation Symposium*, Dellweg, H. Ed., 1976, pp 124.

Chapter 11

New Methods in the Synthesis of Thermal Poly(aspartates)

Robert J. Ross, Grigory Ya. Mazo, and Jacob Mazo

Donlar Corporation, 6502 South Archer Road, Bedford Park, IL 60501

Thermal Polyaspartate (TPA), a useful class of biodegradable, water-soluble polymeric materials is synthesized by the thermal polymerization of aspartic acid. An investigation of the acid catalyzed polymerization of aspartic acid in cyclic carbonate solvents. Propylene carbonate solvent is an attractive choice due to its low environmental impact and good solvating power. The polymerization of aspartic acid in propylene carbonate affords polyaspartate polymers in high yields and when desired, high molecular weights. Preliminary data indicate that the biodegradability of the resulting polyaspartates is not adversely affected by the use of this solvent.

Thermal Polyaspartates (TPA) are biodegradable, water soluble polypeptides obtained via the thermal polymerization of aspartic acid.(1,2) These protein-like polymers are highly biodegradable and non-toxic, and are produced by an environmentally benign manufacturing process. TPA's currently have a variety of commercial uses as mineral scale and corrosion inhibitors in mining (3) and oil production (4), as a lubricant and coolant in metal working (5), and as fertilizer enhancers in agriculture (6).

Polyaspartic acid was first synthesized by Schiff in the late 19th century (7). A number of methods have been used to produce polyaspartates since that time. One method involves polymerization of the N-carboxyanhydride of aspartic acid beta esters (NCA method) in organic solvents (8). The NCA method provides an excellent laboratory procedure for the synthesis of polyaspartate homopolymers and copolymers; however, the process is much too costly for most of the industrial markets for which polyaspartates have utility.

More practical are the so-called thermal methods. The heating of ammonium maleate or fumarate or of maleamic acid at temperatures of above 150 °C produces a polymeric imide known as polysuccinimide (9). Hydrolytic ring opening of the polyimide produces poly-α,β-D,L-aspartate in the molecular weight range of ca. 1000 to 2500 daltons. Recent work by Wolk *et al.* suggests that the polyaspartate produced by polymerization of ammonium maleate is highly branched and less biodegradable than polyaspartates produced from aspartic acid (10).

The thermal polymerization of aspartic acid (I), at temperature in excess of 180 °C also produces polysuccinimide (11). This process is solventless, with both product and reactant remaining in a flowable particulate state and can be done in a continuous manner using a plate drier (1). Treatment of the resulting polysuccinimide (II) with aqueous base, such as sodium hydroxide, affords sodium poly-α,β-D,L-aspartate also known as Thermal Polyaspartate (III) having a molecular weight of ca. 5000 daltons.(See Figure 1)

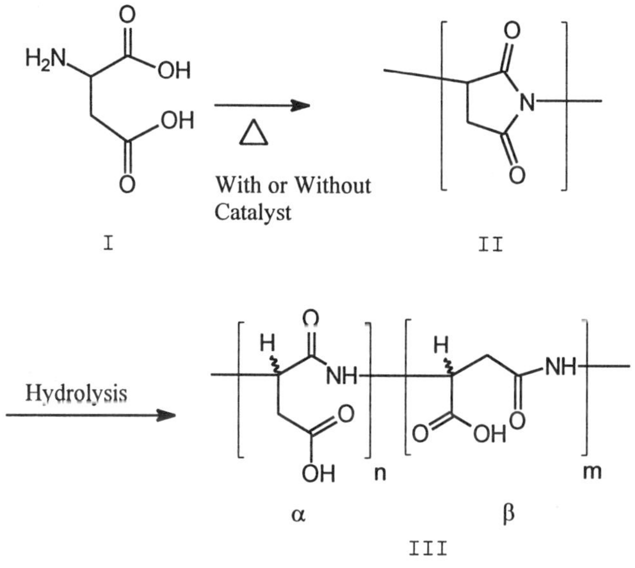

Figure 1. Thermal polymerization of aspartic acid to polysuccinimide and subsequent hydrolysis to polyaspartate.

Acid catalysts, such as phosphoric acid, have been added to the aspartic acid to afford higher molecular weight polysuccinimides than are obtained in the non-catalyzed polymerization (12). One difficulty with the catalyzed polymerization of aspartic acid is that a viscous melt phase is formed during intermediate stages of the polymerization reaction, which is also a problem in the ammonium maleate polymerization (13). In addition, generally very high catalyst loading is required to obtain high molecular weight (e.g. 30,000). The melt then solidifies as the reaction is completed further complicating the handling of the product and reducing the reproducibility of the synthesis (12). The phase changes that occur during the polymerization lead to difficulties in handling and in reproducibility.

The polymerization of aspartic acid in solution provides a means for controlling molecular weight of polysuccinimide without the processing problems associated with the previously described solventless catalyzed methods. Several workers have examined solvent based polymerizations of aspartic acid and ammonium maleate (14,15,16,17,18). Some of these methods involved suspension of the reactants in an inert solvent, such as diphenyl ether (15) or polyalkylene glycols (16) for heat transfer and to produce a fluid slurry that can be handled in conventional stirred reactors. In essence these reactions are still heterogeneous in nature. Homogeneous solution polymerization of aspartic acid has also been investigated. Polar aprotic solvents such as N,N-dimethylformamide (DMF), dimethylsulfoxide (DMSO), and sulfolane have also been investigated (17) as well as mixed solvents of polar aprotics with aromatic solvents such as mesitylene (18). This paper presents results from studies of acid catalyzed polymerization of aspartic acid in cyclic alkylenecarbonate solvents (19). (See Figure 2)

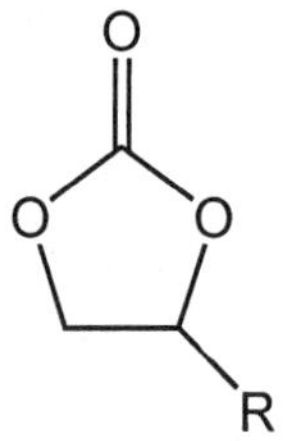

Figure 2. Cyclic carbonate solvents: R = H (ethylene carbonate); R = Me (propylene carbonate); R = Et (butylene carbonate)

Experimental Methods and Results

Materials. L-aspartic acid was obtained from Nanjing Jinke, China. Ethylene carbonate, propylene carbonate and butylene carbonate were obtained from Huntsman Corporation (Austin, TX). All other reagents were obtained from Aldrich Chemical Company (Milwaukee, WI).

Molecular Weight Determinations of molecular weights of thermal polyaspartates were made using gel permeation chromatography (GPC) on 0.1 % solutions that had been adjusted to pH 9.5 before dilution with eluent. GPC was performed on a Shimadzu LC10AD Chromatographic system with UV detection at 220 nm utilizing 2 Synchrom GPC columns in tandem: GPC 500 and GPC 100 (250 mm x 4.6 mm, each). Weight average molecular weight (Mw), number average molecular weight

(Mn), and polydispersity (Mw/Mn) were determined using a Hitachi D-2520 GPC Integrator. The Mw, Mn and Mw/Mn of the experimental samples were determined by comparison to commercial polyacrylate standards with peak molecular weights ranging from 1,200 to 60,000 daltons. Elution was with 0.05 N KH_2PO_4 buffer (pH 7). Values for Mw and Mn typically vary by +/- 10%. Samples of polysuccinimide were hydrolyzed for molecular weight determination by the following procedure: 10 mL of 1 N NaOH solution was added to 1.0 g of polysuccinimide. The resulting mixture was stirred for 1 hour at room temperature to afford a clear solution of sodium polyaspartate (12% w/w). An aliquot of the solution was diluted to 0.1% concentration with the elution buffer and the sample was filtered through a 0.45 micron filter prior to injection.

Biodegradability The biodegradabilities of polyaspartate produced by solventless thermal polymerization of aspartic acid catalyzed by sodium bisulfate (3000 Mw) and polyaspartate produced by polymerization of aspartic acid catalyzed sulfuric acid in propylene carbonate solvent (3100 Mw) were determined by Professor A. P. Wheeler, of Clemson University (Clemson, South Carolina USA) using the OECD 301B protocol (20). In this procedure, the biodegradability is measured by the amount of carbon dioxide produced by microorganisms during metabolism of the test substance. The test substances, at 20 ppm concentration, provide the only exogenous source of organic nutrients available to the microorganisms. The microorganisms used were from sludge obtained from the Clemson waste treatment facility. The percentage of CO_2 evolved, based upon the theoretical amount possible from that amount of polymer, is equaled with the percentage of biodegradation. L-aspartic acid monomer was also tested for comparison purposes. The results are recorded in Table 1.

Table 1. OECD 301B. Modified Sturm Test. Biodegradability of Thermal Polyaspartates

DAY	*% of Theoretical CO_2 Evolution*		
	No Solvent 3000 Mw TPA	*PC Solvent 3000 Mw TPA*	*L-Aspartic acid*
0	0	0	0
2	1.80	0.72	35.6
4	1.62	1.44	69.3
7	5.58	6.47	90.8
10	40.13	46.73	96.8
14	61.01	57.16	98.8
18	66.41	66.86	-----
22	70.01	71.36	101.2
25	72.35	74.41	100.8
28	75.05	75.85	98.7

Effect of solvent hydrophobicity. Aspartic acid was polymerized in three different cyclic carbonate solvents of increasing hydrophobicity: ethylene carbonate (EC), propylene carbonate (PC) and butylene carbonate (BC). (See Figure 2)

Cyclic carbonate (EC, PC or BC) solvent (25 g) was placed in a stirred reactor flask of 50 mL capacity. The reactor was equipped with a magnetic stirring bar, a thermometer, a water removal condenser, and a port for introducing anhydrous nitrogen gas. Next, phosphoric acid catalyst (85 weight % in water, 1.0 g) was added to the solvent with stirring to form an acidic solution. Aspartic acid (3.0 g) was then added to the acidic solution with stirring to form a reaction mixture. Nitrogen gas flow through the flask was begun and the temperature of the reaction mixture was elevated to 170 °C and maintained under stirring for a polymerization period of 2 hours. During this polymerization period, it was noted that all solids dissolved and a homogeneous solution was formed, at which time water of condensation began to collect from the condenser At the end of the polymerization period, the reaction mixture was cooled to ambient temperature. The product was then triturated with 250 mL of acetone to cause precipitation. The precipitate was recovered by filtration and was washed with water then acetone. The solid was dried in a vacuum oven at a temperature of about 60 °C overnight. The yields and molecular weights of the resulting polysuccinimide products are presented in Table 2.

Table 2. Polymerization of Aspartic Acid in Cyclic Carbonate Solvents at 170 C with 85% Phosphoric Acid Catalyst.

Solvent	*Mw*	*Yield*
Ethylene Carbonate	4722	86%
Propylene Carbonate	15020	94%
Butylene Carbonate	5347	85%

Effects of Reduced Atmosphere (Vacuum) on Polymerization. Aspartic acid was polymerized in propylene carbonate solvent using 85% phosphoric acid as catalyst under reduced atmosphere. The effect of vacuum level on the polymerization was determined.

Aspartic acid (170 g) added to propylene carbonate (670 g) containing 85% phosphoric acid (17g). the reaction was heated at 170 °C for 4 hours at atmospheric pressure. The reaction mixture was then cooled to ambient room temperature and the polysuccinimide product was precipitated by pouring the reaction mixture into acetone (6 L). The product was isolated by filtration, was washed with water (3 x 500 g) to remove catalyst, and was dried *in vacuo* at 60 °C overnight. The reaction was repeated under reduced pressure at the pressures indicated in Table 3. The yields and molecular weights of the polysuccinimide products are also recorded in Table 3.

Table 3. Effect of Vacuum on Polymerization of Aspartic Acid in Propylene Carbonate at 170 °C.

Pressure (Torr)	*Mw*	*Yield*
750 (atmospheric)	9134	96%
620	9280	93%
500	11365	94%
435	14729	98%
370	17311	>98%
310	24792	>98%
235	30000	>98%

Effects of Catalyst Concentration on the Polymerization Reaction. Aspartic acid was added to propylene carbonate containing 85% phosphoric acid in amounts indicated in Table 4. The reactions were heated at 170 - 180 °C for 2 hours under reduced pressure (310 Torr). The reaction mixtures were then cooled to ambient room temperature and the polysuccinimide products were precipitated by pouring the reaction mixtures into acetone as above. The products were isolated by filtration, were repeatedly washed with water to remove catalysts, and were dried *in vacuo* at 60 °C overnight. The yields and molecular weights of the polysuccinimide products are also recorded in Table 4.

Table 4. Effects of Phosphoric Acid Catalyst Concentration on the Polymerization of Aspartic Acid.

Propylene Carbonate	*Aspartic Acid*	*H_3PO_4 (85%)*	*Asp:H_3PO_4*	*Mw*	*Yield*
25 g	10	10 g	1:1	6035	95%
25 g	3	1 g	3.5:1	15020	94%
25 g	3	0.63 g	5.6:1	18000	93%
250 g	50	3 g	19.4:1	34157	91%
25 g	3	0.12 g	30:1	5047	97%
25 g	3	0.10 g	35:1	4412	45%

Effects of Selected Catalysts on the Polymerization. Aspartic acid was added to propylene carbonate containing various acid catalysts (85% phosphoric acid, anhydrous phosphoric acid crystals, hypophosphorous acid, sulfuric acid, potassium pyrosulfate, and methanesulfonic acid) in amounts indicated in Table 5. The reactions were heated at 170 - 180 °C for 2 hours under reduced pressure (310 Torr). The reaction mixtures were then cooled to ambient room temperature and the polysuccinimide products were precipitated by pouring the reaction mixtures into acetone as above. The products were isolated by filtration, were repeatedly washed with water to remove catalysts, and were dried *in vacuo* at 60 °C overnight. The yields and molecular weights of the polysuccinimide products are recorded in Table 5.

Table 5. Effects of Selected Catalysts on the Polymerization of Aspartic Acid in Propylene Carbonate.

Catalyst	*Mw*	*Yield*
Aqueous H_3PO_4	30,000	98%
Anhydrous H_3PO_4	150,000	99%
H_3PO_2	60,000	95%
$K_2S_2O_7$	25,000	80%
H_2SO_4	8000	95%
CH_3SO_3H	5000	79%

Discussion

The acid catalyzed polymerization of aspartic acid in solution to form polysuccinimide offers several attractions compared to the polymerization without solvent. Most notably, the reaction takes place homogeneously, rather than heterogeneously, as in the case of the solventless reactions. The solvent also offers better heat transfer, and better mass transfer of water out of the reaction mixture. We have found that cyclic carbonate solvents (particularly propylene carbonate) are environmentally acceptable solvents for the acid catalyzed polymerization of aspartic acid. Despite the reactive nature of the carbonate structure toward amines, the cyclic carbonate solvents are stable under the reaction conditions required for aspartic acid polymerization. In an effort to characterize the effects of this solvent system on the polymerization reaction we investigated several different variables: solvent structure, vacuum effects (water mass transfer), catalyst concentration and catalyst chemical composition.

Solvent hydrophobicity plays an important role in the polymerization reaction in cyclic carbonate solvents. It is clear from the data in Table 2 that propylene carbonate (R= Me) is the most effective of these solvents in the polymerization reaction. This can be related to the solubility of water in the solvent and to the solubility of aspartic acid monomer and polysuccinimide in the solvent. Water solubility decreases with increasing hydrophobicity (R=H > R=Me >R=Et). Decreased water solubility leads to lower residual water concentrations and thus more efficient water removal. More efficient water removal, in turn, leads to higher molecular weights in the condensation polymerization. The solubilities of both aspartic acid monomer and polysuccinimide polymer are decreased with increasing solvent hydrophobicity. Decreased monomer and polymer concentration lead to lower molecular weights. The balance between water solubility and monomer/polymer solubility is optimized using propylene carbonate solvent.

Another factor that affects water removal efficiency, and thus molecular weight, is the use of vacuum. In Table 3, it can be seen that molecular weight increases as the pressure decreases, all other factors being equal. At atmospheric pressure (750 Torr), the polymerization of aspartic acid in propylene carbonate catalyzed by 85% phosphoric acid at 170 °C provided a polysuccinimide of ca 9000 Mw. As the pressure was decreased to 235 Torr, the molecular weight tripled.

The third major factor we investigated was the catalyst concentration. In the solventless acid catalyzed polymerization of aspartic acid, polymer molecular weight is proportional to catalyst concentration. The highest molecular weights (30K to 50 K daltons) are obtained at high catalyst loadings (e.g. 2:1 aspartic acid : phosphoric acid). Reducing the catalyst level leads to reduced molecular weight. This phenomenon is likely due to two factors. The catalyst (phosphoric acid) acts as a solvent, which increases the mobility of the monomer in the melted reaction mixture. Decreasing the acid level decreases the homogeneity of the reaction mixture and decreases the mobility of the monomer. Likewise, the catalyst acts as a dehydrating agent. More efficient dehydration increases molecular weight in a condensation reaction.

In contrast, in the solution phase polymerization of aspartic acid, the acid catalyst no longer plays the role of solvent, *per se*, and thus the benefit of high catalyst levels is not seen. Additionally, water removal is facilitated by the somewhat hydrophobic solvent and reduced pressure . The presence of acid catalysts such as phosphoric acid, in large quantities would tend to increase water levels in the solution and thus is detrimental to the polymerization reaction (decreased Mw). This effect is clear from the data in Table 4 in that decreasing catalyst level leads to increasing molecular weight. This is trend is opposite of the what is seen in the solventless reactions. There is a point, however, wherein the catalyst level becomes too low. At this point (about 30:1 aspartic acid : phosphoric acid, under these reaction conditions) both the yield and the molecular weight decrease. The yield decrease appears to be related to the role that the catalyst plays in monomer solubility. Aspartic acid, in absence of added strong acids, is not soluble in cyclic carbonates. Enough acid catalyst must be present in the reaction mixture to substantially solubilize the monomer at the reaction temperature and concentration. The critical concentration of 85% phosphoric acid catalyst appears to be between 20:1 to 30:1 aspartic acid : phosphoric acid.

Finally, the chemical nature of the catalyst plays a role in the reaction. In general, phosphorus containing catalysts afford the highest molecular weight polyaspartates (see Table 5). In fact anhydrous orthophosphoric acid crystals afforded a molecular weight of 150,000, five times the Mw obtained with 85% phosphoric acid. Lower valence phophorus-containing acids , such as hypophosphorous acid can also lead to high molecular weights. The sulfur based acids tend to produce lower molecular weights, both under the reported reaction conditions and under optimized conditions. In the case of both sulfuric acid and methanesulfonic acid, there appears to be some degradation of polysuccinimide under the reaction conditions, which leads to lower molecular weights.

Conclusions

The acid catalyzed polymerization of aspartic acid to form polysuccinimide and subsequently polyaspartic acid, provides a highly biodegradable water soluble polymer with a large number of current and potential industrial uses. The use of solvents in the polymerization process can facilitate the reaction and lead to higher molecular weight polymers while utilizing lower catalyst levels than non-solvent routes. Propylene carbonate solvent is an attractive choice due to its low environmental impact and good solvating power. The polymerization of aspartic acid in propylene carbonate affords polyaspartate polymers in high yields and when desired, high molecular weights. Preliminary data indicate that the biodegradability of the resulting polyaspartates is not adversely affected by the use of this solvent.

Literature Cited

1. Low, K.C., Wheeler, A.P., and Koskan, L.P. "Commercial Poly(aspartic acid) and Its Uses" **1996**, Chapter 6 in *Hydrophylic Polymers; Performance With Environmental Acceptability*. *Advances in Chemistry Series 248*. E. Glass (Ed.), American Chemical Society, pp. 99.
2. Roweton, S. Huang, S.J. and Swift, G. "Poly(aspartic acid): Synthesis, Biodegradation, and Current Application" *J. Environ. Polymer Degradation* **1997**, Vol 5., pp 175.
3. Hater, W. *Corrosion 98*, Paper No. 213, **1998**, NACE Int'l.
4. Wilson, D., Hepburn, B. J., *Corrosion 97*, Paper No. 172, **1997**, NACE Int'l .
5. Kalota, D.J., Spickard, L.A., and Ramsey, S. *US Patent Nos.* 5,401,428 **1995** and 5,616,544 **1997**.
6. Murphy, L. S. and Sanders, J. L. Paper presented at ACS National Meeting, Div. Of Fertilizer and Soil Chemistry, American Chemical society, Las Vegas, NV September 7 – 11, **1997**.
7. Schiff, H. *Chem. Ber.* **1897,** Vol. 30, pp 2449 ; *Ann.* **1899,** Vol. 307, pp 231.
8. For example see: Rao, V. G., Lapointe, P., and McGregor, D. N. *Macromol. Chem.,* **1993**, Vol. 194, pp 1095.
9. Harada, K. *J. Org. Chem.* **1959**, Vol. 24, pp 1662.
10. Wolk, S.K., Swift, G., Paik, Y.H., Yokum, K. M., Smith, R.L., Simon, E.S., *Macromolecules*, **1994**, Vol. 27, pp 7613.
11. Kovacs, J., Kovacs, H.N., Könyves, I., Csaszar, J., Vajda, T., and Mix, H. *J.Org.Chem.* **1961,** Vol. 26, pp 1084.
12. Knebel, J. and Lehmann, K. *US Patent No.* 5,142,062 **1992**.
13. Fox, S. W. and Harada, K. *J. Am Chem. Soc.* **1960,**Vol 82, pp 3745.
14. Nagatomo, A., Tamatani, H., Ajioka, M., and Yamaguchi, A. *European Patent Application No.* 0640641 **1995**.

15. a) Jacquet, B., Papantoniou, C., Land, G., and Forestier, S. *US Patent No.* 4,363,797 **1982** b) Jacquet, B., Papantoniou, C., Land, G., and Forestier, S. *FR Demande No.* 2,424,292 **1979**.
16. Paik, Y. H., Simon, E. S. and Swift, G. *US Patent No.* 5,380,817 **1995**.
17. Freeman, M.B., Paik, Y. H., Simon, E. S. and Swift, G. *US Patent No.* 5,612,447 **1997** b) Groth, T., Joentgen, W., Müller, N. *US Patent No.* 5,714,558 **1998**.
18. Tomita, Y., Nakato, T., Kuramochi, M. Shibata, M., Matsunami, S., and Kakuchi, T. Polymer, **1996**, Vol. 37, pp 4435
19. Mazo, G. Y., Batzel, D. A., Kneller, J. F., Mazo, J. *US Patent No.* 5,756,595 **1998**.
20. "OECD Method 301B", *OECD Guidelines for Testing of Chemicals*, **1981,** Organization for Economic Cooperation and Development.

Microbial and In-Vitro Synthetic Methods

Chapter 12

Enzyme-Mediated Reactions of Oligosaccharides and Polysaccharides

Qu-Ming Gu

Hercules Incorporated Research Center, 500 Hercules Road, Wilmington, DE 19808–1599

Two types of enzymatic reactions are given here as examples of synthetic problems encountered in industry. In the first case, commercially available β-D-galactosidase from *Escherichia coli* was used as a catalyst to transfer galactose from lactose to oligosaccharides. A preference for galactosyl transfer to the 3 or 4-position of the sugar moiety of the oligosaccharide was observed for the products. As expected, only the β-anomer was produced. A wide variety of sugars, including disaccharides, trisaccharides, cellotetraose as well as maltodextrins have been shown to act as acceptors, yielding oligosaccharides. In the second example, α-galactomannan that has been previously treated to contain cationic groups (cationic guar gum) was subjected to treatment with a series of commercial enzymes such as lipases, protease and cellulases. Some enzyme preparations showed significant changes in the viscosities of 1% cationic guar solution. For example, lipases from *Aspergillus niger* and *Aspergillus saitoi* and Protease XIII from *Rhizopus niveus* produced substantial viscosity reduction (0-20% of original viscosity). These examples demonstrate the utility of low-cost enzymes in manipulating polymer structures.

Polysaccharides are used widely in industry for a variety of applications. These polysaccharides are obtained from natural sources, and many of them are approved for food use, e.g., starch, locust bean gum, guar, gum arabic, pectin, agar, alginate, carrageenan, xanthan, dextran, gellan, and pullulan. These are all commercially important materials with desirable properties (1-4). In some applications, however, the properties of natural polysaccharides may need improvements. In these cases, modifications of the polysaccharide structures may be necessary. Such modifications have been most often done for starch and cellulose (5). As a result, a wide range of starch and cellulosic derivatives are available commercially.

In the past several years a lot of papers have appeared on the use of enzymes and microbes for organic synthesis. Many enzymatic, chemo-enzymatic, and microbial reactions have been developed (6-7). Several of these reactions are applicable toward polysaccharides (8) and in fact some of them were designed specifically for polysaccharides. Through these means, new or improved products or processes may be obtained. In reviewing the literature, there are at least three ways whereby new structures may be generated:

1. Synthesis of polysaccharides, starting with sugars or sugar derivatives. Whereas these processes are likely to result in materials at higher costs than the natural polysaccharides, the resulting materials may have unusual structures or enhanced properties. As an example of this approach, Kobayashi et al. (9) have reported the synthesis of low-molecular-weight cellulose using cellobiosyl fluoride and cellulase. An alternative method to obtain pure cellulose is to use the bacterium *Acetobactor xylinium* which yields high-molecular-weight cellulose with favorable tensile and flexural properties (10).

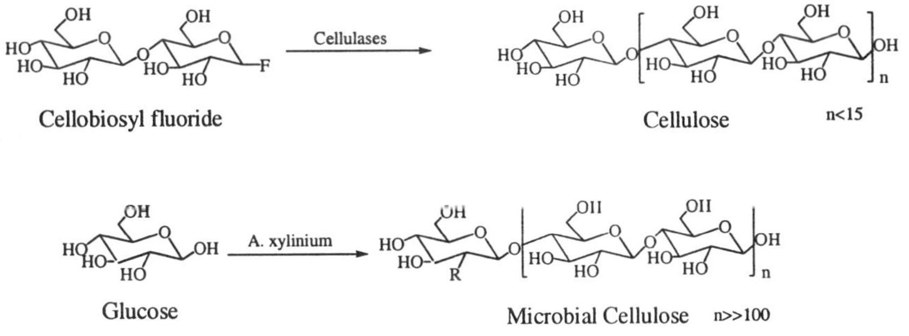

Cellobiosyl fluoride → (Cellulases) → Cellulose n<15

Glucose → (A. xylinium) → Microbial Cellulose n>>100

Figure 1. Synthesis of cellulose oligomers and polymers using isolated enzymes and microbes.

2. Degradation of polysaccharides to produce oligosaccharides and low-molecular-weight materials. As an example, carboxymethylcellulose (CMC) can be degraded by cellulase to give a low-molecular-weight polymer (11).

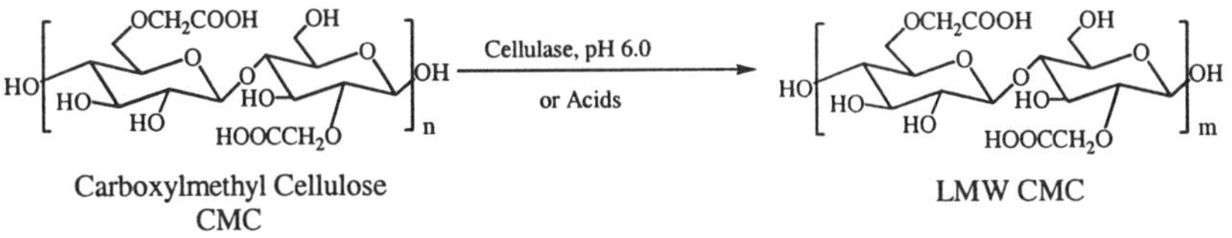

Carboxylmethyl Cellulose CMC → (Cellulase, pH 6.0 or Acids) → LMW CMC

Figure 2. Degradation of CMC to produce low molecular weight CMC.

3. Modification of the polysaccharide structure by adding selected functional groups that impart desirable properties. One possible example is to acylate a polysaccharide using a protease and vinyl alkanoate (12).

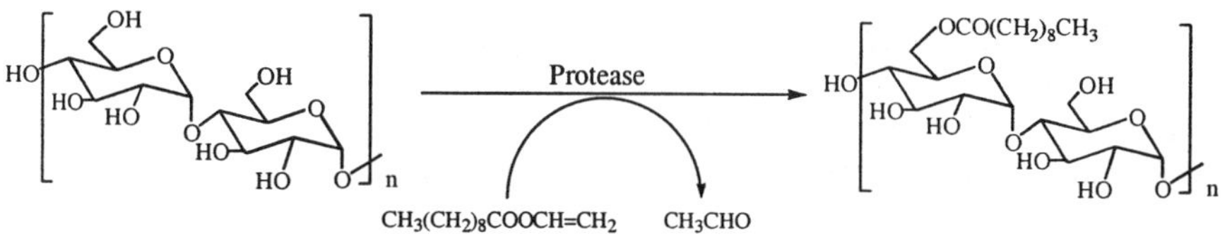

Protease

$CH_3(CH_2)_8COOCH{=}CH_2$ CH_3CHO

Figure 3. Acylation of amylose.

In the industrial perspective, these various reactions are tools in a "tool-box" whereby an industrial chemist can alter the structure of a compound or a material and thereby optimize the activities or properties of a product. Of course, the commercial viability of such a product depends on a large number of factors: potential market size, performance criteria, price constraints, enzyme and processing cost, availability of competitive products, and proprietary position. In many cases, product development requires the synthesis (or screening) or a large number of structures and some effort in structure/activity or structure/property correlations.

In this work two types of reactions will be described as examples of problems encountered in industry. The first is an example of a synthetic reaction (galactosylation) to produce new galactosyl-oligosaccharides. The second example uses degradation by which the polysaccharide properties are modified; cationic guar gum is the example here.

Galactosylation of Oligosaccharides and Polysaccharides

Enzymatic synthesis of oligosaccharides and polysaccharides is of growing industrial interest (13). Glycosidases constitute one group of enzymes which may be used for this purpose (7, 14). Whereas glycosidases usually hydrolyze glycosidic linkages in water, under suitable reaction conditions *in vitro* they can form glycosidic linkages. Hydroxyl-containing molecules other than water, such as alcohols, can serve as acceptors (15-18). The transglycosylation activity of galactosidase has also been used for the synthesis of disaccharides and trisaccharides.

For many studies, lactose has been employed as the galactosyl donor since it is inexpensive. Lactose is involved in the synthesis of both industrial materials and a variety of biologically active molecules. For example, Stevenson et al. (19) reported the large-scale production of alkyl galactosides which can be used as substrates for the lipase-catalyzed synthesis of surfactants and emulsifiers. In other reports, lactose and galactosidases were successfully employed for the galactosidation of several drugs, such as genins (20) and chloramphenicol (21).

Recently, we screened a variety of carbohydrate enzymes for their ability to transfer galactose (from lactose) onto different oligosaccharides and polysaccharides. During the search, we found the β-galactosidase from *E. coli* to be particularly facile in transferring galactose onto maltose and cellobiose in high yields. We have applied this reaction to the synthesis of a number of galactosyl-oligosaccharides (e.g., Figure 4). The regioselectivity as well as the enantioselectivity of the transgalactosylation are also examined.

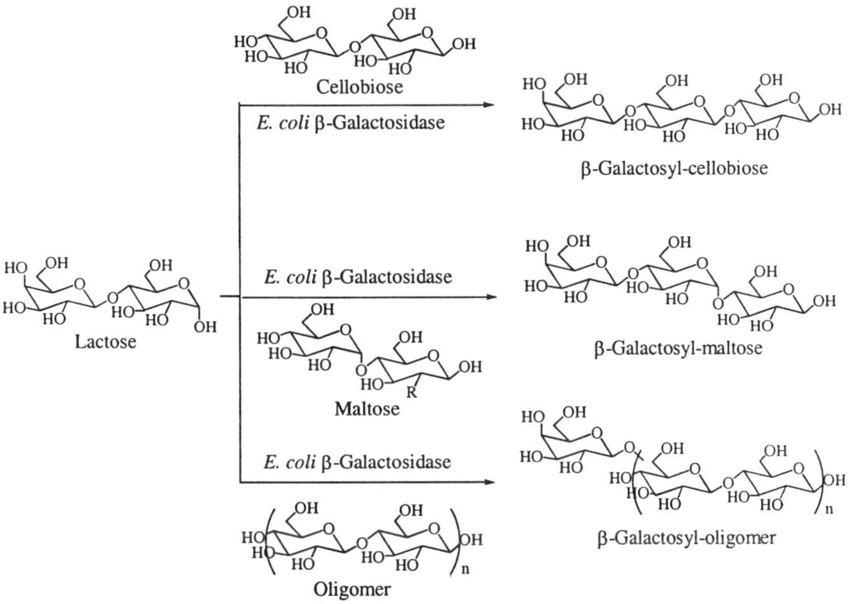

Figure 4. β-Galactosidase catalyzed synthesis of oligosaccharides.

For the synthesis of D-galactosylcellobiose, the concentrations of lactose and maltose were chosen to be 0.8 M and 0.3 M, respectively. The solvent used was aqueous phosphate buffer at pH 7.0. The progress of the reaction was monitored by TLC (isopropanol/H_2O/NH_4OH, 7:4:1) and reverse phase HPLC. Figure 5 shows the time course studies of a β-galactosidase-catalyzed reaction. The overall yield of the trisaccharides rises quickly as lactose is consumed but reaches a plateau as the consumption of lactose levels off.

Clearly, the effective rate of product hydrolysis increases as the lactose concentration in the reaction mixture decreases. The highest trisaccharide yields (50-60%) are obtained if the reaction is stopped when 50-60% of lactose is consumed. Although the reaction is facile, isolation of the desired product from the reaction mixture can be difficult because the incubation solution contains the starting material, reaction product, and hydrolyzed materials. In the case of the synthesis depicted in Figure 5, the crude trisaccharide was obtained by elution with the same solution that was used for the TLC system from the silica column. The fractions containing the trisaccharide were separated from D-galactose, glucose (from lactose hydrolysis), and cellobiose. These fractions were pooled and concentrated to give a syrup, which showed a single spot on a TLC plate. The trisaccharide was identified to be D-galactosyl –(1-4)-β-cellobiose.

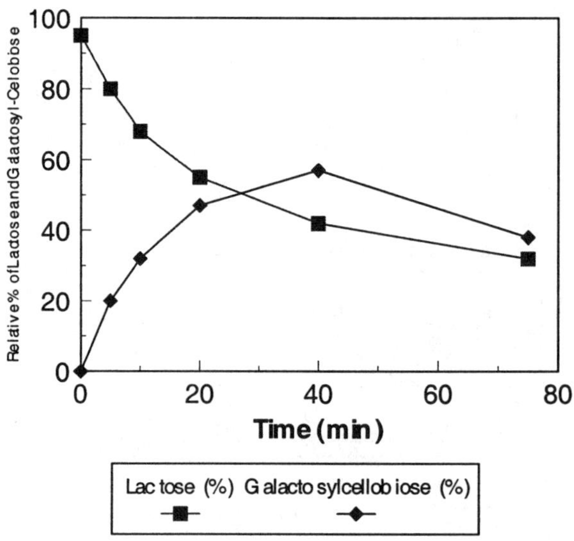

Figure 5. Time-course of β-galactosidase-catalyzed transgalactosylation of 0.8 M lactose with 0.3 M cellobiose. The reaction mixture contained 8 mmol lactose, 3 mmol cellobiose in 10 ml of 0.1 M , pH 7.0 phosphate buffer at 37°C. The reaction mixture was withdrawn at certain time intervals and analyzed by reverse phase HPLC.

Several other enzymes have been screened for the transgalactosylation reaction. The enzymes β-galactosidase from *Aspergillus niger* and *Aspergillus oryzae* only hydrolyzed lactose under the same conditions, and no trisaccharides were observed. Neither hydrolysis nor transgalactosylation have been observed when α-galactosidase from *Aspergilluas niger* and β-amylase were used.

A preference for galactosyl transfer to the 3 or 4-position of the sugar moiety of the oligosaccharide was observed for the products. According to NMR analysis, it is clear that there was no 1,6-β-glycosidic linkage in the trisaccharides. In all cases only the β-anomer was produced, as indicated by the absence of α-anomer signal in the anomeric region of NMR spectrum of the product. As an example, the NMR data for lactose/maltose reaction are shown in Figure 6. The ^{13}C and ^{1}H chemical shifts for the starting materials and for the product (D-galactosylmaltose) are given. The observed C-1 proton signal appears at 4.25 ppm, indicating the β-configuration. (The α-anomer has the proton C-1 signal higher than 5.1 ppm.) The regioselectivity of the reaction can be seen by the ^{13}C NMR analysis. The C-1 carbon of a 1,6-linkage has normally a signal at 93-95 ppm. The newly-formed glycosidic linkage has a C-1 carbon signal appearing at 103 ppm, indicating that the linkage occurs between 1-3 or 1-4. By comparing with the ^{13}C NMR spectra of a number of known oligosaccharides, we could ascertain the linkage to be at the 1,4-position.

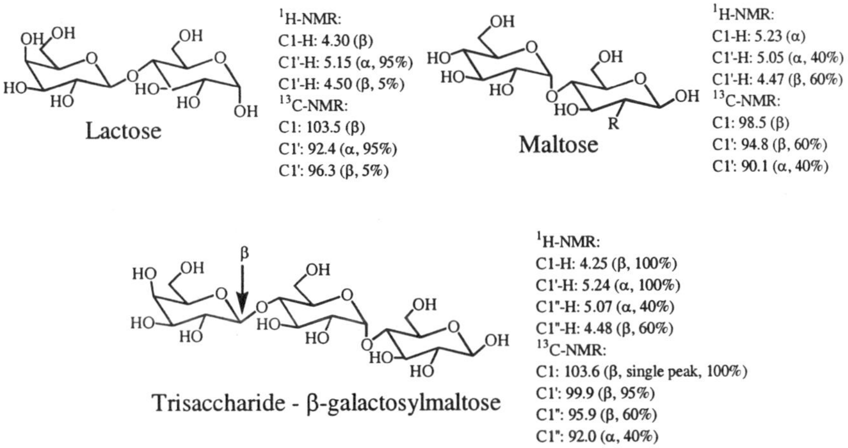

Lactose

^{1}H-NMR:
C1-H: 4.30 (β)
C1'-H: 5.15 (α, 95%)
C1'-H: 4.50 (β, 5%)
^{13}C-NMR:
C1: 103.5 (β)
C1': 92.4 (α, 95%)
C1': 96.3 (β, 5%)

Maltose

^{1}H-NMR:
C1-H: 5.23 (α)
C1'-H: 5.05 (α, 40%)
C1'-H: 4.47 (β, 60%)
^{13}C-NMR:
C1: 98.5 (β)
C1': 94.8 (β, 60%)
C1': 90.1 (α, 40%)

Trisaccharide - β-galactosylmaltose

^{1}H-NMR:
C1-H: 4.25 (β, 100%)
C1'-H: 5.24 (α, 100%)
C1"-H: 5.07 (α, 40%)
C1"-H: 4.48 (β, 60%)
^{13}C-NMR:
C1: 103.6 (β, single peak, 100%)
C1': 99.9 (β, 95%)
C1": 95.9 (β, 60%)
C1": 92.0 (α, 40%)

Figure 6. ^{1}H and ^{13}C-NMR data of β-galactosyl-maltose

One possibility of this reaction is to derivatise higher molecular weight oligosaccharides by attaching galactose onto the non-reducing end of the oligosaccharides. Since the results with cellobiose and maltose were encouraging, we extended this approach to higher oligosaccharides. Using the same enzymatic procedure, we synthesized the tetrasaccharide (D-galactosyl-raffinose) from lactose and raffinose; the yield was 10-15%. We then used the procedure on two maltodextrins (Grain Processing Co., M-250, MW= 720, and M040, MW = 3600) which gave yields of 1-5% based on the HPLC analysis. The results with different types of oligosaccharides are summarized in Table I.

These results indicate that the β-galactosidase-catalyzed transglycosidation is less effective when the molecular weight of the oligosaccharide acceptor increases. This low activity in larger oligosaccharides is of interest although little studied. Further work is ongoing to investigate more complex substrates.

Table I. Summary of Oligosaccharides Synthesis Through β-Galactosidase-Catalyzed Transgalactosylation[a]

Type	*Acceptors*	*Yields[b]*
Disaccharides	Cellobiose	40-60%
	Maltose	40-60% (30%[c])
	Sucrose	<1%
Trisaccharides	Raffinose	25% (12%[c])
	Cellotriose	16%
	Maltriose	20%
Tetrasaccharide	Cellotetraose	6-10%
Oligosaccharides	Maltodextrins (M040, M100)	1-5%

[a]Lactose and saccharide acceptor were reacted under the conditions given in Figure 5 with 0.1 M phosphate buffer at 37 C and pH 7.0. [b]The yields were either determined by HPLC analysis or estimated by TLC analysis. [c]The value shown is the actual isolated yield.

Enzymatic Treatment of Guar Solution

Guar gum is found in the seeds of two annual leguminous plants which grow mostly in semi-arid regions of the world such as India and Pakistan (1). The guar seed is typically made up of 40-46% germ, 38-45% endosperm, and 14-16% husk. The guar gum is prepared by removing the husk and the germ portions and extracting from the endosperm (Table II).

Table II. Composition of Guar Seed Components[a]

Seed Part	*Protein[b] (%)*	*Fat[c] (%)*	*Ash (%)*	*Crude Fiber (%)*	*Moisture (%)*
Hull	5.0	0.3	4.0	36.0	10.0
Endosperm	5.0	0.6	0.6	1.5	10.0
Germ	55.3	5.2	4.6	18.0	10.0

[a]Guar contains 80% soluble dietary fiber and 20% total insoluble material, including insoluble galactomannan, fatty acids and cellulosic materials. [b]Commercial guar contains 7% proteins overall. Cationic guar has been found to have 1.5% proteins. [c]Commercial guar contains 2-3% fatty acids overall.

Guar gum is a carbohydrate polymer containing galactose and mannose as structural building blocks. The ratio of the two components may vary depending on the origin of the seed. It is generally considered to contain one galactose unit for every two mannose units. The molecular weight of guars are between 250,000 to 2,000,000. Guar gum is used as a thickening agent for water. Etherification with cationic reagents has extended the utility of the gum. The need for supplies of guar gum and cationic guar continues to grow. With increasing industrial applications for guar and guar derivatives, the development of new or improved processes for the modification of guar gum is of interest. Enzymatic processes are generally preferable over chemical reactions since the modified products are more acceptable for food uses. As part of our investigations of carbohydrate-based materials, we report here our findings on the susceptibility of cationic guar gum to a variety of enzyme preparation and the viscosity changes after enzymatic treatments.

In a screening experiment, we incubated 1% cationic guar with a range of 15 different enzyme preparations, including 7 proteases, 7 lipases, and a hemicellulase at pH 6.5 to 7. Enzymatic treatment with lipase from *Candida antarctica* and Protease VIII from *Subtilisin Carlsberg* was found to slightly increase the viscosity of the cationic guar solution. Treatment with lipase from *Aspergillus niger*, lipase from *Aspergillus saitoi*, hemicellulase from *Aspergillus niger* and Protease XIII from *Rhizopus niveus* resulted in substantial viscosity reduction from 100% to 0-20% (Table III). Some enzymes also display sensitivity to pH condition for the treatments.

Table III. Viscosity Change of 1% Cationic Guar Solution after Enzymatic Treatments[a]

Enzyme	*Sources (Supplier[b])*	*pH*	*Viscosity change (- : ~20% decrease, +: ~20% increase)*
Lipase	Pseudomonas sp. (A)	6.5	-
Lipase	Porcine Pancreas (S)	6.5	- -
Lipase	Candida cylindracea (A)	6.5	no change
Lipase	Candida antarctica (N)	6.5	+
Lipase	Rhizopus niveus (A)	6.5	- - - -
Lipase	Wheat Germ (S)	6.5	no change
Lipase	Aspergillus niger (A)	6.5	- - - -
Hemicellulase	Aspergillus niger (S)	7.0	- - - -
Chymotrypsin	Bovine Pancreas (S)	6.5	no change
Papain	Papaya Latex (S)	6.5	-
Protease VIII	Subtilisin Carlsberg (S)	6.5	+
Protease IV	Streptomyces caespitosus (S)	6.5	-
Protease X	Bacillus thermo (S).	6.5	-
Protease XIII	Aspergillus saitoi (S)	6.5	- - -
Protease XXVII	Nagarse (S)	6.5	no change

[a]Hercules cationic guar (SP813D, DS 0.13) was used for the treatments. A solution of cationic guar (1%) was prepared and was adjusted to different pH's with NaOH or HCl. All solution concentrations are reported as % (w/v) unless otherwise stated. To the cationic guar solution (1%, 5 ml) was added enzymes at a ratio of enzyme: cationic guar = 1/10 (w/w). The cationic guar solution was tested for viscosity changes (reduction or increase) after incubation at 37 C for 48 hrs. [b]A: Amano; N: Novo; S: Sigma

We subsequently incubated the guar solution with a broader selection of enzymes from various bacterial, fungal, and plant sources at different pHs. The majority of enzymes showed viscosity reduction for the cationic guar solutions. Some of them produced large viscosity reduction. This observation is interesting because the guar backbone structure consists of mannose only. Since the commercial enzymes we used were not necessarily pure, the presence of other enzyme contaminants might be a possible reason for the observed hydrolytic activities. However, the nature of the enzymes responsible for the viscosity reduction still needs to be investigated. It is unknown whether the activities are contributed by a mannanase or other enzymes with specific activities to the guar backbone structure.

The pH effects on the enzymatic treatments have also been investigated. The viscosity change of 1% cationic guar solution was observed under different pH conditions. The activity of these enzymes to galactomannan (guar gum) was found to be sensitive to pH. The optimum pH conditions of each enzyme might be slightly different from those of its natural substrates. As an example, hemicellulase caused 90% viscosity reduction after 16 hours of incubation at room temperature when the starting pH was 7.0. However, the viscosity was reduced to 60% viscosity when the starting pH was 7.5 under the same reaction conditions (Table IV). In contrast, subtilisin treatment enhanced the viscosity of 1% guar solution by 10% at pH 7.5 but showed no viscosity increase when the reaction pH was 7.0.

Table IV. The pH Effects on Viscosity Change with Enzyme-Treated Guar Solution[a]

Enzymes	*Enz. mg/g. guar*	*Start pH*	*Final pH*	*Viscosity (cps)[b]*
None	0	6.4	7.1	1900
None	0	7.5	8.3	1020
Pseudomonas sp. lipase	50	7.0	7.7	2052
Pseudomonas sp. lipase	50	8.0	8.3	480
Hemicellulase	20	7.5	8.3	760
Hemicellulase	20	7.0	7.6	168
Subtilisin	25	7.5	7.6	2040
Subtilisin	25	7.0	7.2	1980

[a]Hercules cationic guar (SP813D, DS 0.13) was used for the treatments. A solution of cationic guar (1%) was prepared and was adjusted to different pH's with NaOH or HCl. After adding the enzymes, the mixtures were incubated at room temperature for 17 hrs. The solution was tested for viscosity change without further treatment. [b]The solution viscosity was determined using a Brookfield spindle-speed combination.

The effect of protease on the viscosity of guar solution indicates the possible effects of proteins on the solution properties of guar. As shown in Table II, a significant amount of the protein exists in guar gum as well its derivatives. It is suspected that the protein is covalently attached to guar molecules. Removal of the protein from guar has also been confirmed through a protein assay using the Bio-Rad protein assay method (22).

Our investigation has demonstrated that the viscosity of cationic guar can be modified by inexpensive enzyme preparations. An interesting discovery is the viscosity increase by the treatment with lipase from *Candida antarctica* and Protease VIII from *Subtilisin Carlsberg*. To the best of our knowledge, there is no previous report of such property change of guar gum by enzymatic treatments.

Acknowledgements

I am grateful to Dr. H. N. Cheng for his advice and encouragement on this work and also for the useful discussion during the work. Thanks are also due to Mrs. Ruth W. Kliment for experimental assistance in the guar work.

References

1. Davidson, R., L. *Handbook of Water-Soluble Gums and Resins*; McGraw-Hill, New York, 1980.
2. Nevell, T. P. and Zeronian, S. H. *Cellulose Chemistry and Its Application*; Ellis Harwood, New York, 1985.
3. Mitchell, J. R. and Ledward, D. A. *Functional Properties of Food Micromolecules*; Elsevier Applied Science Publishers, London, 1986.
4. Cheng, H. N. *Polymeric Materials Encyclopedia* **1996,** *4*, 2568-2574.
5. Yalpani, M. *Carbohydrates and Carbohydrate Polymers*; ATL Press, Inc, Mount Prospect, Il, 1992.
6. Wong, C.-H. and Whitesides, G. M. *Enzymes in Synthetic Organic Chemistry*; Elsevier Science Inc., New York, 1994
7. Gijsen, H. J. M., Qiao, L., Fitz, W. and Wong, C.-H. *Chem Rev*. **1996**, *96*, 443-473.
8. Trilsbach, G. P., Pielken, P., Hamacher, K. and Sahm, H. *Proc. 3rd Europ. Congress Biotechnol.*, Verlag Chemie, Weinheim, 1984, Vol. 2, pp 65-70.
9. Kobayashi, S. and Shoda, R. *Chemistry and Biology* **1993,** *31*, 385-391.
10. Byrom, D. US Patent 4,929,550 (1990); US Patent 5,273,891 (1993).
11. Timonen, M. et al. US Patent 5,569,483 (1996).
12. Bruno, F. F., Akkara, J.A., Ayyagari, M., Kaplan, D. A., Gross, R. A., Swift, G., and Dordick, J.S. *Macromolecules* **1995**, *28*, 8881-8883.
13. Hodgson, J. *Bio/Technology* **1995**, *13*, 38-41.
14. Nilsson, K. G. I. *Enzymatic Synthesis of Complex Carbohydrates and Their Glycosides. In Applied Biocatalysis* (Blanch, H. W. Ed.), Dekker Inc., New York, 1991, Vol 1, 117-118.
15. Ajisaka, K. and Fujimoto, H. *Carbohydrate Research* **1989**, *185*, 139-146.
16. Crout, D. H. G. and MacManus, D. A. *J. Chem. Soc. Perkin Trans. I* **1990**, 1865-1868.

17. Hedbys, L., larsson, P. O., Mosbach, K. and Svensson, S. *Biochem. Biophys. Res. Commun.* **1984**, *123*, 8-15.
18. Suyama, K., Adachi, S., Toba., Sohma, T., Hwang, C. and Itoh, T. *Agric. Biol. Chem.* **1986**, *50*, 2069-2075.
19. Stevenson, D.E., Stanley. R A. and Furneaux, R H. *Biotechnol. Bioeng*. **1993**, *42,* 657-666.
20. Scheckermann C., Wagner F. and Fisher L. *Enzyme and Microbial Technol.* **1997**, *20*, 629-634.
21. Ooi, Y, Hashimoto T, Mitsuo, N and Satoh T. *Chem. Pharm. Bull* **1985**, *33*, 1808-1814.
22. Bradford, M. *Anal. Biochem*. **1976**, *72*, 248-252.

Chapter 13

Enzyme-Catalyzed Lactone Ring-Opening Polymerizations: Regioselectivity and the Preparation of a Star-Shaped Polymer

Fang Deng[1] and Richard A. Gross[2,*]

[1]Hercules Incorporated, 500 Hercules Road, Wilmington, DE 19808–1599
[2]Department of Chemistry, Polytechnic University, Six Metro Tech Center, Brooklyn, NY 11201

The one-pot biocatalytic synthesis of novel amphiphilic products consisting of an ethylglucopyranoside (EGP) head group and a hydrophobic polyester chain is described. The porcine pancreatic lipase (PPL) catalyzed ring-opening polymerization of ε-caprolactone (ε-CL) by the multifunctional initiator EGP was carried out at 70 ^{0}C in bulk. Products of variable oligo(ε-CL) chain length (M_n = 450, 2200 g/mol) were formed by variation of the ε-CL/EGP ratio. Extension of this approach using Novozym-435 (*Candida antarctica* lipase B) and EGP as the initiator for ε-CL and trimethylene carbonate (TMC) ring opening polymerizations also resulted in the formation of different EGP polyester conjugates. Structural analysis by ^{1}H, ^{13}C, DEPT-135, COSY (^{1}H-^{1}H), HETCOR (^{1}H-^{13}C) NMR experiments showed that the reactions were highly regiospecific i.e. the polyester chains formed were attached by an ester/carbonate link exclusively to the primary hydroxyl moiety of EGP. The resulted EGP polyester conjugates were found to be surface active and the surface activity of the conjugates were related to their molecular weights. The EGP oligo(ε-CL) conjugate was also used as a macromonomer to prepare a novel multiarm block copolymer of PCL and PLA with a well-defined spatial architecture. For this purpose, 1-ethyl 6-oligo(ε-CL) glucopyranoside (EGP) with M_n = 1,120 g/mol (M_w/M_n = 2.16) was first regioselectively end-capped through a transesterification reaction catalyzed by lipase PS30 (from *Pseudomonas cepacia*).

Vinyl acetate was used as an irreversible capping agent. This provided a macromer with an ω-acetyl terminated PCL chain segment linked to the 6-position of EGP. The ring-opening polymerization of L-lactide at the EGP secondary hydroxyl groups was then carried out in bulk using the non-selective chemical catalyst stannous octanoate. From NMR structural analysis and molecular weight measurement, the resulted block copolymer was found to consist of three arms of PLA and one arm of PCL.

Introduction

The fatty acid esters of carbohydrates are now known as an important class of nonionic surfactants. Chemical synthesis of these compounds with defined structures is a difficult task due to the presence of several hydroxyl groups of similar reactivity (*1*). Tedious introduction and removal of protecting groups are normally required to achieve the desired regioselectivity through conventional chemical preparation methods (*2*). On the other hand, the high selectivity displayed by many hydrolases (lipases and proteases) under mild reaction conditions makes them promising candidates for the catalysis of these transformations. Since the first demonstration of using lipases for the selective acylation of monosaccharides by Klibanov and coworkers (*3*,*4*), numerous excellent reports have appeared in the literature regarding the application of enzymatic catalysis in selective carbohydrate acylations and deacylations (*5,6*).

However, the intrinsic polarity of carbohydrate compounds has posed difficulties in developing commercially viable enzyme-mediated reactions. Very few organic solvents (such as dimethyl sulphoxide DMSO, dimethylformamide DMF and pyridine) can solubilize sugars at reasonably high concentrations (*7*). Many enzymes loose their activity in these highly polar reaction media due to the essential water being stripped away. Consequently, only a small number of enzymes remain active under these harsh environments. Even so, the activity of these enzymes in DMSO, DMF and pyridine is dramatically reduced. Moreover, the ultimate removal of these high-boiling aprotic solvents is difficult. Alternative methods to improve the miscibility between carbohydrates and hydrophobic organic species were pursued to enhance both the reaction kinetics and product yield. The procedure reported by Adelhorst *et al* (*8*) involving the prior modification of the sugar moiety followed by solvent-free esterification with molten fatty acids was particularly attractive.

Generally, in the majority of reports in the literature regarding the preparation of carbohydrate esters, the acylating agents were simple aliphatic acids or their activated esters. However, lipases can catalyze transesterification reactions even with different acyl moieties such as anhydrides (*9*), amino acids (*10*) and lactone compounds. Our interest in lipase-catalyzed ring-opening polymerization of lactones (*11*) led us to develop a new approach to use lactones as acyl donors for acylation of carbohydrates.

From previous studies carried out by MacDonald *et al* (*12*) and Hendersen *et al* (*13*) on ε-CL ring-opening polymerizations, it is clear that nucleophiles other than water molecules can serve as effective initiators to modulate the end group structure of poly(ε-caprolactone). Multifunctional compounds such as hydrophilic mono-, di- and oligosaccharides, in particular, are of interest to us. By utilizing the hydroxyl groups in these compounds as initiating sites for the ring-opening polymerization of lactone compounds, while exploiting the exceptional regioselectivity of enzyme-catalysis under mild reaction conditions, novel amphiphilic structures can be constructed. The hydrophobic segments of the resulted oligomers or polymers can be composed of different biodegradable polyesters accessible from ring-opening polymerization routes, while the hydrophilic parts could consist of regioselectively esterified carbohydrate species served as multifunctional initiators during the polymerization. These products would provide numerous opportunities for further modification. For example, it is envisioned that modification of the carbohydrate free hydroxyl groups with charged moieties could be used to 'tailor' product hydrophilic-hydrophobic balance to develop biodegradable 'tunable' surfactants. Alternatively, they could also be used as macromers to prepare various multiarm heteroblock copolymer structures of well-defined spatial architecture. Furthermore, by combining traditional chemical and enzyme-catalyzed transformations, it may be possible to control the repeat unit composition of each arm.

In this report, the above approaches were illustrated by using 1-ethyl glucopyranoside (EGP) as a multifunctional initiator for the ring-opening polymerizations of ε-caprolactone (ε-CL) and trimethylene carbonate (TMC) (*14*). The resulted macromer 1-ethyl 6-oligo(ε-CL) glucopyranoside was further used to construct a novel star shaped heteroarm block copolymer of poly(L-Lactide) (PLA) and poly(ε-caprolactone) (PCL) with a well-defined spatial architecture (*15*).

Experimental Methods

Materials

All chemicals and solvents were of analytical grades and were used as received unless otherwise noted. ε-CL was distilled at 97-98 ^{0}C over CaH_2 at 10 mm of Hg. Trimethylene carbonate (TMC) was synthesized following a procedure described previously (*16*). Prior to its use TMC was dried over P_2O_5 in a dessicator (0.1 mm Hg, 38 h, room temperature). Vinyl acetate (BP 72 ^{0}C) was used as obtained from Aldrich. (3S)-*cis*-3,6-Dimethyl-1,4-dioxane-2,5-dione (L-Lactide) (Aldrich) was recrystallized three times from fresh distilled ethyl acetate and dried in a vacuum oven at 25 ^{0}C. Tin (II) (2-ethylhexanoate)$_2$ (Sn(Oct)$_2$, Sigma) was used as 1.0 M solution in toluene. Ethylglucopyranoside (EGP) was synthesized and characterized by detailed NMR analysis as described elsewhere (*17*). It is a mixture of α- and β-anomers. EGP was dried over P_2O_5 in a vacuum oven (0.1 mm Hg, 38 h, 50 ^{0}C).

Porcine pancreatic lipase (PPL) Type II Crude (activity=61 units/mg protein) and *Candida rugosa* lipase (CCL) Type VII (activity = 4570 u/mg protein) was obtained from Sigma Chemical Co. The lipases PS-30, AK and MAP-10 from *Pseudomonas cepacia*, *Pseudomonas fluorescens* and *Mucor javanicus*, respectively, were obtained from Amano enzymes (USA) Co., Ltd. (specified activities at pH 7.0 were 30,000 u/g, 20,000 u/g and 10,000 u/g, respectively. Immobilized lipases from *Candida antarctica* (Novozym-435) and *Mucor miehei* (Lipozyme IM) were gifts from Novo Nordisk Bioindustrials, Inc. All enzymes, prior to their use, were dried over P_2O_5 (0.1 mm Hg, 25 ^{0}C, 16 h).

Lipase-catalyzed EGP initiated ring-opening polymerizations of ε-caprolactone and trimethylene carbonate

The polymerization of ε-CL or TMC using different lipase catalyst was carried out in a similar manner. The sample procedure is as below. Using a glove bag and dry argon to maintain an inert atmosphere, 200 mg of EGP and 500 mg of PPL (type II, Sigma) which had previously been dried in a vacuum desiccator (0.1 mm Hg, 25 ^{0}C, 16 h) were transferred to an oven dried 20 mL reaction vial. The vial was immediately stoppered with a rubber septum, purged with argon and 0.8 mL of ε-CL (distilled at 97-98 ^{0}C over CaH_2 at 10 mm of Hg) was added via syringe under argon. The reaction vial was then placed in a constant temperature oil bath maintained at 70 ^{0}C for 96 h. A control reaction was set up as described above except PPL was not added. More than 95% EGP (determined by TLC) was consumed in 96 h. The reaction was quenched by removing the enzyme by vacuum filtration (glass fritted filter, medium-pore porosity), the enzyme was washed 3-4 times with 5 mL portions of chloroform, and the filtrates were combined and solvent removed *in vacuo* to give 950 mg of product. To remove unreacted EGP, the biotransformation product (500 mg) was purified by column chromatography over silica gel (10g, 130-270 mesh, 60 Å, Aldrich). A gradient solvent system of hexane-ethylacetate (2 mL/minute) with increasing order of polarity was used as eluent to give 450 mg of purified product (M_n = 2200 g/mole and M_w/M_n = 1.2). The absence of unreacted EGP in the final purified product (R_f = 0.20) was confirmed by TLC.

Polymer chain end derivatization with diazomethane

After its ^{13}C NMR analyse, poly(ε-caprolactone) sample in deuterated chloroform was transferred from the NMR tube into a 7 mL glass vial. Excess amount of CH_2N_2 in ether was then added into the vial until no gas bubble was generated. The solvent in the reaction mixture was then removed by purging with dry N_2 gas. The dried residue was then dissolved in deuterated chloroform and subjected to ^{13}C NMR analysis again.

Regio-selective end capping of EGP oligo(ε-CL)

In a 20 mL oven dried reaction vial, 135 mg (0.12 mmol) of EGP oligo(ε-CL) which had previously been dried in a vacuum desiccator (0.1 mmHg, 25 ^0C, 16 h) was dissolved in 5 mL of freshly distilled THF. Lipase PS30 (Amano) (200 mg) dried through the same procedure as EGP oligo(ε-CL) described above was transferred to the same reaction vial in a glove bag maintained by argon atmosphere. The vial was immediately stoppered with a rubber septum, purged with argon and sealed by Teflon tape. 55 μL of vinyl acetate (0.60 mmol) (Aldrich) was added via a syringe under argon. The reaction mixture was then stirred at 25 ^0C under argon for 8 hours. To terminate the reaction, the enzyme was removed by vacuum filtration (glass-fritted filter, medium-pore porosity) and washed with 3 x 2 mL portions of THF. The filtrates were combined and solvent was removed *in vacuo* to give 98 mg of product (93.2% yield).

Synthesis of a four-armed block copolymer of poly(LA-co-CL)

662.7 mg (4.60 mmol) of L-Lactide (Aldrich, recrystallized three times from fresh distilled ethyl acetate, dried in a vacuum oven at 25 ^0C) and 158.7 mg (0.12 mmol) of EGP oligo(ε-CL) (previously dried in a vacuum desiccator at 25 ^0C for 16 h, 0.1 mmHg) were transferred into a previously silanized, flame-dried, and argon-purged Schlenk tube using a dry argon purged glovebag. 23 μL tin (II) (2-ethylhexanoate)$_2$ ($Sn(Oct)_2$, Sigma) solution, 1.0 M in toluene, (Monomer/Catalyst = 200/1) was then added into the tube by syringe under an argon atmosphere. The solvent residue was removed at reduced pressure and the tube was sealed under dry argon. The reaction was then allowed to proceed for 6 hours at 120 ^0C maintained by a thermostat oil bath. The reaction melt was then cooled to room temperature, dissolved in chloroform, precipitated in cold methanol, separated by filtration, and finally dried *in vacuo* at room temperature. Mn = 11,500 g/mol, M_w/M_n = 1.10 (GPC result).

Instrumental methods

^1H-NMR spectra were recorded on a Bruker ARX-250 spectrometer at 250 MHz. ^1H-NMR chemical shifts in parts per millions are reported downfield from 0.00 ppm using trimethylsilane (TMS) as the internal reference. The concentration used was 4% w/v in $CDCl_3$. The instrumental parameters were as follows: temperature 300 K, pulse width 7.8 μs (30^0), 32K data points, 3.178-s acquisition time, 1-s relaxation delay, and 16 transients. ^{13}C-NMR spectra were recorded at 62.9 MHz on a Bruker

ARX-250 spectrometer with chemical shifts in ppm referenced relative to DMSO-d_6 or TMS at 39.7 or 0.00 ppm, respectively. Spectral acquisitions were conducted with 10% w/v DMSO-d_6 or $CDCl_3$ solutions using the following parameters: 300 K, pulse width 6.3 μs (30^0), 64K data points, 1.638-s acquisition time, 1-s relaxation delay, and 15000-18000 transients. DEPT-135 spectra were recorded at 62.9 MHz using a Bruker ARX-250 spectrometer with 10% w/v DMSO-d_6 solutions at 300 K, 64 K data points, 1.638-s acquisition time, 1-s relaxation delay and 10000-15000 transients. For the (^1H-^1H) COSY experiment (4% w/v product in DMSO-d_6) the data were collected in a 1024 x 128 data matrix and zero filled to 512 x 512 using 8 scans per increment, a 1915 Hz sweep width, and 1.93-s delay between transients.

Molecular weights were measured by gel permeation chromatography (GPC) using a Waters HPLC system equipped with model 510 HPLC pump, Waters model 717 autosampler, Waters model 410 refractive index detector (RI), Viscotek model T60 viscosity detector and 500, 10^3, 10^4, 10^5 Å ultrastyragel columns in series. Chloroform (HPLC grade) was used as eluent at a flow rate of 1.0 ml/min. The sample concentration and injection volumes were 0.5 % (w/v) and 100 μL, respectively. Molecular weights were determined based on a universal calibration curve generated by narrow molecular weight distribution polystyrene standards (3.00 x 10^2, 1.00 x 10^3, 2.50 x 10^3, 4.00 x 10^3, 1.40 x 10^4, 9.00 x 10^4 and 2.07 x 10^5, Polysciences). RI and viscosity data were processed by Viscotek TriSEC GPC Software (version 3).

The surface tension measurements were conducted at 25 °C, using a Fisher Scientific Surface Tensiomat 21 equipped with a 6 cm Platinum-Iridium ring (Du Nouy method).

Results and Discussion

Ethyl glucopyranoside (EGP) as a multifunctional initiator for lipase-catalyzed ring-opening polymerization of ε-caprolactone

α,β-ethylglucopyranoside (EGP) was used as the multifunctional initiator for ε-caprolactone (ε-CL) ring-opening polymerization (Scheme I). This provided a novel route for the one-pot synthesis of oligomers or polymers with ethyl glucose head groups.

Screening experiment

Commercially available lipases were evaluated for EGP initiated ε-CL ring-opening polymerization (ε-CL/EGP = 7:1 mol/mol; 70 ^{0}C, 96 h). The result of screening experiment is given in Table 1. The lipases from *Candida rugosa* (formerly known as *Candida cylindracea*, CCL), *Psuedomonas cepacia* (PS-30), *Mucor miehii* (lipozyme IM), *Mucor javanicus* (MAP-10) and *Psuedomonas fluorescens* (AK) gave

Scheme I

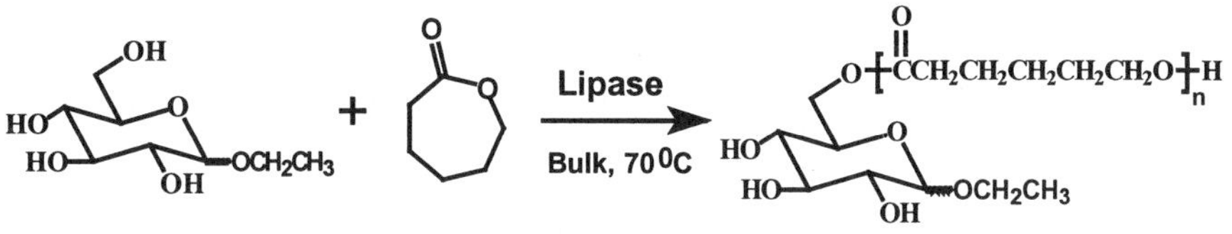

$$\text{EGP} + \varepsilon\text{-CL} \xrightarrow[\text{Bulk, } 70\,^{0}\text{C}]{\text{Lipase}} \text{EGP-}O\text{-}[\overset{O}{\overset{\|}{C}}CH_2CH_2CH_2CH_2CH_2O]_n\text{-}H$$

low conversions of ε-CL to ring-opened product (2, 9, 15, 29 and 54 %, respectively). In contrast, the lipase from *Candida antarctica* (Novozym-435) and porcine pancreatic lipase (PPL) gave relatively higher ε-CL conversion (>80 %) under identical reaction conditions. Therefore, lipase PPL and Novozym 435 were used as biocatalysts in the following ring-opening polymerization of ε-CL.

Table 1. Lipase-catalyzed Ring-opening Polymerization of ε-CL in Bulk with the Presence of Ethyl Glucopyranoside (EGP) at 70 ^{0}C for 96 Hours

Entry	Enzyme[a]	Molar ratio of ε-CL/EGP	ε-CL conversion (%)	Mn[b]	Mw/Mn[b]
1	PPL	7/1	81.1	775	1.8
2	Novozym 435	7/1	92.0	5,180	2.8
3	Lipase AK	7/1	54.0	700	1.6
4	Lipase MAP10	7/1	29.4	500	2.2
5	Lipozyme IM	7/1	15.1	350	1.1
6	PS30	7/1	9.2	440	1.2
7	CCL	7/1	2.0	395	1.2

a). The amount of enzymes used for each entry was 500 mg.
b). Determined by GPC method.

Lipase-catalyzed regioselective ring-opening polymerization of ε-caprolactone

Initially, lipase PPL was used to catalyze the ring-opening polymerization of ε-CL by EGP (ε-CL/EGP ratio of 7:1) at 70 ^{0}C for 96 h in bulk. The resulted crude product was then purified by a silica-gel column chromatography to remove unreacted EGP. The resulting end-group 'tailored' oligopolyester had a number average molecular weight (M_n) and polydispersity (M_w/M_n) of 2200 g/mol and 1.26, respectively. The reaction was highly regioselective and resulted in oligo(ε-CL) chains which were attached by an ester group exclusively to the primary hydroxyl moiety of EGP as discussed below. The structure of the product was confirmed by nuclear magnetic resonance (NMR) experiments as is elaborated below.

The ^{1}H-NMR spectrum of the EGP-oligo(ε-CL) conjugate, due to poorly resolved signals, was not useful in determining the position at which oligo(ε-CL) was linked to the sugar head group. Well resolved ^{13}C NMR spectra were therefore used in the following experiments to establish the regioselectivity of the polymerization products. DEPT (distortionless enhancement by polarlization transfer) experiments were found to be particularly useful. By editing the spectrum, carbon can be categorized according to the number of protons. A comparison of the DEPT-135 spectrum of EGP and the ring-opened products from PPL-catalyzed reactions between EGP and ε-CL showed that for the latter, signals corresponding to the EGP C-6 α- and β- anomers were shifted downfield by 2.4 ppm (Figure 1b). Furthermore, signals in Figure 1a observed at 61.9 and 62.0 ppm were no longer detected in the product (Figure 1b). These results indicated that the C-6 primary hydroxyl position of EGP served as sites for initiation of ε-CL polymerization. This conclusion was supported by an upfield shift for EGP C-5 of 3.0 ppm subsequent to ε-CL polymerization (Figure 1b). The upfield shift is consistent with substitution at the ε-position relative to C-5 (*18*). Moreover, all other carbon resonances assigned to EGP showed no substantial change in chemical shift subsequent to ε-CL polymerization (Figures 1a and 1b).

To give an unambigious assignment of all NMR signals of EGP oligo(ε-CL) conjugates, low molecular weight fractions are needed. In this respect, an EGP oligo(ε-CL) dimer in which two ε-CL units were linked to EGP was successfully fractioned by column chromatography. The structure of this EGP oligo(ε-CL) dimer was confirmed by detailed NMR analyses (^{1}H-^{1}H COSY and ^{1}H-^{13}C HETCOR) in deuterated chloroform. The ^{13}C-NMR spectrum of this dimer showed two signals at 180 to 160 ppm regions (spectrum not shown). By comparing the literature data of ^{13}C NMR spectrum of poly(ε-caprolactone) (*19*), the signal at 174.0 ppm was assigned to the ester carbonyl carbon linked to EGP head group and the signal at 173.8 ppm was ascribed to the ester carbonyl carbon between two ε-caprolactone units. This assignment was further strengthened in a comparison with the ^{13}C NMR spectrum of EGP oligo(ε-CL) monoadduct in which the EGP head group was linked to one ε-CL unit. The DEPT-135 NMR spectrum of this diadduct conjugate was then compared to that of EGP-oligo(ε-CL) in Figure 2. Identical carbon signals were observed for the two EGP-oligo(ε-CL) conjugates with the different chain lengths of oligo(ε-CL) segments. Therefore, it was concluded that PPL-catalysis resulted in regioselective initiation by EGP at C-6 to form well defined macromers containing oligo(ε-CL) with an EGP head group. Controlled experiments without the presence of enzyme in the reaction mixture showed no ε-CL monomer conversion.

It is noteworthy to mention that there was no observed preference between the α- and β-anomers of EGP for PPL-catalyzed reaction with ε-CL. Compared with the report by Adelhorst *et al* (*8*) that ethyl β-D-glucopyranoside exhibited about twice the reactivity of the ethyl α-D-glucopyranoside, this may be due to the fact that ε-caprolactone is an activated acyl donor compared with the fatty acid used in their study.

Similar reactions catalyzed by lipase Novozym 435 using 1:2 and 1:30 ratios of the EGP to ε-CL resulted in the synthesis of EGP-oligo/poly(ε-CL) conjugates, M_n =

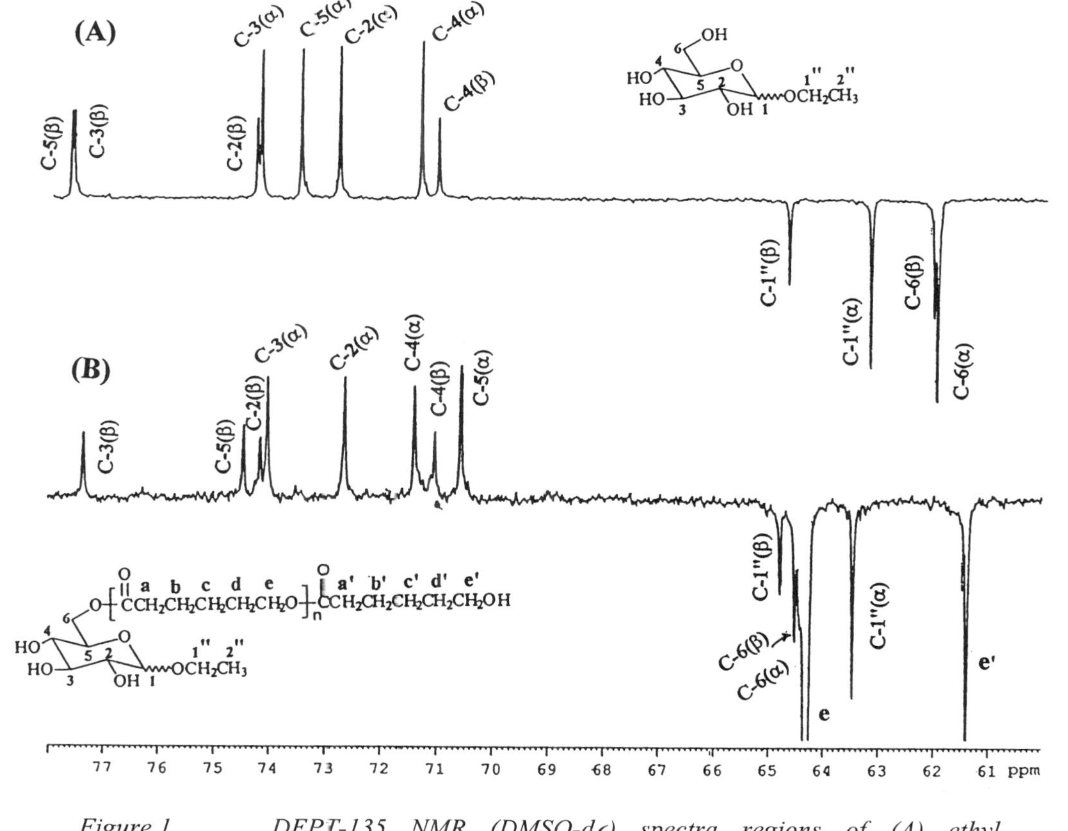

Figure 1. *DEPT-135 NMR (DMSO-d_6) spectra regions of (A) ethyl glucopyranoside (EGP), (B) oligo(ε-CL)-EGP conjugate (Mn= 2,200 g/mol).*

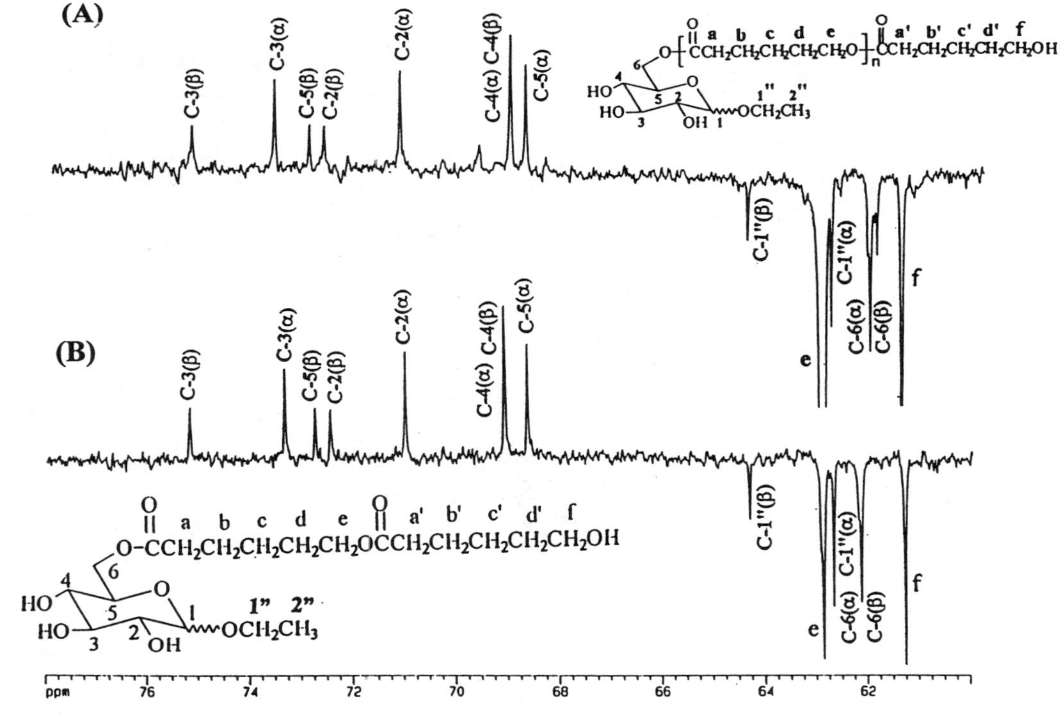

Figure 2. DEPT-135 NMR ($CDCl_3$) spectra regions of (A) oligo(ε-CL)-EGP conjugate (Mn= 2,200 g/mol), (B) oligo(ε-CL)-EGP diadduct.

486 and 12,500 g/mol, respectively. Fractionating of the EGP-oligo(ε-CL) conjugates (M_n = 486 g/mole) by column chromatography afforded an EGP oligo(ε-CL) monoadduct in which one ε-CL unit was linked to EGP head group. The structure of this EGP oligo(ε-CL) monoadduct was confirmed by detailed NMR analyses in deuterated chloroform. The ^{13}C-NMR spectrum of this monoadduct at carbonyl carbon region showed one signal at 174.0 ppm (spectrum not shown), and it was assigned to the ester carbonyl carbon linked to EGP head group. The DEPT-135 NMR spectrum of this monoadduct was also compared to that of EGP-oligo(ε-CL) diadduct in Figure 3. Similar results were observed as that in lipase PPL catalysis. This suggested that lipase Novozym 435-catalyzed ring-opening polymerization of ε-CL also proceeded in a regioselective fashion with the exclusive esterification of primary hydroxyl group of EGP head group. Controlled experiments with deactivated Novozym 435 and the absence of enzyme in the reaction mixture showed no ε-CL monomer conversion.

During the lipase-catalyzed bulk polymerization of lactones and cyclic carbonates, water is known to act as an initiator *(20,12,13)*. Therefore, there is the possibility of competitive initiation by EGP and water. However, if a fraction of the oligo(ε-CL) chains were initiated by water, this would result in carboxyl terminal chain ends having a carbonyl signal at 177.3 ppm (Figure 4a). After derivatization with diazomethane, the carboxyl chain ends converted to their methyl ester derivative resulting in a corresponding upfield shift of the carbonyl signals by about 4 ppm (Figure 4b). Comparison of the spectra region of EGP oligo(ε-CL) to that of PCL showed that the signal at 173.8 ppm is due to the intra-chain carbonyl groups of oligo(ε-CL). From previous NMR analysis on EGP oligo(ε-CL) monoadduct, the signal at 174.0 ppm was ascribed to the ester carbonyl groups connecting the ε-CL repeat units with the primary hydroxyl groups at EGP terminal. The absence of any resonances in the 176 to 178 ppm region suggested that the initiation of the oligo(ε-CL) chains occurred exclusively by EGP (Fig. 5a). This conclusion was further strengthened by comparing the ^{13}C-NMR spectra of these products prior to and after diazomethane derivatization. Since there was no notable change in the spectra displayed in Figures 5a and 5b, this supports that if both water and EGP functioned as competitive initiators, the initiation of chains by water occurred infrequently.

Ethyl glucopyranoside (EGP) as a multifunctional initiator for lipase-catalyzed regioselective ring-opening polymerization of trimethylene carbonate (TMC)

The same lipases and protocol as above were used to establish conditions for EGP initiated TMC ring-opening polymerization (TMC/EGP = 7:1 mol/mol; 70 ^{0}C, 96 h) (Scheme II). There were significant differences in the activity of the lipases for EGP initiated TMC ring-opening polymerization reaction. Low conversions were observed with lipases PS-30, CCL, Lipozyme IM and MAP-10 (20, 5, < 2, and 15 %, respectively). However, the lipases PPL, Novozym 435 and Lipase-AK gave much higher conversions for the same reaction (>78 %). Of all the lipases screened,

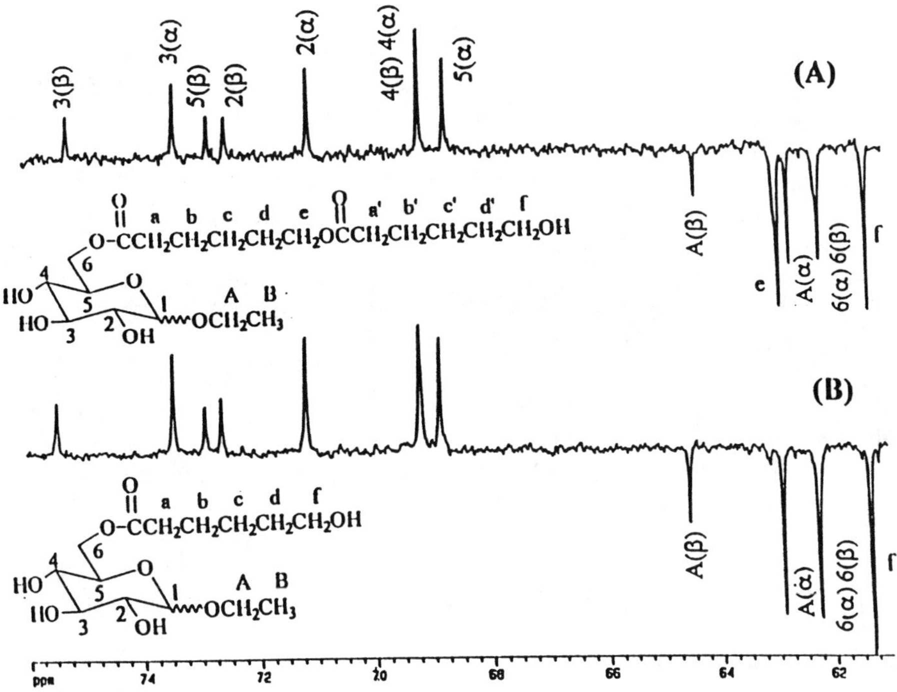

Figure 3. DEPT-135 NMR ($CDCl_3$) spectra regions (61-76 ppm) of (A) oligo(ε-CL)-EGP diadduct, (B) oligo(ε-CL)-EGP monoadduct.

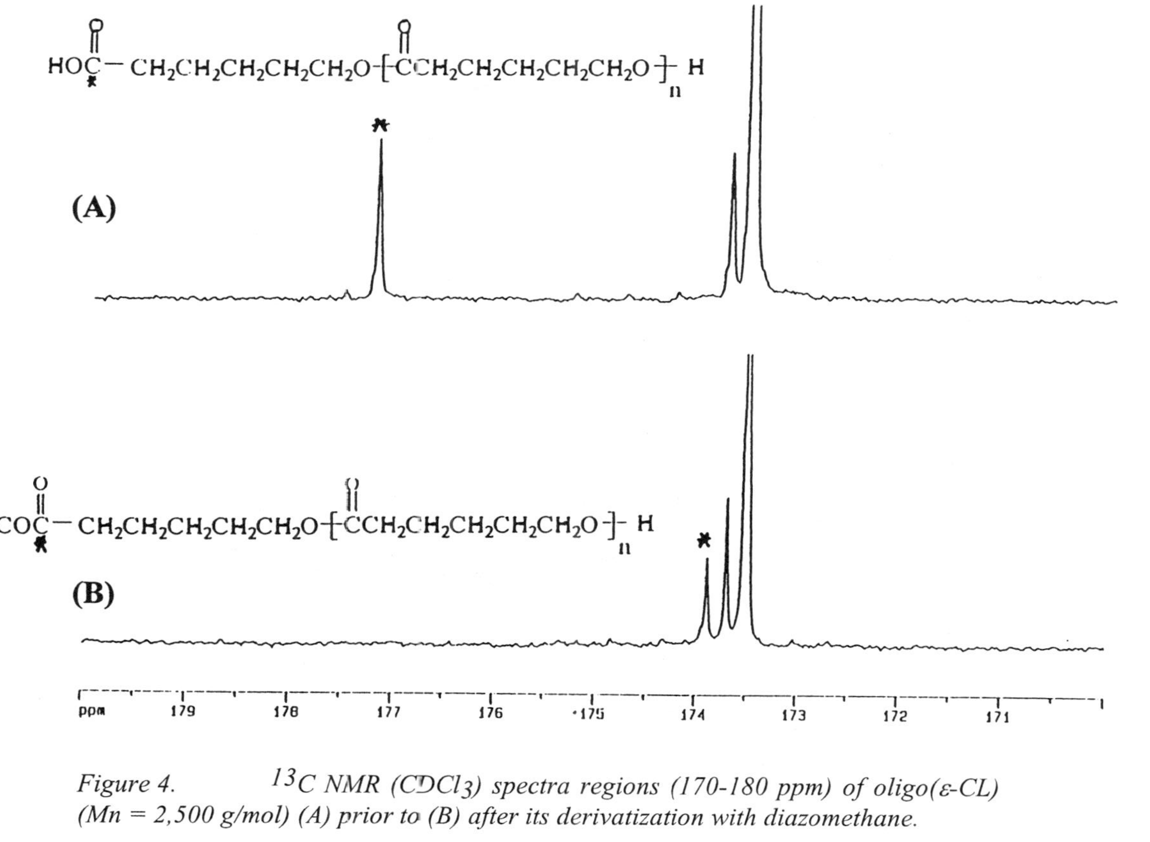

Figure 4. ^{13}C *NMR ($CDCl_3$) spectra regions (170-180 ppm) of oligo(ε-CL) (Mn = 2,500 g/mol) (A) prior to (B) after its derivatization with diazomethane.*

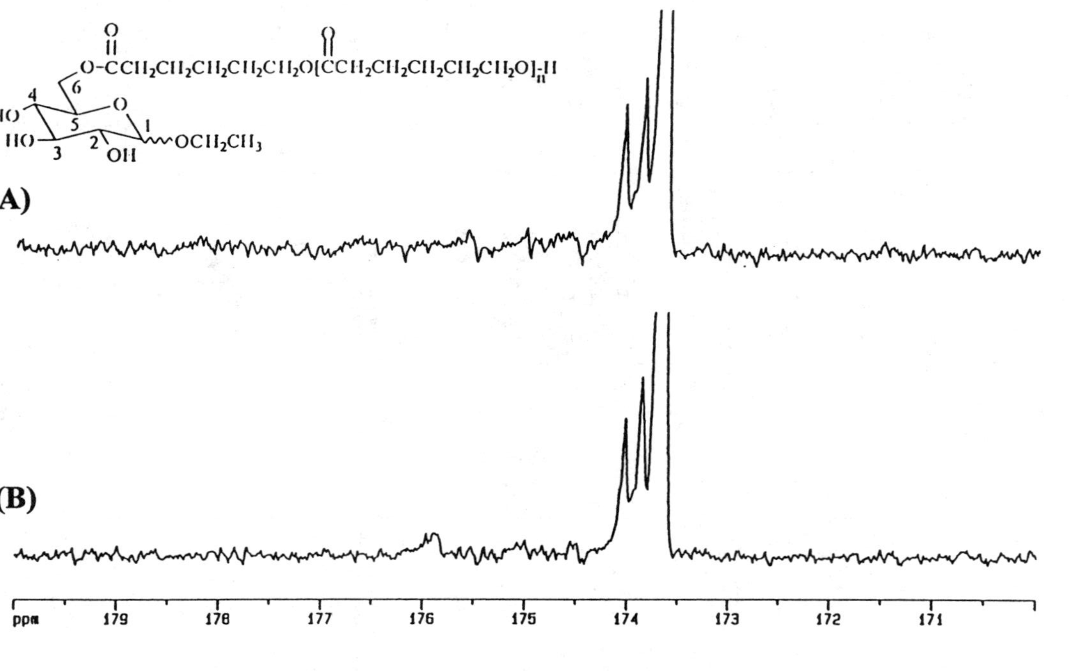

Figure 5. ^{13}C *NMR ($CDCl_3$) spectra regions (170-180 ppm) of oligo(ε-CL)-EGP conjugate (Mn = 2,200 g/mol) (A) prior to (B) after its derivatization with diazomethane.*

Novozym-435 gave the highest TMC conversion (97%). Work was then performed to establish the chain end structure of the products.

Scheme II

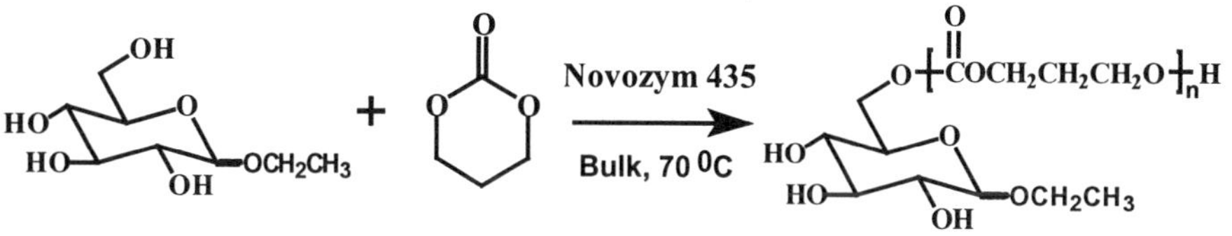

Novozym 435

Bulk, 70 ^{0}C

The Novozym-435 catalyzed formation of a EGP-oligo(TMC) conjugate was highly regiospecific and only the primary hydroxyl moiety of EGP was attached to the oligo(TMC) chains. After column chromatographic purification to remove unreacted EGP, an EGP-oligo(TMC) having an M_n and M_w/M_n of 7200 g/mol and 2.5, respectively, was obtained. An analysis by NMR was then undertaken to establish the site of reaction in the EGP-oligo(TMC) conjugate. A similar reaction with a 1:2 molar ratio of EGP to TMC followed by column chromatographic purification gave the EGP-TMC diadduct. The structure of this EGP-TMC diadduct was confirmed by detailed NMR (^{1}H-^{1}H COSY and ^{1}H-^{13}C HETCOR) analyses in deuterated chloroform. To conclusively determine the site of substitution at EGP where initiation occurs, the same product was then subjected to NMR analysis in dimethyl sulfoxide-*d$_6$* in which EGP is soluble also. The DEPT 135 carbon spectral region of EGP and EGP-TMC diadduct in dimethyl sulfoxide-*d$_6$* is shown in the Figure 6. On comparison with the DEPT 135 carbon spectra of EGP, the EGP-oligo(TMC) conjugate showed a downfield shift (~ 6 ppm) in the resonance positions of carbon-6 from ~62.1 to ~68.0 ppm (Figure 6b). Also, the carbon-5 signals were shifted upfield (~ 3 ppm) due to the γ-shift (C-5β from 77.5 to 74.4 ppm, C-5α from 73.8 to 70.5 ppm). Furthermore, all other signals assigned to EGP showed no substantial shift subsequent to TMC polymerization. Therefore, it was concluded that the Novozym 435 catalyzed ring-opening polymerization initiated by EGP led to the exclusive participation of EGP C-6 hydroxyl groups. Again, there was no observed preference between the α- and β-anomers of EGP for the Novozym 435-catalyzed reaction between EGP and TMC. Controlled experiments with deactivated Novozym 435 under the same reaction condition showed no monomer conversion of TMC.

It is noteworthy to point out that when methyl D-glucopyranoside (MGP) was used instead of EGP as the initiator under similar reaction conditions, no reactions were observed for both monomer (ε-CL, TMC) systems.

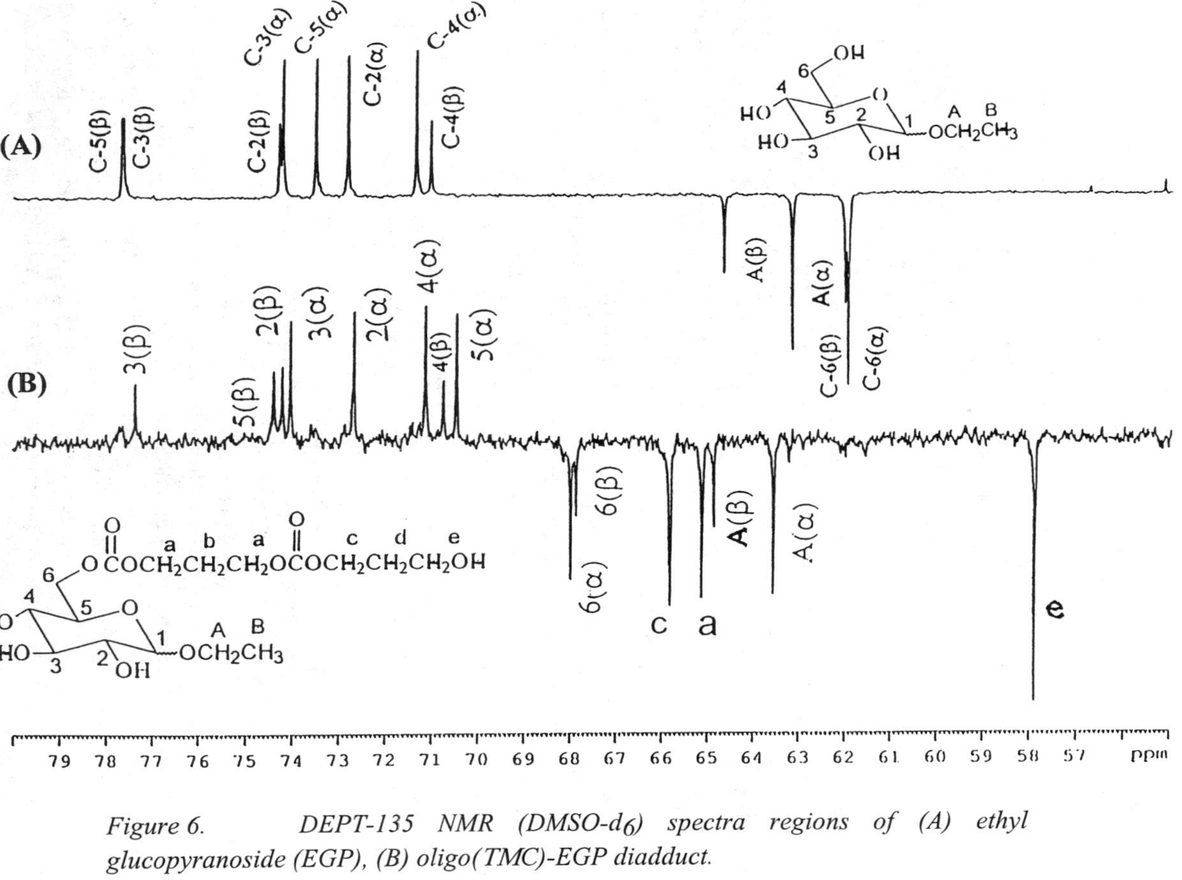

Figure 6. *DEPT-135 NMR (DMSO-d_6) spectra regions of (A) ethyl glucopyranoside (EGP), (B) oligo(TMC)-EGP diadduct.*

Proposed reaction mechanism

Based on the defined structure of the products synthesized herein and related work which is further elaborated below, a general mechanism for the lipase-catalyzed EGP initiated ring-opening polymerization reaction was proposed (Scheme III). The monomer ε-CL was selected for the purpose of illustration. Initiation is believed to involve: *i*) reaction between the lipase and ε-CL to form an enzyme-activated monomer (EAM) complex and *ii*) reaction of EGP with the EAM complex to form ethyl 6-(6-hydroxyhexanoyl) glucopyranoside (*21*). The regioselectivity of chain initiation must be controlled by the lipase catalysis. Propagation than occurs by reaction of either **1** or **2** with the EAM complex. This reaction scheme is consistent with the general mechanism of lipase-catalyzed ring-opening polymerization of lactones proposed by MacDonald *et al* (*12*), Henderson *et al* (*13*) and Uyama *et al* (*22*).

Scheme III

Initiation

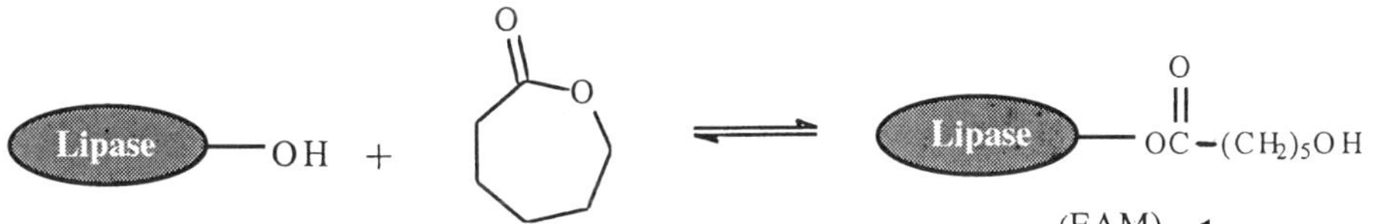

Lipase—OH + ε-CL ⇌ Lipase—$OC(=O)-(CH_2)_5OH$ (EAM) **1**

Propagation

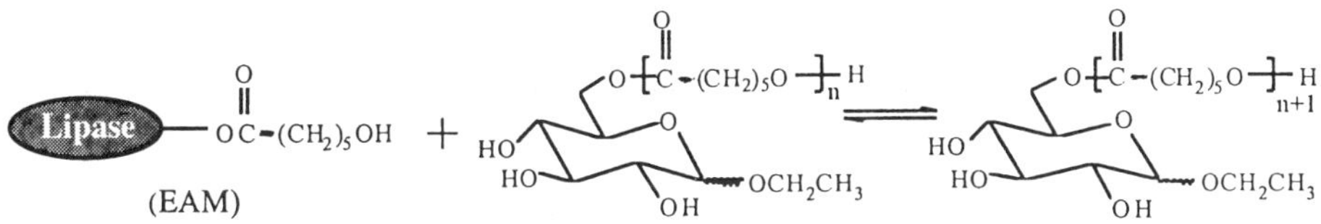

Lipase—$OC(=O)-(CH_2)_5OH$ (EAM) + **2** [6-O-$[C(=O)-(CH_2)_5O]_n$H ethyl glucopyranoside, OCH_2CH_3] ⇌ 6-O-$[C(=O)-(CH_2)_5O]_{n+1}$H ethyl glucopyranoside (OCH_2CH_3)

Surface activity of EGP oligo(ε-CL) conjugates

It was possible to control the length of the polyester chain by variation in the molar ratio of EGP to lactone monomer. The surface activities of two EGP oligo(ε-CL) conjugates with number-average molecular weight Mn = 486 and 2200 g/mole were depicted in Figure 7. As expected, these amphiphilic oligomeric conjugates are surface active based on tensiometric measurements at the air-water interface (Figure 7). The efficiency and effectiveness of these conjugates for surface tension reduction was a function of the oligo-ester chain length (Figure 7). The higher the molecular weight of the hydrophobic segment of the conjugate, the more efficient it is to reduce the surface tension. Therefore, the potential exists to enhance the surface activity of these conjugates by careful modulation of their hydrophilic/hydrophobic balance. Also, numerous reports in the literature attest to the biodegradability and biocompatibility of poly(ε-CL) and poly(TMC) (*23,24*). Therefore, amphiphilic structures generated in this work are likely to be biodegradable

Chemoenzymatic synthesis of a multiarm poly(lactide-co-ε-caprolactone)

As mentioned earlier, another useful application of EGP oligomer or polymer conjugates is that they can be used as macromers to construct multiarm heteroblock copolymers with well-defined spatial architecture. As an illustration of this approach, EGP oligo(ε-CL) obtained from the lipase-catalyzed ring-opening polymerization process as described previously was used as a macromer to further prepare a star shaped copolymer with poly(L-Lactide).

Copolymers of L-Lactide and ε-caprolactone [poly(LA-co-CL)] are of interest for use as biodegradable and biocompatible materials. By variation in copolymer structure, products may be designed that offer 'tailored' physico-mechanical properties and controlled degradation rates. In contrast to the extensive efforts made to prepare linear block copolymers from L-LA and other monomers such as caprolactone, glycolide and ethylene oxide (*25-27*), there are few reports on the preparation of their counterparts with alternative architectural structures (*28,29*). Our enzymatic approach to prepare EGP oligo- or poly(ε–CL) conjugates provides us a novel route to a star shaped heteroarm block copolymer of PLA and PCL with a well-defined spatial architecture. For this purpose, the macromer 1-ethyl 6-oligo(ε-CL) glucopyranoside was prepared so that the ω-hydroxyl group of oligo(ε-CL) was protected by acetylation. Subsequently, the polymerization of L-LA was initiated from the remaining free hydroxyl groups at the carbohydrate core to form the heteroarm block copolymer.

Regioselective end-capping of EGP oligo(ε-caprolactone)

1-ethyl 6-oligo(ε-CL) glucopyranoside was prepared and purified as described previously. The resulting 1-ethyl 6-oligo(ε-CL) glucopyranoside was separated from unreacted EGP and fractioned by column chromatography. It had a number average molecular weight (Mn) of 1120 g/mol (Mw/Mn = 2.07) measured by gel permeation chromatography (GPC). The extent to which the chain ends were initiated with EGP was analyzed by ^{13}C NMR studies. These analyses involved comparing NMR spectra before and after derivatization with diazomethane. In the ^{13}C-NMR spectrum, the

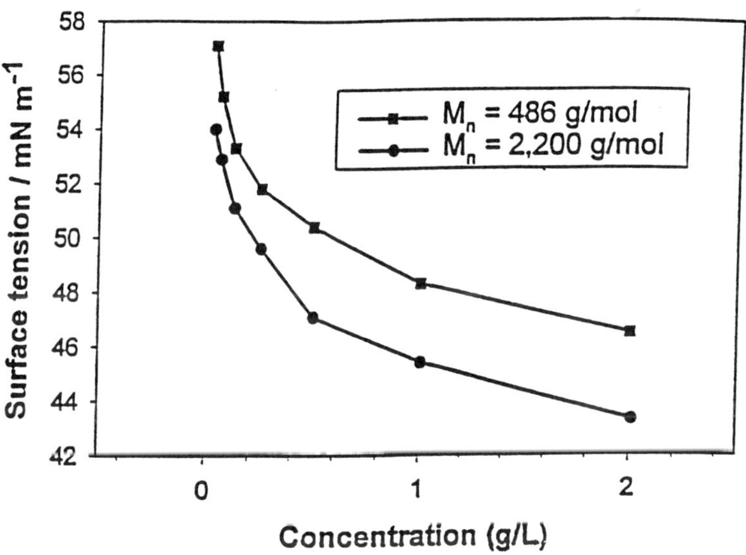

Figure 7. Surface tension versus concentration of oligo(ε-CL)-EGP conjugates with different molecular weights at 25 ^{0}C.

presence of signals corresponding to carbonyl carbons between 170 to 178 ppm would provide information on end-group structures of oligo(ε-CL) chain ends. A shift in the carboxyl carbonyl carbon of the oligo(ε-CL) chain due to reaction with diazomethane from 178 to 174 ppm would indicate the presence of carboxyl chain ends. By comparing the above ^{13}C NMR spectra region of the purified product used in the following experiment, prior to and after diazomethane derivatization, no notable change was observed. This suggests that all the oligo(ε-CL) chains are linked to EGP groups at one end.

To exclusively perform the initiation of L-LA polymerization at the three sugar-ring secondary hydroxyl groups, it was first necessary to protect the terminal ω-hydroxyl group of the oligo(ε-CL) chain. The lipase PS30 (from *Pseudomonas cepacia*) was evaluated for this transformation using vinyl acetate as an irreversible acetylating agent (*30*). The reaction was performed in a closed vial that contained anhydrous THF (5 mL), 1-ethyl-6-oligo(ε-CL) glucopyranoside (135 mg) and lipase PS30 (200 mg) (Scheme IV).

Scheme IV

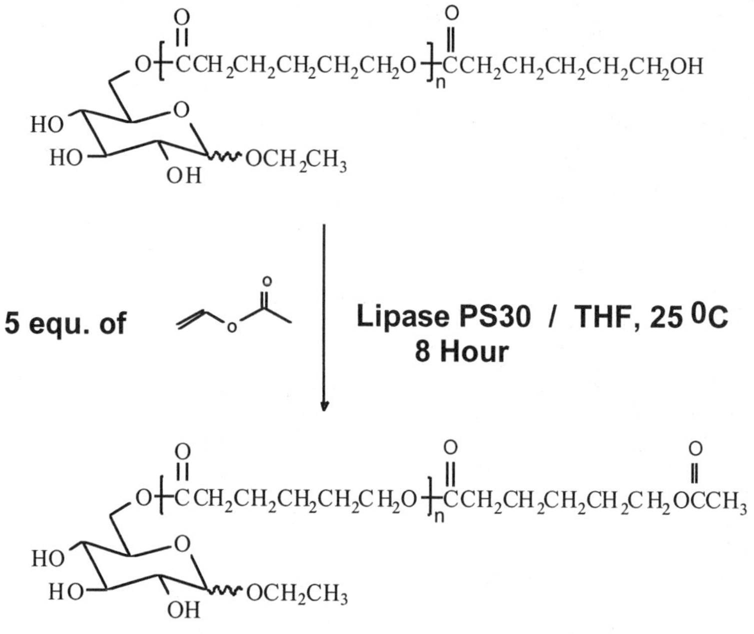

O—[C(=O)CH₂CH₂CH₂CH₂CH₂O]ₙ—C(=O)CH₂CH₂CH₂CH₂CH₂OH

HO, HO, OH, O, OCH_2CH_3

5 equ. of (vinyl acetate) **Lipase PS30 / THF, 25 ^{0}C 8 Hour**

O—[C(=O)CH₂CH₂CH₂CH₂CH₂O]ₙ—C(=O)CH₂CH₂CH₂CH₂CH₂OC(=O)CH₃

HO, HO, OH, O, OCH_2CH_3

The course of the reaction was monitored by ^{1}H-NMR. When the molar ratio of vinyl acetate to EGP oligo(ε-CL) glucopyranoside was 5 to 1, the lipase PS30 catalyzed selective acetylation of the ω-hydroxyl oligo(ε-CL) terminal groups was complete within 8 hours. Furthermore, under these reaction conditions, acetylation at the sugar secondary hydroxyl positions was not observed. The ^{1}H-NMR spectra of EGP oligo(ε-CL) before and after the reaction are displayed in Figure 8. The signal at 3.66 ppm characteristic of the ω-hydroxylmethylene protons of the PCL chains disappeared and a new singlet at 2.05 ppm due to the corresponding acetyl chain-end group protons was observed. Since unreacted vinyl acetate and the by-product acetaldehyde are volatile compounds, they were easily removed *in-vacuo* during product isolation. A comparison of the EGP oligo(ε-CL) conjugate before and after the reaction by ^{13}C-NMR showed a downfield shift of 1.70 ppm for the signal due to the ω-hydroxylmethylene carbon of the PCL segment (spectrum not shown). The new signal from the acetyl group appeared at 21.06 ppm. All other carbon resonances assigned to the EGP oligo(ε-CL) conjugate showed no substantial change which excluded any possible transesterification between EGP secondary hydroxyl groups and main chain esters. GPC analysis showed that no substantial chain degradation occurred during chain-end acetylation.

Synthesis of PLA-co-PCL

Due to its low toxicity and high efficiency, stannous octanoate has been the most frequently used catalyst for synthesizing high molecular weight PLAs. Previous work showed that the bulk polymerization of L-LA, conducted in the presence of a multifunctional-hydroxyl initiator such as inositol, gave star-shaped products (*31*). Han *et al.* prepared a lactide-based poly(ethylene glycol) polymer network for tissue engineering scaffolds by using glycerol as an initiator (*32*). In these stannous octanoate catalyzed polymerizations, both primary and secondary hydroxyl groups were effective for the initiation of L-LA polymerization. These polymerizations were believed to proceed via an insertion coordination mechanism (*33*). Based on these reports, the spatially organized EGP oligo(ε-CL) was used as the multifunctional hydroxyl initiator to prepare a novel heteroarm multi-block PLA-co-PCL copolymer (Scheme V). The L-LA polymerization was performed solventless in the melt at 120 ^{0}C for 6 hours. The residual monomer was removed by precipitation of the product in cold methanol which showed that the monomer conversion was 80%. The ^{1}H NMR spectrum of purified copolymer is shown in Figure 9. The assignments of the main signals were made by comparison to literature data (*28*) and our own previous work (*12*). Proton-^{1}H COSY45 2D NMR experiment (spectrum not shown) further confirmed the above assignments. By comparison of the relative signal intensities of the PCL methylene protons H_e (4.07 ppm) to protons H_f the acetyl end-capped groups (2.05 ppm), the M_n of the PCL chain segment was determined ($I_e/2 \div I_f/3$, 1270 g/mol). Also, the relative intensities of protons H_h (5.17 ppm) and $H_{h'}$ (4.35 ppm) of oligo-PLA chain segments were used to determine the M_n value of the PLA arms (3,107 g/mol for each arm). The values of M_n for oligo-PCL determined by ^{1}H-NMR and GPC were in close agreement (1270 and 1301 g/mol, respectively). Furthermore, the signal intensities of protons $H_{h'}$ and H_f were almost equivalent (1.00 and 1.10, respectively) which suggests that the product has a ratio of oligo-PLA to oligo-PCL

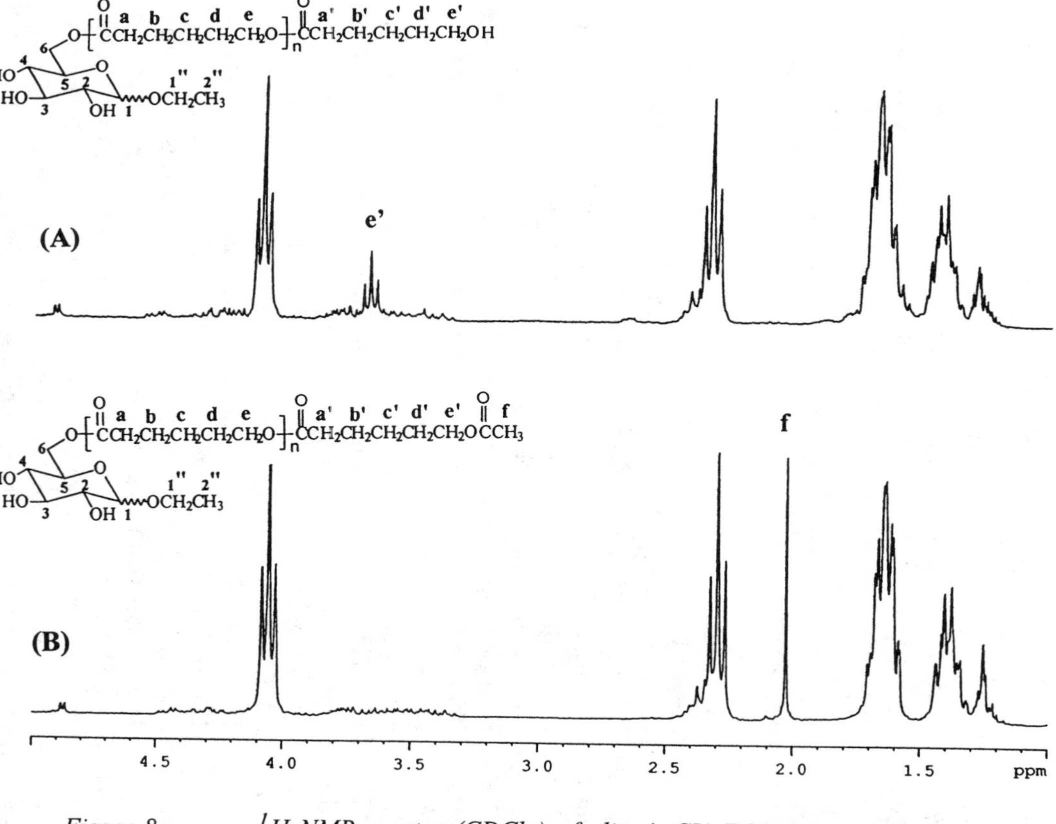

Figure 8. ^1H *NMR spectra ($CDCl_3$) of oligo(ε-CL)-EGP (Mn=1,120 g/mol, Mw/Mn=2.07) (A) prior to (B) after end-capping.*

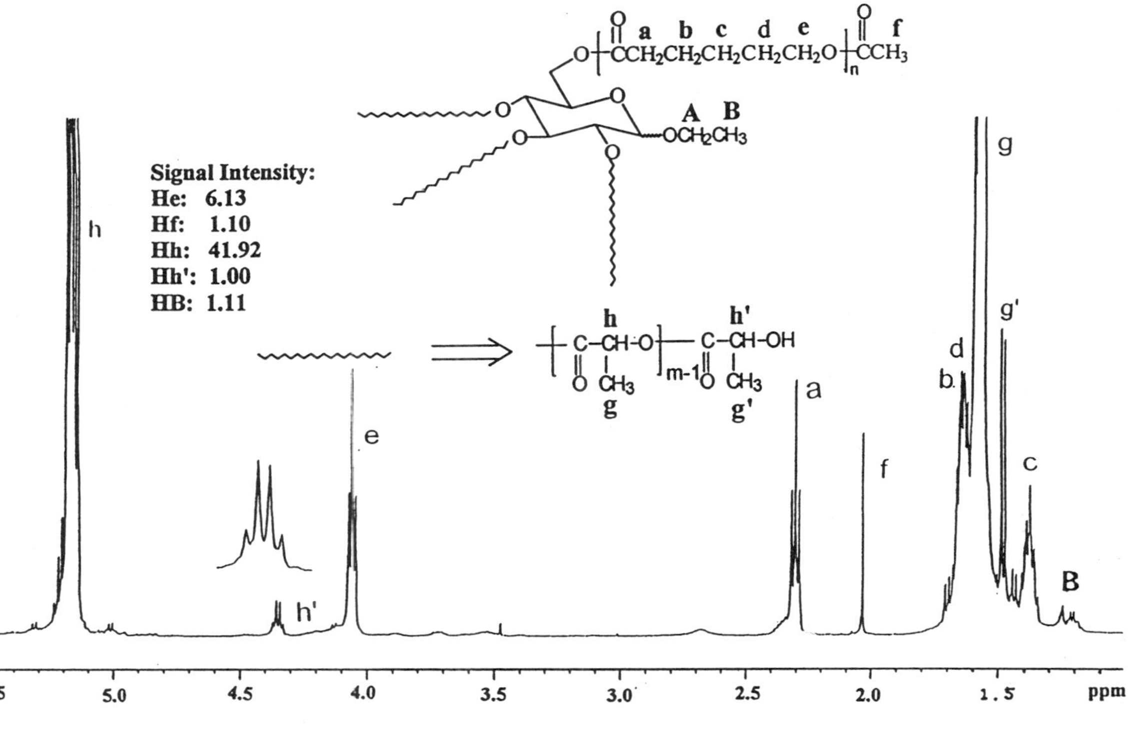

Figure 9. ^{1}H *NMR spectra ($CDCl_3$) of four-armed block copolymer of PLA-co-PCL (Mn= 11,500 g/mol, Mw/Mn=1.10).*

chain segments of 3:1. This conclusion was further strengthened by the observation that the signal intensities of protons $H_{h'}$ and H_B were of similar value (1.00 and 1.11, respectively). Comparing the GPC chromatographs of the heteroarm multi-block PLA-co-PCL copolymer to the starting material (EGP-oligo[ε-CL]) showed an increase in M_n from 1,301 to 11,500, a narrowed dispersity (2.07 to 1.10) and a unimodal distribution with no evidence of remaining unreacted EGP-oligo(ε–CL).

Based on the above ^{1}H NMR experiments, the product *Mn* based on a 3:1 ratio of PLA to PCL chain segments and chain segment lengths for PLA and PCL of 3,107 and 1,270 g/mol, respectively, was calculated and found to be 10,591 g/mol. The excellent agreement between the NMR calculated and GPC experimental values of M_n is consistent with the proposed four-armed heteroblock structure.

The absence of homo-PLA in the product was confirmed by a similar end group analysis experiment as that described preciously. For this purpose, low molecular weight oligo(lactide) was subjected to derivatization with diazomethane. After the derivatization, the carboxylic acid end group was converted to methyl carboxylate. Correspondingly, the carbonyl signal at 177.5 ppm due to the carboxyl chain end disappeared, and a new signal at 175.9 ppm ascribed to methyl carboxylate ester was

Scheme V

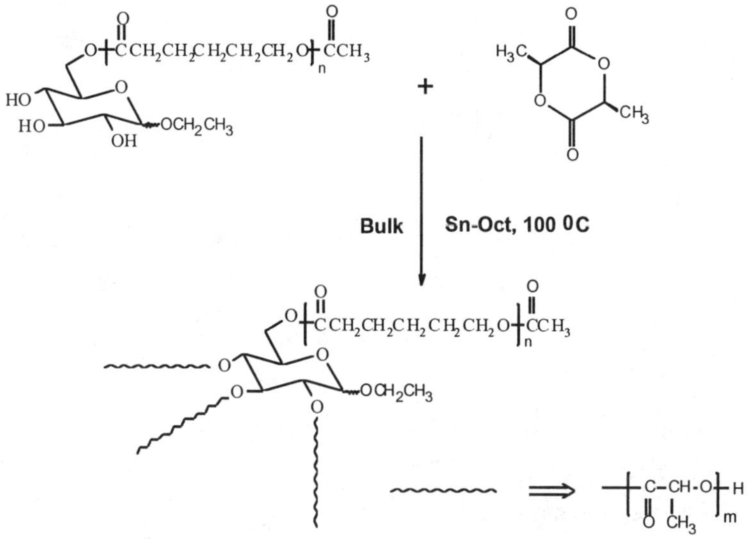

observed. The same procedure was applied to the above block copolymer of PCL and PLA. No signals were observed between 176 to 178 ppm regions. Furthermore, the ^{13}C-NMR spectra of the copolymer prior to and after diazomethane derivatization showed no observable change (spectra not shown). Therefore, it is concluded that no homo-PLA is present in the product that is due to the possible chain initiation by trace amount of moisture.

Conclusions

A convenient one-pot biocatalytic synthesis of novel biodegradable amphiphilic oligomers/polymers is described. The selectivity of different lipases and general applicability of the method was also demonstrated by screening a number of commercial lipases and several different monomer/enzyme systems i.e. ε-CL/PPL, ε-CL/Novozym-435, and TMC/Novozym-435. Lipase Novozym-435 was found to be the preferred catalyst for both monomers (ε-CL, TMC) ring-opening polymerization systems in terms of monomer conversions and product molecular weights. Thus, by this approach, polymer chains with different compositions were formed having sugar head groups without using conventional protection-deprotection chemistry. Reactions conducted using denatured lipases which were inactive for ε-CL and TMC polymerizations showed that the EGP initiated oligomerizations were indeed enzyme-catalyzed.

The resulted sugar polyester conjugates were found to be surface active, and the surface activities of these products were related to their molecular weights. Therefore, biodegradable surfactants with tunable hydrophobic hydrophilic balance can be prepared by further manipulating either the carbohydrate components or the polyester segments.

Further reactions of these macromonomers at positions 2, 3 and 4 of the EGP ring were used to develop a new family of well-defined multi-arm heteroblock copolymers with well-defined spatial architectures. For this purpose, an enzyme-mediated process was first designed to prepare a regioselective end-capped oligo(ε-CL) EGP conjugate with an ω-acetyl terminated PCL chain segment linked to the 6-position of EGP. The non-selective catalysis of lactide ring-opening at all of the EGP secondary positions was then carried out using the traditional chemical catalyst stannous octanoate. This resulted in a spatially well-defined heteroblock copolymer organized around a carbohydrate core. It is expected that by better defining the spatial organization of block copolymer chain segments, it will be possible to derive unique properties from block copolymer systems.

Acknowledgment We are grateful for the financial support received from the Center for Enzymes in Polymer Science (EPS). EPS is a collaborative initiative between the Polytechnic University and Tufts. The current industrial partners of this Center include Novo Nordisk and Rohm & Hass Co.

References

1. Haines, A. H. **1976**. *Adv. Carbohydrate Chem. Biochem.* 33, 11.
2. Bollenback, G. N., Parrish, F. W. **1971**. *Carbohyd. Res.* 17, 431.
3. Therisod, M. and Klibanov, A. M. **1986**. *J. Am. Chem. Soc.*, 108, 5638.
4. Therisod, M. and Klibanov, A. M. **1987**. *J. Am. Chem. Soc.*, 109,3977.
5. Riva, S.; Secundo, F. **1990**. *Chimica Oggi*, 6, 9.
6. Drueckhammer, D. G.; Hennen, W. J.; Pederson, R.L.; Barbas, C. F. III; Gautheron, C. M.; Krach, T.; and Wong, C. H. **1991**, *Synthesis*, 499.
7. Moye, C. J. **1972**. *Adv. Carbohydrate Chem. Biochem.* 27, 85.
8. Adelhorst, K.; Bjorkling, F.; Godtfredsen, S. E.; and Kirk, O. **1990**. *Synthesis*, 112.
9. Nicotra, F.; Riva, S.; Secundo, F.; Zucchelli, L. **1990**. *Synthetic Communications*. 20(5), 679.
10. Fabre, J.; Paul, F.; Monsan, P.; Blonski, C.; Perie, J. **1994**. *Tetrahedron Lett.* 35(21), 3535.
11. *Monomer and polymer synthesis by lipase-catalyzed ring-opening reactions.* Gross, R. A.; Kapaln, D. L.; Swift, G. (Eds.) Enzymes in polymer synthesis. *ACS symposium series 684*. American Chemical Society, Washington, DC. **1998**, p. 90.
12. MacDonald, R. T.; Pulapura, S. K.; Svirkin, Y. Y.; Gross, R. A.; Kaplan, D. L.; Akkara, J.; Swift, G.; Wolk, S. **1995**. *Macromolecules*. 28, 73.
13. Henderson, L. A.; Svirkin, Y. Y.; Gross, R. A.; Kapaln, D. L.; Swift, G. **1996**. *Macromolecules*. 29, 7759.
14. Bisht, K. S.; Deng, F.; Gross, R. A.; Kapaln, D. L.; Swift, G. *J. Am. Chem. Soc.*, **1998**, *120*, 1363.
15. Deng, F.; Bisht, K. S.; Gross, R. A.; Kapaln, D. L.; Swift, G. **1999.** *Macromolecules,* 32, 5159.
16. Ariga, T.; Takata, T.; Endo, T. **1993**. *J. Polym. Sci., Part A: Polym. Chem.* 31, 581.
17. Bisht, K. S.; Gross, R. A.; Cholli, A. L. *Appl. Spectroscopy.* **1998**, *52(11)*, 1472.
18. Breitmaier, E.; Bauer, G. **1984.** *^{13}C NMR spectroscopy: a working manual with exercises.* Harwood Academic Publishers, New York, pp. 54.
19. Vanhoorne, P.; Dubois, Ph., Jerome, R., Teyssie, Ph. **1992**. *Macromolecules*. 25, 37.
20. Bisht, K. S.; Henderson, L. A.; Gross, R. A.; Kapaln, D. L.; Swift, G. **1997**. *Macromolecules*. 30, 2705.
21. It is assumed that the serine residue present at the active site of PPL activates the lactone for subsequent polymerization by forming the EAM complex. This model of enzyme-substrate reaction and substrate activation is based on studies with triglycerides (Brockerhoff, H., Jensen, R. G., **1974**. *Lipolytic Enzymes*. Academic Press, New York).
22. Uyama, H., Namekawa, S., Kobayashi, S. **1997**. *Polymer J.* 29(3), 299.

23. Benedict, C. V.; Cameron, J. A.; Huang, S. J. **1983**. *J. Appl. Polym. Sci.* 28, 335.
24. Bucholtz, B. **1993**. *J. Mater. Sci., Mater. Med.* 4, 381.
25. Jacobs, C., Dubois, Ph., Jerome, R., Teyssie, Ph. **1991**. *Macromolecules* 24(11), 3027.
26. Kricheldorf, H. R.; Jonte, J. M.; Berl, M. **1985**. *Makromol. Chem., Suppl.*, 12, 25.
27. Chen, X.; McCarthy, S. P.; Gross, R. A. **1997**. *Macromolecules*. 30(15), 4295.
28. Tian, D., Dubois, Ph., Jerome, R., Teyssie, Ph. **1994**. *Macromolecules*. 27, 4134.
29. Kim, S. H.; Han, Y.; Kim, Y. H.; Hong, S. **1992**. *Makromol. Chem.* 193, 1623.
30. Wang, Y. F.; Lalonde, J. J.; Momongan, M. *et al.* **1988.** *J. Am. Chem. Soc.* 110, 7200.
31. Spinu, M. **1993.** US patent 5,225,521.
32. Han, D. K.; Hubbell, J. A. **1996**. *Macromolecules*. 29, 5233.
33. Zhang, X.; MacDonald, D. A.; Goosen, M. F. A.; McAuley, K. B. **1994**. *J. Polym. Sci. Polym. Chem.* 32, 2965.

Chapter 14

The Synthesis and Polymerization of Glycolipid-Based Monomers

Kirpal S. Bisht[1] and Richard A. Gross*

Polytechnic University, Center for Biocatalysis and Bioprocessing of Macromolecules, Six Metrotech Center, Brooklyn, NY 11202

Abstract: Enzymatic-synthesis of well-defined sophorolipid-analogs for evaluation of their bioactivities and as new building blocks for the preparation of glycolipid-based amphiphilic polymers is described. This work describes a route to prepare a novel glycolipid containing polymers. Specifically, the synthesis of a 6-O-acryl sophorolipid derivative, its homopolymerization, and copolymerization with acrylic acid and acrylamide are presented. The alkyl esters of sophorolipids, produced by *Torulopsis bombicola* were subjected to Novozym 435 catalyzed acylation in dry tetrahydrofuran (THF) with vinyl acrylate and vinyl acetate to diacyl derivatives. A chemo-enzymatic synthetic strategy was used that exploits advantageous characteristics of both chemical and enzymatic transformations. The regioselective synthesis of 6'-O-acryl sophorolipid derivative was accomplished by a lipase-catalyzed acrylation reaction in dry organic solvent. Of the lipases PPL, CCL, PS-30, AK, MAP-10, Novozym-435 and Lipozyme IM screened to catalyze this transformation, Novozym-435 was found to be the biocatalyst of choice. Subsequent homopolymerization of the C-6' monoacryl sophorolipid derivative and its copolymerization with acrylamide and acrylic acid by AIBN afforded glycolipid containing acrylate polymers. This is the first report of any attempt towards incorporation of these biosurfactant analogs into polymers.

Introduction

Sophorolipids are microbial extracellular surface-active glycolipids. Cells of *Torulopsis bombicola* produce them when they are grown on sugars, hydrocarbons, vegetable oils or mixtures thereof. Gorin et al.[1] described them as a mixture of

[1]Current address: Department of Chemistry, University of South Florida, 4202 East Fowler Avenue, SCA 400, Tampa, FL 33620.

macrolactone and free acid structures that are acetylated to various extents at the sophorose ring primary hydroxyl positions (Figure 1). High performance chromatographic analysis have revealed at least eight structurally different components in sophorolipids.[2] The major components have been identified as 17-L-([2′-O-β-D-glucopyranosyl-β-D-glucopyranosyl]-oxy)-cis-9-octadecenoic acid and its corresponding C4"lactone.[3] Their biogenetic role is still a subject of debate, however, it is widely accepted that these lipids are not only essential for the growth of *T. bombicola* on water insoluble alkanes, but that they also inhibit the growth of other alkane utilizing yeasts.[4] Recent developments on the use of glycolipids to treat very severe immune disorders are certainly promising. In summary, the glycolipids have been reported to be of interest for the following: *i*) *in vivo* cancer treatment/anti tumor cell activity by cytokine upregulation[5]; *ii*) treatment of autoimmune disorders (putatively via regulation of cytokine profile)[6]; *iii*) *in vivo* and *in vitro* anti endotoxic (septic) shock activity (via cytokine downregulation)[7]; *iv*) regulation of angiogenesis (via modification of cytokine profile)[8]; *v*) apoptosis induction by modification of cytokine profiles.[9] Importantly, recent advances in microbial fermentation methods have made it possible to generate up to 700 g/L of sophorolipids[10].

Because of their potential for application in a wide variety of fields, interest in biosurfactants continues to grow. The biocompatiability of biosurfactants is advantageous as problems of toxicity and their accumulation in ecosystem are eliminated. Work has been carried out to tailor-make improved sophorolipids during *in vivo* formation. These studies have mainly involved the selective-feeding of different lipophilic substrates. For example, changing the co-substrate from sunflower to canola oil resulted in a large increase (50 to 73%) of the lactonic portion of SLs.[11] Also, unsaturated C-18 fatty acids of oleic acid may be transferred unchanged into sophorolipids.[12] Thus, to date, physiological variables during fermentations have provided routes to the variation of sophorolipid composition but have not lead to well-defined pure compounds.

Our interest in sophorolipids arises due to their intriguing structures and potential application in wide variety of fields.[13] Specifically, development of a family of new glycolipid-based building blocks for the preparation of amphiphilic polymers has tremendous potential for use in a wide range of industrial and pharmaceutical applications.[14] These polymers will be of interest as an alternative route to regioselective polysaccharide modifications. It is important to realize that polymers bearing sugar residues have been reported to be of great importance for pharmacological applications where the sugar groups play an important role.[14d, 14e] The need to generate well defined sohorolipids monomers that can be subsequently polymerized to generate amphiphilic polymers is clearly evident. An alternative strategy to the 'tailoring' of sophorolipid structure during the *in vivo* processing step is to develop methods for the regioselective modification of sophorolipids after microbial synthesis. Surprisingly, only one report exists that describes the enzyme-catalyzed synthesis of a monoacetylated sophorolipid derivative. This involved deacetylation in a biphasic environment using acetylesterase as the catalyst. Low yields (25-30 %) were reported after long reaction times.[15] Enzyme-catalyzed transesterifications using activated esters for preparation of monoacylated sugars have

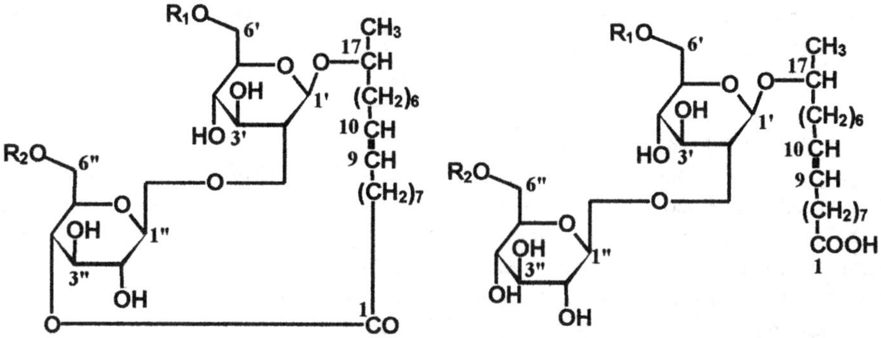

Figure 1: Sophorolipids (SL)

been reported in DMF and pyridine. Vinyl monomers with sugar groups, such as glucosylethyl methacrylate, alkyl or aryl 6-O-acryl-α-D-glucopyranosides that can be subsequently polymerized have been prepared.[16] However, yields reported from these reactions are low.[17] Moreover, using these polar solvents prevents widespread use of such reaction schemes. Also, unfortunately, many enzymes are catalytically inactive in these solvent media.

One of our research goals was the development of suitable synthetic methods so that well-defined sophorolipid-analogs would be in-hand for subsequent studies. In this paper we describe highly regioselective syntheses by the acylation of sophorolipids. These reactions were catalyzed by the lipase from *Candida antarctica* (Novozym 435) in anhydrous tetrahydrofuran at room temperature. The aim was to develop new sophorolipid-analogs that would be polymerizable using conventional polymerization chemistries.

Experimental Section

General Chemicals and Procedures. All chemicals and solvents were of analytical grades and were used as received unless otherwise noted. Sophorolipids were synthesized by fermentation of *Torulopsis bombicola* on glucose/oleic acid mixtures following a literature procedure.[15] Prior to their use, the sophorolipids were dried over P_2O_5 in a vacuum desiccator (0.1 mm Hg; 38 h; room temperature).

Porcine pancreatic lipase (PPL) Type II Crude (activity = 61 units/mg protein) and *Candida rugosa* lipase (CCL) Type VII (activity = 4570 u/mg protein) were obtained from the Sigma Chemical Co. The lipases PS-30, AK and MAP-10 from *Pseudomonas cepacia, Pseudomonas fluorescens* and *Mucor javanicus*, respectively, were gifts from Amano enzymes (USA) Co., Ltd. (specified activities at pH 7.0 were 30000 u/g, 20000 u/g and 10000 u/g, respectively. Immobilized lipases from *Candida antarctica* (Novozym-435) and *Mucor miehei* (Lipozyme IM) were gifts from Novo Nordisk Inc. All enzymes, prior to their use, were dried over P_2O_5 (0.1 mm Hg, 25° C, 16 h).

Column Chromatographic separations were performed over silica gel 60 (Aldrich Chemical Company, USA).

Nuclear Magnetic Resonance. ^{1}H and ^{13}C NMR spectra were recorded using Bruker ARX-250 and DRX-500 spectrometers. Chemical shifts in parts per millions are reported downfield from 0.00 ppm using trimethylsilane (TMS) as the internal reference. Unambiguous assignments were derived from COSY and HETCOR spectra. The following abbreviations are used to present the spectral data: s= singlet, brs =broad singlet, d = doublet, dd= doublet of doublet, t = triplet, dt =doublet of triplet, q = quartet, and p= quintet.

Mass Spectrometry Instrumentation. Mass analyses were performed using a Bruker Biflex™ MALDI-TOF spectrometer.

Other Instrumental Methods. Optical rotations were measured using a Perkin-Elmer 241 digital polarimeter. Infrared Spectra were recorded using a Perkin-Elmer FT-IR Spectrometer model 1760 X. The samples for infrared spectroscopy were dissolved in THF and deposited as thin films on NaCl optical disks.

Synthesis of methyl 17-L-([2′-O-β-D-glucopyranosyl-β-D-glucopyranosyl]-oxy)-cis-9-octadecenoate (SL-Me, 1): In a typical reaction, to a 100 mL round bottomed flask equipped with a reflux condensor were added 10 g of dry crude sophorolipid and 10.0 mL 0.022 N sodium methoxide in methanol solution. The reaction assembly was protected from atmospheric moisture by a $CaCl_2$ guard tube. The reaction mixture was refluxed for 3 h, cooled to room temperature (30 °C), and acidified using glacial acetic acid. The reaction mixture was concentrated by rotoevaporation and poured with stirring into 100 mL of ice cold water that resulted in the precipitation of the sophorolipid methylester as a white solid. The precipitate was filtered, washed with ice-water and lyophilized (8.77 g, yield 95.0 %) : $[\alpha]^{25}_{D}$ -9.77 (c = 0.0145 g/mL, THF); IR [cm^{-1} (%T)] 3354 (1.7%), 2929 (2.0%), 2860 (3.4 %), 1741 (4.4 %), 1165 (4.8 %), 1074 (2.0 %), 1030 (3.2 %); ^{1}H NMR (500 MHz, CD_3OD) δ 1.28 (3H, d, J = 7.0 Hz, H-18), 1.38 (14H, brs, H- 4-7 & -12-14), 1.48 (2H, m, H-15), 1.62 (4H, p, J = 7.0 Hz, H-3 &-16), 2.04 (4H, dt, J = 7.0 Hz, H-8 &-11), 2.34 (2H, t, J = 7.5 Hz, H-2), 3.23 - 3.38 (4H, m, H-2″,-4″,-4′, &-5′), 3.40 (1H, t, J = 8.4 Hz, H-3″), 3.48 (1H, t, J = 8.4 Hz, H-2′), 3.58 (1H, t, J = 8.4 Hz, H-3′), 3.65-3.73 (3H, m, H-6′a,-6″a, & -5″), 3.68 (3H, s, OCH_3), 3.80-3.96 (3H, m, H-6′b, -6″b & -17), 4.47 (1H, d, J = 7.0 Hz, H-1′), 4.66 (1H, J = 6.7 Hz, H1″) and 5.38 (2H, m, H-9 &-10); ^{13}C NMR (125.8 MHz) δ 20.89, 25.02, 25.26, 27.12, 27.17, 29.13, 29.16, 29.22, 29.39, 29.86, 29.90, 29.95, 33.85, 36.80, 50.99, 61.82, 62.13, 70.56, 70.85, 74.89, 76.76, 76.83, 77.24, 77.31, 77.76, 81.12, 101.69, 103.75, 130.45, 130.60, 175.04; MALDI-TOF *m/z* 659.84 (M + Na)$^{+}$.

Methyl 17-L-([2′-O-β-D-glucopyranosyl-β-D-glucopyranosyl]-oxy)-cis-9-octadecenoate 6′,6″-diacrylate (2). To a 50 mL round bottom flask under dry argon were added 1 g of Novozym 435 and a solution of **1** (500 mg) in 20 mL of dry THF. Excess of vinyl acrylate (2 mL) was then added to the reaction mixture and the contents were stirred magnetically at 35 °C for 96 h. The reaction setup was secluded from the light with black paper. The usual work-up procedure led to the isolation of 0.51 g of the crude product. Product purification by column chromatography yielded 0.48 g (yield 87%) of **2**: $[\alpha]^{25}_{D}$ -1.38 (c = 0.0125 g/mL, THF); IR [cm^{-1} (%T)] 3394 (26.9 %), 2928 (16.6 %), 2856 (26.8 %), 1729 (9.8 %), 1636 (54.9 %), 1618 (61.2 %),

1438 (42.1 %), 1408 (28.4 %), 1375 (41.3 %), 1297 (25.5 %), 1276 (31.2 %), 1198 (13.9 %), 1082 (9.8 %); ^{1}H NMR (250 MHz, CD_3OD) δ 1.20 (3H, d, J = 6.5 Hz, H-18), 1.38 (16H, brs, H- 4-7 & -12-15), 1.62 (4H, p, J = 7.0 Hz, H-3 &-16), 2.04 (4H, dt, J = 7.0 Hz, H-8 &-11), 2.34 (2H, t, J = 7.5 Hz, H-2), 3.23 - 3.65 (8H, m, H-2″-5″ & -2′-5′), 3.68 (3H, s, OCH_3), 3.76 (1H, m, H-17), 4.25-4.94 (4H, m, H-6′ & 6″), 4.49 (1H, d, J = 7.0 Hz, H-1′), 4.58 (1H, J = 6.7 Hz, H1″), 5.38 (2H, m, H-9 &-10), 5.92 (2H, 2dd, J = 10.1 & 2.0 Hz, $COCH{=}CH_{2cis}$), 5.92 (2H, 2dd, J = 17.0 & 10.0 Hz, $COCH{=}CH_2$), and 6.44 (1H, 2dd, 17.0 & 2.0 Hz, $COCH{=}CH_{2\,trans}$); ^{13}C NMR (62.9 MHz, CD_3OD) δ 20.81, 25.01, 25.27, 27.11, 27.15, 29.10, 29.13, 29.18, 29.36, 29.73 (double), 29.85, 33.85, 36.81, 50.98, 63.91, 64.02, 70.46, 70.67, 73.96, 74.69, 75.11, 76.64, 76.89, 77.41, 82.97, 101.49, 104.74, 128.44 (double), 129.79, 129.95, 130.63, 130.75, 166.51, 166.57, 175.04; MALDI-TOF m/z 767.55 $(M + Na)^+$.

Methyl 17-L-([2′-O-β-D-glucopyranosyl-β-D-glucopyranosyl]-oxy)-cis-9-octadecenoate 6′,6″-disuccinate (3). Compound **1** (500 mg) was dissolved in 20 mL of dry THF. To this solution, succinic anhydride (2 mL) and Novozym 435 (1 g) were added, and the suspension was stirred magnetically at 35 °C for 96 h. The enzyme was filtered off, the solvent evaporated, and the product purified by column chromatography (eluent $CHCl_3$-MeOH, 8:2) to give 492 mg (yield 75%) of **3**: $[\alpha]^{25}_D$ -3.57 (c = 0.0110 g/mL, THF); IR [cm^{-1} (%T)] 3373 (56.9 %), 2927 (50.9 %), 2855 (58.4 %), 1735 (40.7 %), 1374 (63.9 %), 1168 (51.1 %), 1078 (49.7 %); ^{1}H NMR (250 MHz, CD_3OD) δ 1.12 (3H, d, J = 7.0 Hz, H-18), 1.25 (16H, brs, H- 4-7 & -12-15), 1.52 (4H, p, J = 7.0 Hz, H-3 &-16), 1.98 (4H, dt, 7.0 Hz, H-8 &-11), 2.20 (2H, t, J = 7.5 Hz, H-2), 2.50 (8H, m, $COCH_2CH_2COOH$), 3.10 - 3.60 (8H, m, H-2″-5″ & -2′-5′), 3.56 (3H, s, OCH_3), 3.68 (1H, m, H-17), 4.15-4.30 (4H, m, H-6′ & 6″), 4.35 (1H, d, J = 7.0 Hz, H-1′), 4.50 (1H, d, J = 6.7 Hz, H1″), 5.28 (2H, m, H-9 &-10); ^{13}C NMR (62.9 MHz, CD_3OD) δ 20.86, 25.01, 25.27, 27.11, 27.17, 28.85, 28.90, 28.98, 29.12, 29.18 (double), 29.37 (double), 29.76, 29.86, 33.85, 36.84, 51.00, 63.97, 63.99, 70.48, 70.59, 73.98, 74.68, 74.90, 76.59, 76.85, 77.47, 82.36, 101.48, 104.35, 129.81, 129.97, 171.75, 171.95, 174.33; MALDI-TOF m/z 861.29 $(M + Na)^+$.

Novozym-435 catalyzed synthesis of 17-L-([2′-O-β-D-glucopyranosyl-β-D-glucopyranosyl]-oxy)-cis-9-octadecenoic acid 1′, 6″-lactone (Sophorolactone, 4). For the following inert atmosphere was maintained using a glove bag and dry argon. To an oven dried 100 mL round bottomed flask were transferred 1.5g of methyl 17-L-([2′-*O*-β-D-glucopyranosyl-β-D-glucopyranosyl]-oxy)-*cis*-9-octadecenoate (**1**), 2.2 g of zeolite, 2 g of Novozym 435 (dried in a vacuum dessicator: 0.1 mm Hg, 25 °C, 16 h) and dry THF (50 mL) and the round-bottomed flask was immediately stoppered. The flask was then placed in a constant temperature oil bath maintained at 35 °C for 96 h and the contents were stirred magnetically. A control reaction was set up as described above except Novozym 435 was not added. TLC ($CHCl_3$: MeOH, 7:3) was used to follow the progress of the reaction. The reaction was quenched by removing the enzyme and zeolite by vacuum filtration (glass fritted filter, medium-pore porosity), the enzyme was washed 3-4 times with 5 mL portions of THF, the filtrates were combined and solvent removed *in vacuo* to give 1.45 g of the product. The crude product (1.45 g) was purified by column chromatography over silica gel (100g, 130-270 mesh, 60 Å, Aldrich) using a gradient solvent system of chloroform-methanol (2 mL/minute) with increasing order of polarity to give 1.2 g (yield 84%) of purified product: $[\alpha]^{25}_D$ -4.25 (c = 0.0141 g/mL, THF); IR [cm^{-1} (%T)] 3360 (50.1

%), 2927 (44.0 %), 2855 (52.0 %), 1735 (53.1 %), 1458 (65.5 %), 1375 (64.4 %), 1174 (58.9 %), 1076 (39.3 %); ^{1}H NMR (500 MHz, CD_3OD) δ 1.24 (3H, d, J = 7.0 Hz, H-18), 1.38 (14H, brs, H- 4-7 & -12-14), 1.42 (2H, m, H-15), 1.62 (4H, p, J = 7.0 Hz, H-3 &-16), 2.08 (4H, dt, J = 7.0 Hz, H-8 &-11), 2.36 (2H, t, J = 7.5 Hz, H-2), 3.23 - 3.48 (4H, m, H-2″,-4″,-2′, &-5′), 3.48 (2H, m, H-3′ & -5″), 3.58 (2H, m, H-3″ & -4′), 3.7 (1H, m, H-6′a), 3.85 (2H, m, H-6′b & -17), 4.20 (1H, m, H-6″a), 4.42 (1H, m, H-6″b), 4.49 (1H, d, J = 7.0 Hz, H-1′), 4.52 (1H, d, J = 6.7 Hz, H1″) and 5.38 (2H, m, H-9 &-10); ^{13}C NMR (125.8 MHz) δ 20.71, 24.70, 25.47, 26.15, 27.21, 27.88, 28.21, 28.66, 28.97, 29.57, 29.72, 29.99, 33.99, 36.49, 61.76, 63.38, 70.12, 70.40, 74.69, 75.24, 76.58, 76.67(double), 77.05, 83.87, 101.11, 105.41, 129.78, 130.29, 174.30; MALDI-TOF *m/z* 627.95 $(M + Na)^+$.

17-L-([2′-*O*-β-D-glucopyranosyl-β-D-glucopyranosyl]-oxy)-*cis*-9-octadecenoic acid 1′, 6″-lactone 6′-acrylate (5). To an oven-dried 100 mL round-bottomed flask were added Compound **9** (1.0 g), 2.0 g of zeolite, and 3 g of Novozym 435 that was dried (0.1 mm Hg, 25 °C, 16 h). Then, dry THF (30 mL) was added and the round-bottomed flask was immediately stoppered. Vinyl acrylate (2.5 mL) was added and the round-bottomed flask was wrapped with black paper to seclude it from the light. This flask was then placed in a constant temperature oil bath (35° C) with magnetic stirring for 96 h. A control reaction was set-up as was described above except that Novozym 435 was not added. Progress of the reaction was followed by TLC ($CHCl_3$:MeOH, 9:1). The reaction was quenched by removing the enzyme and zeolite by vacuum filtration (glass fritted filter, medium-pore porosity), the enzyme was washed 3-4 times with 5 mL portions of THF, and the filtrates were combined and solvent removed *in vacuo* to give 1.5 g of the product. The crude product (1.05 g) was purified by column chromatography over silica gel (50g, 130-270 mesh, 60 Å, Aldrich) using a gradient solvent system of chloroform-methanol (2 mL/minute) with increasing order of polarity to give 0.9 g (yield 83%) of purified product: $[\alpha]^{25}_D$ -2.81 (c = 0.0116 g/mL, THF); IR [cm^{-1} (%T)] 3378 (49.6 %), 2928 (42.8 %), 2854 (50.1 %), 1735 (39.8 %), 1710 (52.0 %), 1457 (63.4 %), 1410 (61.9 %), 1354 (60.3 %), 1295 (55.3 %), 1185 (49.6 %), 1077 (29.7 %); ^{1}H NMR (250 MHz, $CDCl_3$) δ 1.22 (3H, d, J = 7.0 Hz, H-18), 1.38 (14H, brs, H- 4-7 & -12-14), 1.42 (2H, m, H-15), 1.66 (4H, p, J = 7.0 Hz, H-3 &-16), 2.08 (4H, dt, J = 7.0 Hz, H-8 &-11), 2.36 (2H, t, 7.5 Hz, H-2), 3.23 - 3.70 (8H, m, H-2″- 5″& -2′-5′), 3.78 (1H, q, H-17), 4.10-4.64 (6H, m, H-6′, -6″, -1′ & -1″), 5.38 (2H, m, H-9 &-10), 5.92 (1H, dd, J = 10.1 & 2.0 Hz, $COCH{=}CH_{2cis}$), 5.92 (1H, dd, J = 17.0 & 10.0 Hz, $COCH{=}CH_2$), and 6.44 (1H, dd, J = 17.0 & 2.0 Hz, $COCH{=}CH_{2trans}$); ^{13}C NMR (125.8 MHz) δ 20.66, 24.67, 25.48, 26.12, 27.15, 27.88, 28.18, 28.67, 28.92, 29.47, 29.61, 29.92, 33.97, 36.64, 63.41, 63.88, 70.18, 70.67, 73.99, 74.69, 75.34, 76.69, 76.92, 77.00, 83.57, 101.24, 105.28, 128.42, 129.77, 130.25, 130.58, 166.58, 174.80; MALDI-TOF *m/z* 681.90 $(M + Na)^+$.

Synthesis of Poly (17-L-([2′-*O*-β-D-glucopyranosyl-β-D-glucopyranosyl]-oxy)-*cis*-9-octadecenoic acid 1′, 6″-lactone 6′-acrylate). In a small flask, a solution of 250 mg of **5** in 3 mL of DMF was prepared (freeze/pump/thaw cycles), and 0.1 % (w/v) AIBN was added. The flask was then placed in an oil bath maintained at 65 °C. The polymerization was continued for 16 h. The reaction was terminated by precipitating the polymer in ethyl acetate, and the white precipitate was washed with acetone to yield 200 mg (80% yield) of the polymer.

Copolymerization of Acrylic acid and 5. The polymerization reaction was conducted under similar conditions as was described above for homo-polymerization of **5**. Compositional copolymers were generated by varying the monomer feed ratio, Acrylic acid : **5**, 1:1; 1:9; 1:19; 1:32.3; and 1:0.

Copolymerization of Acrylamide and 5. The polymerization reaction was conducted under similar conditions as was described above. Compositional copolymers were generated by varying the monomer feed ratio, Acrylamide : **3**; 1:1; 1:19; 1:32.3; 1:99 and 1:0. The compositions of the copolymers were calculated, from the percent nitrogen determined by their CHN analysis, to be 46.5, 4.2, 2.1, 3.2, 0.7, 0.9 mole percent of **5**.

Results and Discussion

Sophorolipids produced by the yeast *T. bombicola* grown on oleic acid and glucose give a mixture of at least eight different compounds. The products exist mainly in the lactonic and acidic forms with variable degree of acetylation at position(s) 6′ and 6″ (Figure 1). Scheme 1 shows the strategy that was conceived for the incorporation of acryl group in sophorolipids. Esterification was required to obtain pure compounds from a complex mixture of sophorolipids. This route was considered as an alternative to the isolation and purification of individual pure components from the sophorolipid mixture.

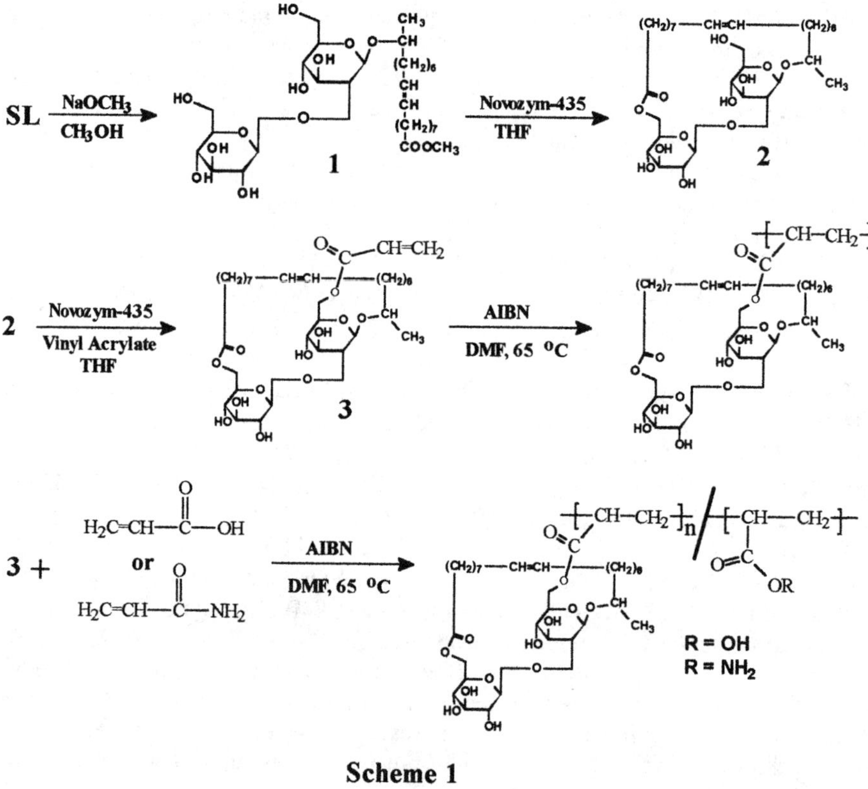

Scheme 1

Synthesis of ester sophorolipid derivatives

The sophorolipid mixture was refluxed with an alcoholic solution of sodium methoxide to yield the methyl ester (SL-Me, **1**). The optically active compound, $[\alpha]^{25}_{D}$ -9.77, was isolated in 95 % yield. The product was soluble in anhydrous THF and was identified based on the results of its MALDI-TOF mass spectrum (*m/z* 659.85 $(M+Na)^+$) as well as detailed spectral data. The ^1H NMR spectrum of **1** was complex and chemical shifts were spread over the 1.2 to 5.5 ppm range. The acetyl group resonances observed around 2.2 ppm for the natural sophorolipid mixture were no longer seen and the methyl ester group was a singlet at 3.68 ppm. The assignments shown were derived from a ^1H-^1H COSY 45 NMR spectrum. The ^{13}C NMR spectrum of **1** showed a signal corresponding to the methyl group at 50.99 ppm. The carbonyl group (C-1, Scheme 1) resonance was observed at 175.04 ppm. Assignments of the resonances in the ^{13}C NMR spectrum were made based on a ^1H-^{13}C HETCOR experiment.

Enzyme Screening

The SL-Me was then subjected to lipase–catalyzed esterification with vinyl acrylate in dry THF (Scheme 1). A study to determine the efficiency of different lipases (PPL, CCL, PS-30, AK, MAP-10, Novozym-435 and Lipozyme IM) for the acylation of **1** in dry organic solvent at room temperature was performed. In 48 h, though, all lipase showed conversion to certain extent there were considerable differences with the source (PPL ~ 40%, CCL < 30%, PS-30 ~ 50%, AK ~ 50%, MAP-10 < 10% and Novozym-435 > 70%). Hence, Novozym-435 was found to be the lipase of choice for the acylation with vinyl acrylate.

Lipase-Catalyzed Regioselective Acylations

Initial attempts at the synthesis of the mono acryl sophorolipid derivative were conducted using excess vinyl acrylate and Novozym 435 as the catalyst (Scheme 1). The resulting product was purified by column chromatography over silica gel and was identified from its proton and carbon NMR spectra. The compound had $(M + Na)^+$ ion peak at *m/z* 767.55 from MALDI-TOF which is 43 mass units higher than that calculated for the mono acryl derivative. The proton NMR of the compound showed resonances at 5.92 - 6.44 ppm due to two acryl groups in the molecule. The proton NMR spectrum of the compound showed a downfield shift in the resonance positions of both the C-6′ and C-6″ protons. However, conclusive determination of the position of the acryl groups in the molecule was not possible from the proton NMR spectrum due to its complexity.

The ^{13}C NMR spectrum of this product was edited using a DEPT 135 pulse sequence to separate out the resonances due to the methine and methyl carbons from those due to the methylene carbons. Expanded regions of the DEPT 135 ^{13}C NMR spectrum of compounds **1** and **2** are shown in Figures 2 and 3. The DEPT 135 spectrum of the product clearly showed peaks due to two acryl groups ($COCH{=}CH_2$) at 129.79, 129.95, 130.63 and 130.75 (Figure 2b). The peaks due the methine carbons appear inverted. Additionally, the carbons C-9 and C-10 in the lipid chain of the molecule appeared at 129.79 and 129.95 ppm. Importantly, the resonances at 63.91

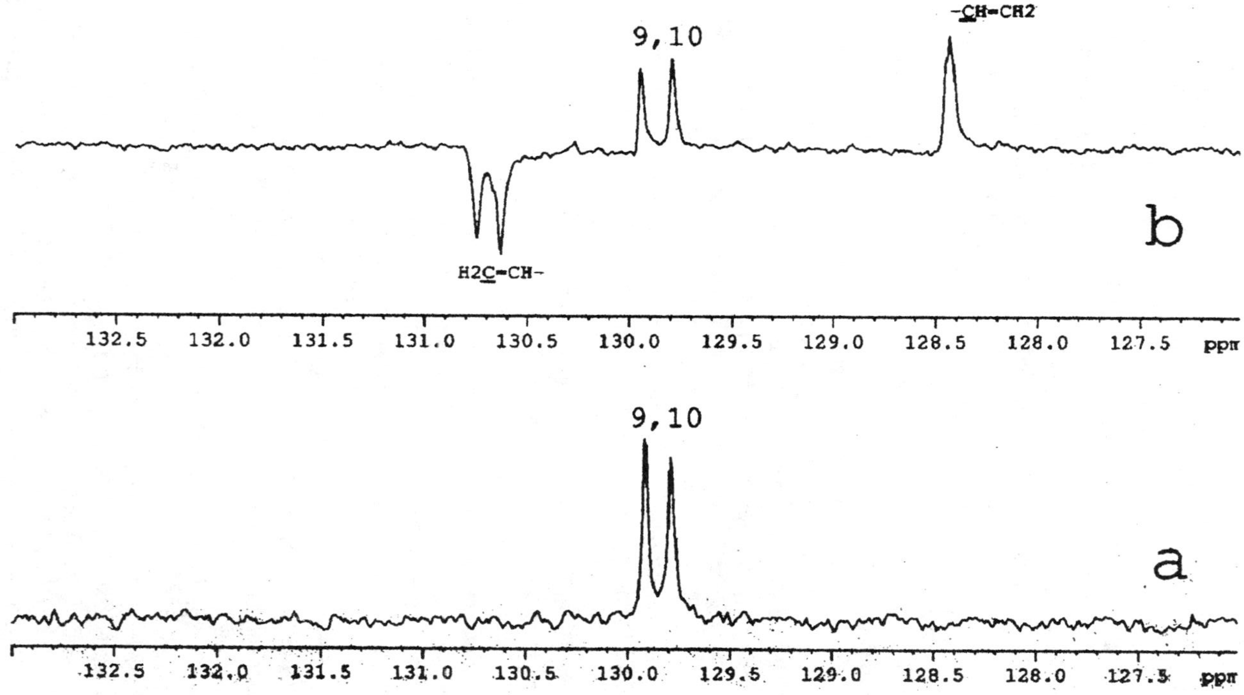

Figure 2: Regions 126-134 ppm of the DEPT-135 spectra of a) the SL-Me, **1** and b) the 6′, 6″-diacrylate derivative, **2**.

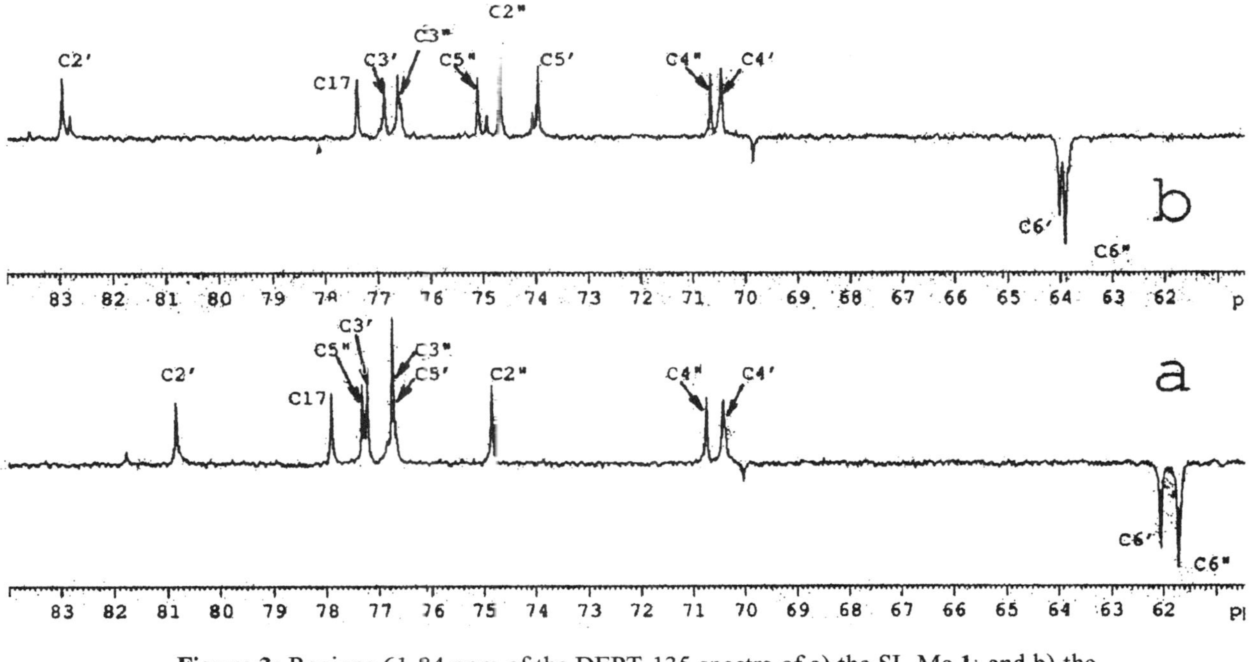

Figure 3: Regions 61-84 ppm of the DEPT-135 spectra of a) the SL-Me,**1**; and b) the 6′, 6″-diacrylate derivative, **2**.

and 64.02 ppm for C-6′ and C-6″, respectively, were 2.0 ppm downfield when compared to the corresponding methyl ester (Figure 3b). Hence, these results suggest the formation of a diacryl product that is acylated at positions C-6′ and C-6″. Interestingly, it was observed that the resonance position of C-2′ (carbon bearing no free hydroxyl group) was shifted downfield by ~ 2 ppm upon formation of the acyl ester at C-6′ and C-6″. This down field shift in the resonance position of C-2′ might very well be due to the conformational changes in the lipid structure upon acylation. Furthermore, an upfield shift of 2.5 ppm for the resonances corresponding to C-5′ and C-5″ and no significant changes in the resonance positions of the other carbons in the molecule was found (Figure 3b). The upfield shift in the resonance positions of C-5′ and C-5″ is consistent with the γ-effect that is caused by the attachment of the acryl groups at the C-6′ and C-6″ hydroxyls. Therefore, these NMR results showed that the Novozym-435 catalyzed acylation proceeded in a highly regioselective fashion where only the C-6′ and C-6″ hydroxyls in the molecule were acrylated. This product (Compound **2**) was therefore conclusively identified as methyl 17-L-([2′-*O*-*β*-D-glucopyranosyl-*β*-D-glucopyranosyl]-oxy)-*cis*-9-octadecen-oate 6′,6″-diacrylate. Compound **2** is currently under investigation for polymerization with diamines by Michael addition at the acryl groups.

The preparation of regioselectively modified sophorolipids that contain carboxylic acid functionalities is of interest. The generated dicarboxylic acid could be condensed with an activated diol to give glycolipid-containing polyester. Hence, the acylation of **1** as above was conducted using succinic anhydride as the acylating agent to prepare methyl 17-L-([2′-*O*-β-D-glucopyranosyl-β-D-glucopyranosyl]-oxy)-*cis*-9-octadecenoate 6′,6″-disuccinate (**3**, *m/z* 861.29 $(M + Na)^+$, $[\alpha]^{25}_D$ -3.57). From its detailed spectral analysis, the positions of the succinate groups were established at C-6′ and C-6″.

Synthesis of Sophorolactone

One of the objectives of this work was to accomplish site-selective synthesis of a monoacryl derivative of **1**. Therefore, reactions of **1** were conducted where the ratio of **1** to vinyl acrylate was varied. Such a product could be used as a glycolipid monomer that is polymerizable to linear chains through well-established free radical methods. When the ratio of **1** to vinyl acrylate was 1:1 or less, Compound **4** ($[\alpha]^{25}_D$ -4.25, *m/z* 627.95 $(M + Na)^+$) was formed (see Scheme 1). Compound **4** was separated from the unreacted SL-Me by column chromatography. Comparison of the ^{1}H NMR spectra for Compounds **4** and **1** showed substantial differences in the appearance of the proton signals in the 3.25 to 4.5 ppm region that are due to carbohydrate protons. Assignments to resonances were made from a ^{1}H-^{1}H COSY NMR spectrum (Figure 4). Interestingly, the ^{1}H NMR spectrum of **4** did not indicate that acrylation occurred and had no resonances corresponding to the methyl ester. These anomalous features of the ^{1}H NMR spectrum of **4** suggested the formation of a lactone where the ester linkage was between the carboxylic acid end group of the fatty acid chain and one of the hydroxyl groups of the sophorose ring.

The ^{1}H NMR spectrum of **4** also showed a 0.5 ppm downfield shift in the resonance position of the C-6″ protons suggesting participation of the C-6″ hydroxyl group in the formation of the lactone ring. However, conclusive determination of the hydroxyl group(s) on the sophorose moiety that took part in lactone formation required additional studies by ^{13}C NMR. Specifically, the ^{13}C NMR of **4** was recorded after editing by the DEPT-135 pulse sequence. This permitted resolution of the resonances due to the methyl, methylene, and methine carbons. The DEPT-135 spectra of **1** and **4** were compared (Figures 3a and 5a). A downfield shift in the resonance position of the C-6″ by 1.5 ppm was accompanied by an upfield shift of 1.3 ppm in the resonance position of the C-5″ (Figure 5a). This observation confirmed that a bond between the hydroxyl of C-6″ and the lipid carbonyl carbon (C-1, Scheme 1) formed the lactone ester.

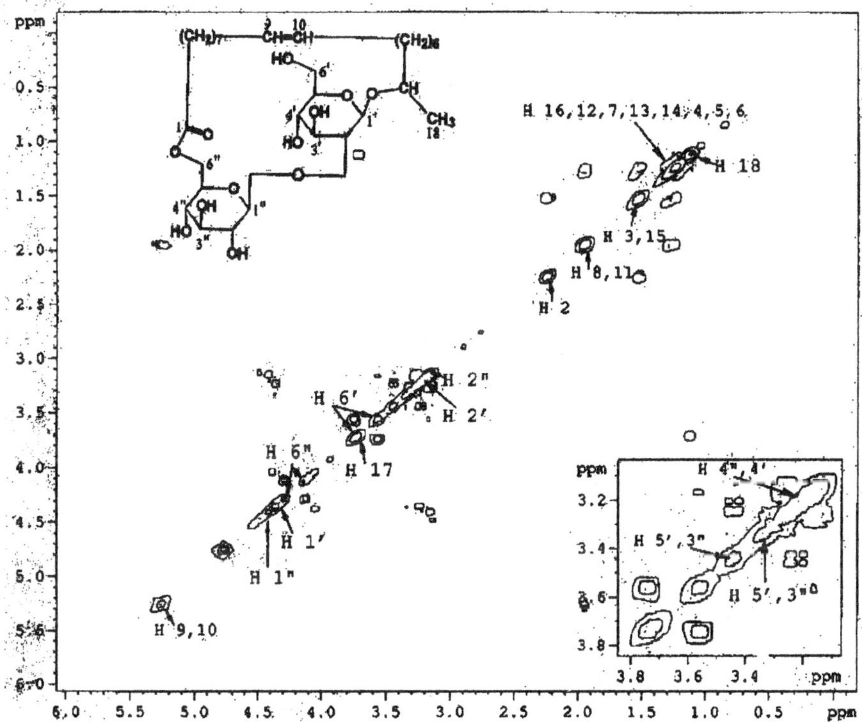

Figure 4: ^{1}H-^{1}H COSY NMR spectrum of Sophorolactone, **4** (250 MHz, CD_3OD).

The structure of the lactone **4** is very interesting, as it is an unnatural analog of the microbially produced macrolactone. The lactone **4** differs in the site at which the sophorose ring is attached to the fatty acid. Specifically, in **4**, unlike the natural sophorolipids, the fatty acid carboxyl carbon (C-1) is linked to the C-6″ hydroxyl, not to the C-4″ hydroxyl.

Synthesis of monoacryl sophorolipid monomer

The successful synthesis of **4** provided a sophorolipid-analog that had only one primary hydroxyl group. Hence, this compound was an excellent candidate for the regioselective conversion of **4** to the corresponding monoacryl derivative linked only

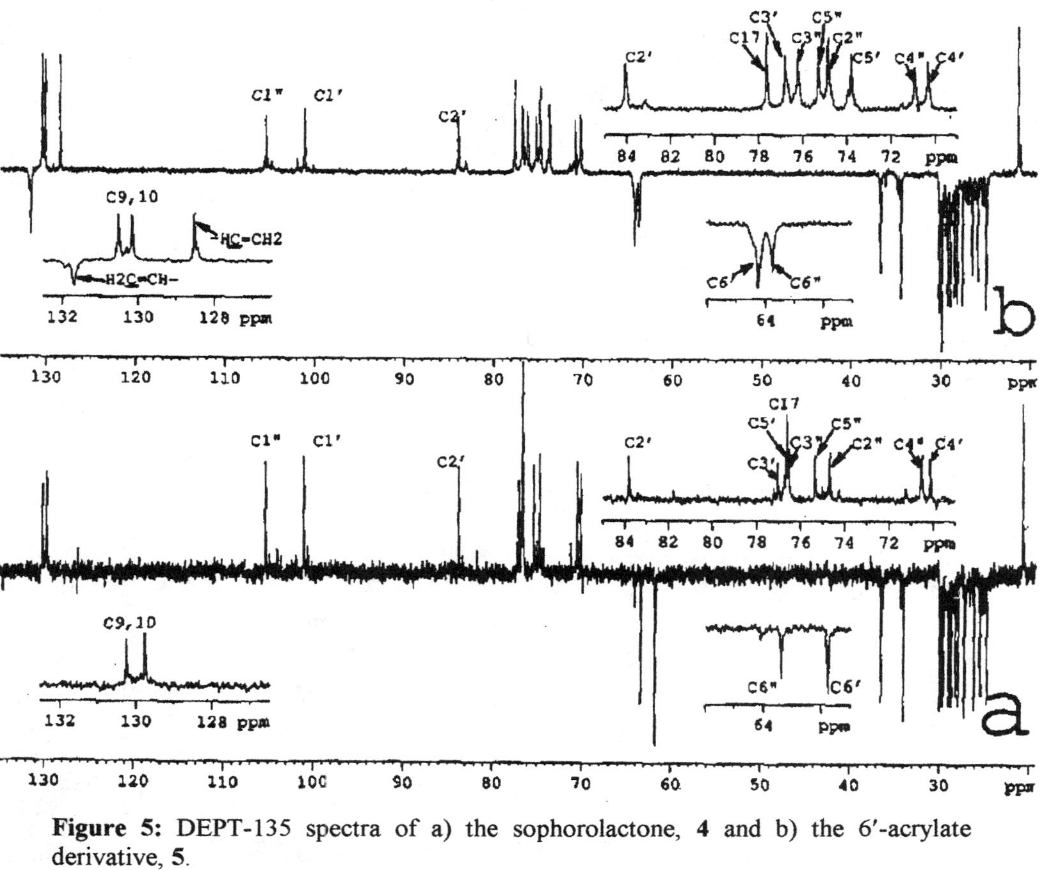

Figure 5: DEPT-135 spectra of a) the sophorolactone, **4** and b) the 6'-acrylate derivative, **5**.

to the one remaining primary site. Indeed, the reaction of **4** with vinyl acrylate catalyzed by Novozym-435 in dry THF gave compound **5** ($[\alpha]^{25}_D$ -2.81, *m/z* 681.90 $(M + Na)^+$). The ^1H NMR spectrum of **5** confirmed that the monoacrylation did indeed take place, 5.92 (1H, dd), 6.25 (1H, dd), and 6.44 (1H, dd). In addition, when compared to the ^1H NMR spectrum of **4**, the resonance position of the methylene on carbon 6′ for **5** showed a 0.7 ppm downfield shift. Further proof of the acrylation position for **5** was obtained by comparison of the ^{13}C DEPT-135 NMR spectra for Compounds **4** and **5** (Figure 5). The spectrum of **5** showed a downfield shift of 2.0 ppm in the resonance position of the C-6′ carbon and an upfield shift of 2.5 ppm in the resonance position of the C-5′ (Figure 5b).

Polymerization of the acryl monomer

The homopolymerization of **5** was conducted by using AIBN as the initiator in dry DMF. In representative example, to a small flame dried flask, a solution of 250 mg of **5** in 3 mL of DMF was added. The solution was degassed (freeze/pump/thaw cycles), and 0.1 % (w/v) AIBN was added. The flask was then placed in an oil bath maintained at 65 °C. The polymerization was continued for 16 hours. Precipitating the polymer in ethyl acetate terminated the reaction, and the white precipitate was washed with acetone to yield 200 mg (80% yield) of the polymer. The ^1H NMR spectrum of the homopolymer did not show resonances due to the acryl protons (see Figure 6). In addition, inspection of the corresponding resonances showed that the integrity of the sophorolipid moiety was maintained under the reaction conditions used. The ^{13}C-NMR spectrum of the polymer further supported that the sugar moiety was not disturbed and that the [-(-CH-CH_2-)-] carbon backbone was formed. The polymer was soluble in DMF and DMSO but was insoluble in water. The lack of solubility of this polymer in water is explained by the presence of the large 1′, 6″-lactone ring that introduces substantial hydrophobic character. It is noteworthy to mention that the unusual amphiphilic character of the polymer side groups may provide unique inter- and intramolecular interactions.

Copolymers of **5** with acrylic acid and acrylamide were prepared to generate products having diverse solution properties and increased hydrophilic character. In other words, for some copolymer compositions, it was anticipated that the products would be water-soluble. The copolymerizations were conducted by procedures similar to the one described above for homopolymerization using AIBN as the initiator in dry DMF. In addition, adjusting the monomer feed ratio controlled the copolymer composition. The compositions of acrylamide/**5** copolymers were determined from the percent nitrogen of the copolymers. Table 1 shows the copolymer composition in the acrylamide/**5** copolymers along with the percent nitrogen. Interestingly, only copolymers with < 3 mole percent of **5** showed solubility in water.

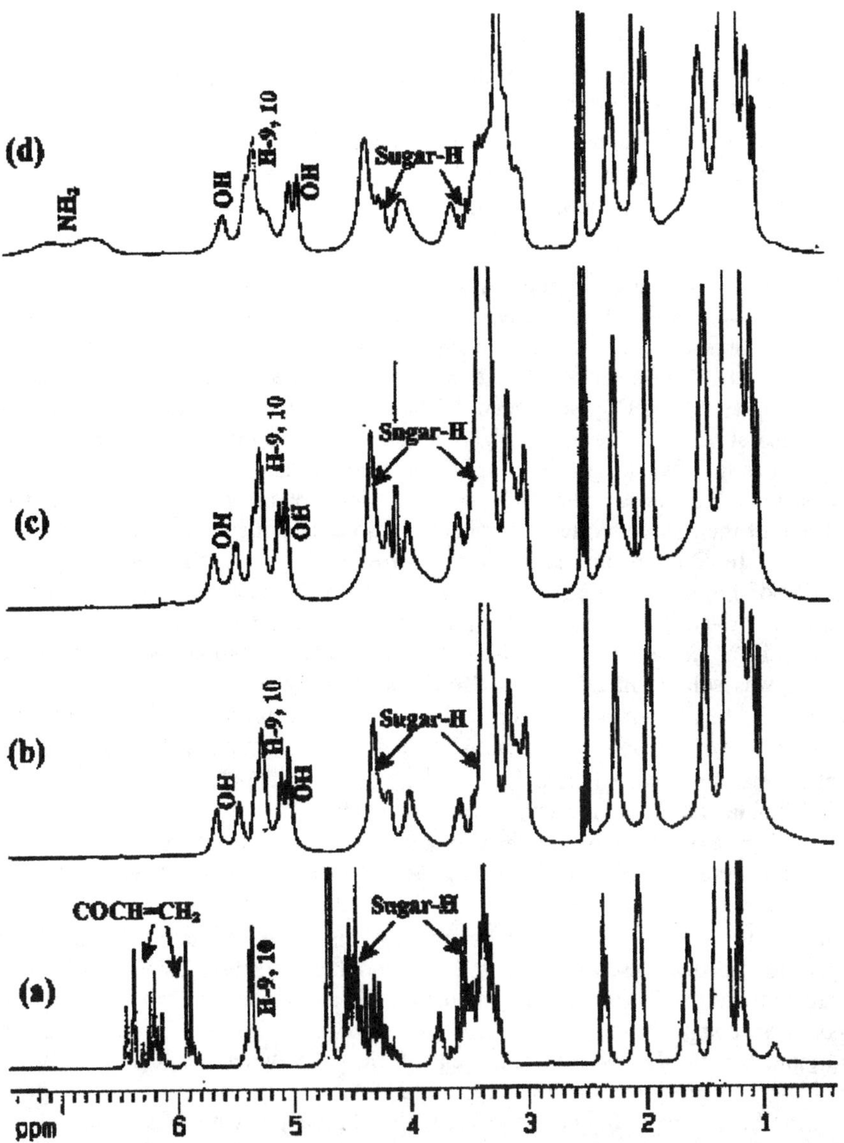

Figure 6. ^{1}H NMR spectra of **5** ($CDCl_3$, **a**); poly **5** (DMSO d_6, **b**); 1:1 copolymer **5**/acrylic acid (DMSO d_6, **c**) and 1:1 copolymer **5**/acrylamide (DMSO d_6, **d**).

Table 1. Composition for acrylamide/ 5 copolymers.

no.	mole percent of 3 in the feed (%)	mole percent of 3 in the copolymer (%)#	Nitrogen in the copolymer (%)#	yield (%)
1	50	46.5	2.17	78
2	5	4.2	14.01	92
3	3	2.1	16.40	87
4	2	3.2	15.12	97
5	1	0.7	18.55	90
6	0.5	0.9	18.09	94

fromCHN analysis

Following the above-described procedure, copolymers of **5** and acrylic acid were prepared. Copolymers were generated with 3, 5, 10, and 50 mole % of **5** in the feed. The solution properties of these novel polymers are currently under evaluation in our laboratories. Analysis by ^{13}C-NMR of the diad and triad sequences for acrylic/**5** and acrylamide/**5** copolymers was attempted. NMR spectra were recorded in DMSO-d_6, at 300K on a bruker DRX 500 MHz spectrometer. Observation of the carbonyl regions for the acrylamide /3 copolymer (46.5 mole % of 3, entry 1 in table 1) showed signals at 176.02 and 173.93 ppm. These peak positions are identical to that of the respective homopolymers. In other words, the signals at 176.02 and 173.93 ppm correspond to poly(acrylamide) and poly(3), respectively. Similarly, the carbonyl region for the 50/50 (feed ratio) copolymer of acrylic acid/3 showed signals at 175.68 and 173.81 ppm. Again, these peaks occur at identical positions as the corresponding homopolymers. These results suggest that the copolymers formed are blocky in nature. However, we cannot exclude the possibility that diads of acrylamide-3 and acrylic acid-3 are present but are not resolved.

Conclusions

In the present study we have established an enzymatic strategy that could provide a wide range of new sophorolipid analogs. Specifically, regioselective synthesis of a 6'-monoacryl sophorolipid derivative was achieved via an interesting unnatural sophorolactone. This sophorolactone was found to have the lactone ester link between C-1 of the 17-hydroxy-cis-9 octadecenoic acid chain and the C-6" hydroxyl group of the sophorose moiety. Subsequent polymerization and copolymerization of the monoacryl derivative with acrylic acid and acryl amide resulted in formation of sophorolipids containing linear polyacrylates. An interesting outcome of this work was that attempts to prepare a monoacryl derivative lead to certain analogs, such as the disuccinyl and acryl sophorolipid derivatives, that would be useful as monomers for the preparation of unusual amphiphilic polymers by well established condensation polymerization chemistries. Important concepts that motivated this work were the following: *i*) it is relatively simple to take the mixture of sophorolipids and convert them to a pure intermediate, *ii*) the choice of a sophorolipid intermediate that is readily soluble in moderate polarity solvents facilitates further lipase-catalyzed reactions that do not proceed well in polar aprotic media, *iii*) sophorolipids are produced in large amounts by fermentation methods. It is envisioned that the sophorolipid containing polymers might behave as important immuno-modulators.

References and Notes

1. Gorin, P.A.J.; Spencer, J.F.T.; Tulloch, A.P. *Can. J. Chem.* **1961**, *39*, 846.
2. Davila, A.M.; Marchal, R.; Monin, N.; Vandecasteele, J.P. *J. Chromatography* **1993**, *648*, 139.
3. a) Tulloch, A.P.; Spencer, J.F.T.; Gorin, P.A.J. *Can. J. Chem.* **1962**, *40*, 1326. b) Tulloch, A.P.; Hill, A.; Spencer, J.F.T. *Can. J. Chem.* **1968**, *46*, 3337.
4. Ito, S.; Inoue, S. *Appl. Environ. Microbiol.* **1982**, *43*, 1278. Ito, S.; Kinta, M.; Inoue, S. *Agric. Biol.Chem.* **1980**, *44*, 2221.
5. Isoda, H.; Kitamoto, D.; Shinmoto, H.; Matsumura, M.; Nakahara, T. *Biosci. Biotechnol. Biochem.* **1997**, *61*, 609.; Satoh, M.; Inagawa, H.; Shimada, Y.; Soma, G.; Oshima, H.; Mizuno, D. *J. Biol. Response Mod.* **1987**, *6*, 512.; Yamazaki, M.; Okutomi, T. *Cancer Res.* **1989**, *49*, 352.; Yang, D.; Satoh, M.; Ueda, H.; Tsukagoshi, S.; Yamazaki, M. *Cancer Immunol. Immunother.* **1994**, *38*, 287.; Kumazawa, E.; Akimoto, T.; Kita, Y.; Jimbo, T.; Joto, N.; Tohgo, A. *J. Immunother. Emphasis Tumor. Immunol.* **1995**, *17*, 141.; Akimoto, T.; Kumazawa, E.; Jimbo, T.; Joto, N.; Tohgo, A. *Int. J. Immunopharmacol.* **1994**, *16*, 887.; Nguyen, N.M.; Lehr, J.E.; Shelley, C.I.; Andersen, J.C.; Pienta, K.J. *Anticancer Res.* **1993**, *13*, 2053.; Nigam *et al.*, 1997.; Kamireddy, B.; Darsley, M.; Simpson, D.; Massey, R. US Patent 5,5975.73, **1997**.
6. Piljac, G.; Piljac, V. US Patent 5,514,661, **1996**.; Piljac, G.; Piljac, V. US Patent 5,455,232, **1995**.
7. Carlson, R.W. US Patent 5,648,343, **1997**.; Kawata, T.; Bristol, J.R.; Rose, J.R.; Rossignol, D.P.; Christ, W.J.; Asano, O.; Dubuc, G.R.; Gavin, W.E.; Hawkins, L.D.; Kishi, Y. *Prog. Clin. Biol. Res.* **1995**, *392*, 499.; Kiani, A.; Tschiersch, A.; Gaboriau, E.; Otto, F.; Seiz, A.; Knopf, H.P.; Stutz, P.; Farber, L.; Haus, U.; Galanos, C.; Mertelsmann, R.; Engelhardt, R. *Blood* **1997**, *90*, 1673.; Nakamura, H.; Ishizaka, A.; Urano, T.; Sayama, K.; Sakamaki, F.; Terashima, T.; Waki, Y.; Soejima, K.; Tasaka, S.; Hasegawa, N. *Clin. Diagn. Lab. Immunol.* **1995**, *2*, 672.; Matsuura, M.; Kiso, M.; Hasegawa, A.; Nakano, M. *Eur. J. Biochem.* **1994**, *221*, 335-341.; Kamireddy, B.; Darsley, M.; Simpson, D.; Massey, R. US Patent

5,5975.73, **1997**., Schutze, E.; Hildebrandt, J.; Liehl, E.; Lam, C. *Circ. Shock* **1994**, *42*, 121.

8. Sato, K.; Yoo, Y.C.; Mochizuki, M.; Saiki, I.; Takahashi, T.A.; Azuma, I. *Jpn. J. Cancer Res.* **1995**, *86*, 374-382.; Madarnas, P.; Benrezzak, O.; Nigam, V.N.. *Anticancer. Res.* **1989**, *9*, 897.; Benrezzak, O.; Bissonnette, E.; Madarnas, P.; Nigam, V.N. *Anticancer Res.* **1989**, *9*, 1815.; Gastl, G.; Hermann, T.; Steurer, M.; Zmija, J.; Gunsilius, E.; Unger, C.; Kraft, A.. *Oncology* **1997**, *54*, 177.; Saiki, I.; Sato, K.; Yoo, Y.C.; Murata, J.; Yoneda, J.; Kiso, M.; Hasegawa, A.; Azuma, I. *Int. J. Cancer.* **1992**, *51*, 641.; Mattsby-Baltzer, I.; Jakobsson, A.; Sorbo, J.; Norrby, K.. *Int. J. Exp. Pathol.* **1994**, *75*, 191.
9. Norimatsu, M.; Ono, T.; Aoki, A.; Ohishi, K.; Tamura, Y.J. *Med. Microbiol.* **1995**, *43*, 251.; Bingisser, R.; Stey, C.; Weller, M.; Groscurth, P.; Russi, E.; Frei, K. *Am. J. Respir. Cell Mol. Biol.* **1996**, *15*, 64-70.; Haendeler, J.; Messmer, U.K.; Brune, B.; Neugebauer, E.; Dimmeler, S. *Shock* **1996**, *6*, 405-409.)
10. Marchal, R.; Lemal, J.; Sulzer; C. US Patent 5,616,479, **1997**.
11. Zhou, Q.-H.; Kosaric, N. *J. Am. Oil Chem. Soc.* **1995**, *72*: 67.; Zhou, Q.-H.; Kelekner, V.; Kosaric, N.. *J. Am. Oil Chem. Soc.* **1992**, *69*, 89.; Asmer, H.J.; Slegmund, L.; Fritz, W.; Wray, V. *J. Am. Oil Chem. Soc.* **1988**, *65*, 1460.; Davila, A.M.; Marchal, R.; Vandecasteele, J.P. *Appl. Microbiol. Biotech.* **1992**, *38*, 6.; Tulloch A.P.; Spencer J.F.T.; Gorin, P.A.J. *Can. J. Chem.* **1962**, *40*, 1326.
12. Rau, U.; Manzke, C.; Wagner, F. *Biotechnol. Lett.* **1996**, *18*, 149.
13. Bisht, K. S.; Gross, R. A.; Kaplan, D. L. *J. Org. Chem.* **1999**, 64,780. Bisht, K. S.; Waterson, A. C.; Gross, R. A. *Proc. Am. Chem. Soc.: PMSE*, **1998**, *78*, 246. Bisht, K. S.; Gross, R. A. *Macromolecules* **1999**, communicated.
14. a) Malm, C.J.; Meneh, J.W.; Kendall, D.L.; Hiatt, G.D. *Ind. Eng. Chem.* **1951**, *43*, 684. b) Matsumura, S. *J. Environ. Polym. Deg.* **1994**, *2*, 89. c) Wang, P.; Tao, B.Y. *J. Appl. Polym. Sci.* **1994**, *52*, 755. d) Nishio, K.; nakaya, T.; Imoto, M. *Makromol.Chem.* **1974**,*175*, 3319. e) Carpino, L. A.; Ringsdrof, H.; Ritter, H. *Mackromol. Chem.* **1976**, *177*, 1631.
15. Asmer, H.J.; Lang, S.; Wagner, F.; Vray, V. *J. Am. Oil Chem. Soc.* **1988**, *65*, 1460.
16. Martin, B. D.; Ampofo, S. A.; Linhardt, R. J.; Dordick, J. S. *Macromolecules* **1992**, *25*, 7081. Ikeda, I.; Klibanov, A. M. *Biotechnol. Bioengg.* **1993**, *42*, 788.
17. Therisod, M.; Klibanov, A. M. *J. Am. Chem. Soc.* **1986**, *108*, 5638. Sweers, H. M.; Wong, C. H. *Ibid.* **1986**, *108*, 6421. Klibanov, A. M.; *Biotechnol. Bioeng.* **1987**, *29*, 648. Therisod, M.; Klibanov, A. M. *J. Am. Chem. Soc.* **1987**, *109*, 3977. Kloosterman, M.; Mosmuller, E.W.J.; Schoemaker, H.E.; Meijer, E.M. *Tetrahedron Lett.* **1987**, *28*, 2989. Chopineau, J.; McCafferty, F.D.; Therisod, M.; Klibanov, A.M. *Biotechnol. Bioeng.* **1988**, *31*, 208. Riva, S.; Chopineau, J.; Kieboom, A.P.G.; Klibanov, A.M. *J. Am. Chem. Soc.* **1988**, *110*, 584. Carrea, G.; Riva, S.; Secundo, F.; Danieli, B. *J. Chem. Soc., Chem. Commun.* **1989**, 1057. Shaw. J.F.; Gotor, V.; Pulido, R.J. *J. Chem. Soc., Perkin Trans. 1* **1991**, 491.

Chapter 15

Bioengineered Emulsans

Arthur Trapotsis[1], Bruce Panilaitis[2], Vladimir Guilmanov[2], Juliet Fuhrman[2], Richard Gross[3], and David Kaplan[1]

Departments of [1]Chemical and Biological Engineering and [2]Biology, Tufts University, Medford, MA 02155
[3]Department of Materials Science and Engineering, Polytechnic University, Brooklyn, NY 11201

Emulsans are a family of lipopolysaccharides produced by the bacterium, *Acinetobacter calcoaceticus*. A series of studies have demonstrated that the structural features of these polymers can be manipulated by selective feeding of exogenous fatty acids or through the generation of transposon mutants deficient in fatty acid metabolic pathways. The results suggest that major shifts in fatty acids decorating the polysaccharide main chain can be achieved, leading to a family of structurally-related polymers. These changes result in significant alteration in the solution properties of the polymers, such as in emulsification properties and critical micelle formation. In addition, these structures can be used to explore important biomedical applications, such as vaccine adjuvants. This application was explored by macrophage activation *in vitro* and immunomodulation *in vivo*.

Introduction

Bioemulsifiers (biosurfactants) are produced by bacteria, yeast, and fungi during cultivation on various carbon sources, especially hydrophobic substrates such as hydrocarbons. These compounds are amphipathic molecules that congregate at oil/water or air/water interfaces, reducing surface and interfacial tension and forming emulsions with hydrocarbons. Most surfactants currently in commercial use are chemically derived from petroleum; however, in recent years there has been an increase in interest in the use of biologically-derived surfactants. There are a number of advantages to using naturally produced surfactants in comparison to those which are synthetically produced: (1) there is no need for petroleum feedstocks and organic solvents for synthesis, (2) the products are biodegradable and non-toxic, (3) the compounds are more selective, therefore smaller amounts are required in emulsification application (4) the potential exists to control the structural composition and therefore the emulsification properties, and (5) the opportunity to utilize recombinant DNA technology as well as classical techniques of mutation and selection, to modify and improve strains of microorganisms and thus production.

Emulsans are a family of anionic polysaccharides produced by the Gram negative bacterium, *Acinetobacter calcoaceticus* strain RAG-1 when grown on a variety of hydrocarbon sources including fatty acids and methanol [1,2]. Emulsans form an extracellular cell-associated capsule which is subsequently released into the fermentation broth. The polysaccharide main chain contains three aminosugars, D-galactosamine, D-galactosaminouronic acid, and a dideoxydiamino hexose in the ratio 1:1:1 [3]. The polymer has O-acyl and N-acyl bound side chain fatty acids ranging in chain length from C10 to C18. These fatty acid substituents constitute about 5 to 23% (wt/wt) of the polymer. The structure of emulsan is shown in Figure 1. Approximately 15% by weight of protein is tightly associated with emulsan.

A variety of possible applications for emulsans have been explored. For example, a pilot study was conducted that utilized emulsan to form an oil-in-water emulsion (70:30 ratio) which reduced the viscosity of a highly viscous crude oil [4]. Lowering the viscosity can be accomplished by forming an emulsion, which is an attractive alternative (reduced cost) to diluting the crude oil with light oil stock or heating the oil. The addition of emulsan improved the handling characteristics (e.g., lowered viscosity) of the crude oil while maintaining the same combustion qualities. These results were not observed with synthetic commercial surfactants. In addition, the emulsion was shown to be stable over several days of pumping with no apparent coalescence of the oil or inversions to water-in-oil emulsions [5,6].

There exist other potential applications for emulsan which have been studied and demonstrated with other bioemulsifiers. For example, a novel bioemulsifier from *Candida utilis* has shown potential use in the food industry, specifically in salad dressing [7]. Biosurfactants have also been used in the health care and cosmetic industries [8]. For example, a product containing a sophorolipid and propylene glycol was shown to be skin compatible and is used in skin moisturizers, lipstick, and hair products [9,10]. Sophorolipids are mainly produced by yeasts and consist of a dimeric carbohydrate sophorose linked to a long-chain hydroxy fatty acid. Other important biologically-derived glycolipids that exhibit surfactant properties include rhamnolipids and trehalolipids. It is assumed that there is also a wide range of biosurfactants yet to be identified from the vast array of microorganisms present in natural ecosystems. Since hydrophobic-hydrophilic interfaces predominate in many environments, such as hydrocarbon/water or air/water surfaces, these compounds play a critical role in the nutrient flow and the dynamics of metabolism. It is also conceivable that these compounds play an important role in adhesion/biofilm formation and in gene transfer in the environment.

Biosynthesis

The biosynthetic pathway for emsulan formation has not been characterized. Glycosyl synthases and acyl transferases are presumably involved in control of polymerization of the main chain sugars and the decoration of the sugars with fatty acids, respectively. An esterase putatively involved in the release of emulsan from the cell membrane was reported [11]. It is probable, based on studies with other biological systems such as *Escherichia coli* and mammalian cells, that the fatty acids come from at least two sources: (1) fatty acid biosynthesis mediated via acetyl-CoA, and (2) the β-oxidation pathway.

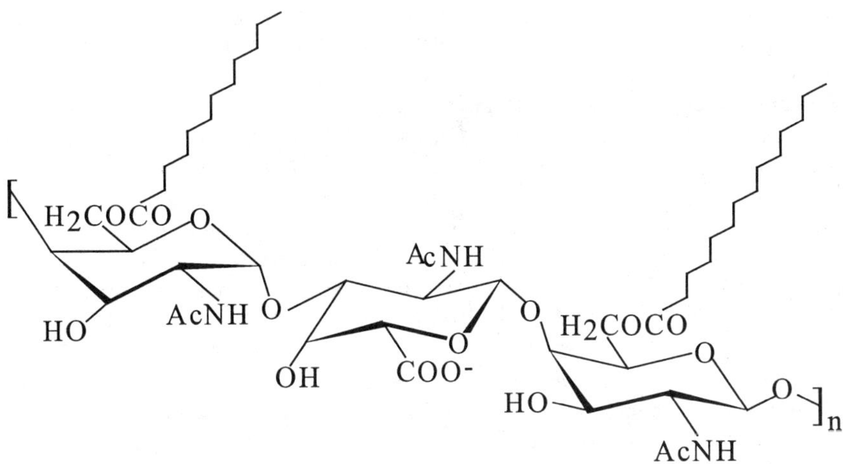

Figure 1. *Structure of emulsan from A. calcoaceticus RAG-1.*

The biological synthesis of lipopolysaccharides provides some important advantages from a structure and function viewpoint: Emulsans are directly biologically synthesized with a high level of control over structural features (molecular weight, polydispersity, regioselectivity, stereoselectivity). This control is considered critical since the formation of, and biological responses to, materials are directly tied to structural features. This level of structural control would otherwise be extremely difficult to achieve using more traditional synthetic organic chemistry approaches and the associated elaborate chemical blocking and deblocking reactions. This level of control is essential to the interrogation of biological function and response mechanisms related to biomedical applications for these compounds, while perhaps less critical to commodity chemical applications such as in lowering the viscosity of crude oils, environmental clean up or as an additive in various cosmetic formulations. In addition, emulsans are secondary metabolites, thus synthesis is not growth associated, and high productivity (up to 25 g/L) can be achieved. Thus, once suitable structural analogs are identified the specific polymers can be generated in amounts sufficient to carry out environmental studies or clinical trials, depending on the target application, and the costs to do so would be nominal. An additional benefit to this system is that purification of emulsans is relatively simple since the polymer is an extracellular product. The purification process involves an ammonium sulfate precipitation from the extracellular culture medium. Further purification requirements depend on the intended applications. For environmental and commodity needs, no further purification may be required. For biomedical applications, more exhaustive purification with precipitation, column separations, and hot phenol extraction to remove associated protein may be required.

Structural Control

We have addressed the structural 'tailorability' of emulsans through a combination of selective feeding strategies and the use of genetic manipulation. Our initial studies were directed at exploring the metabolic flexibility of *A. calcoaceticus* RAG-1 to incorporate exogenous fatty acids under a variety of batch culture strategies. These studies led to the observation that significant manipulation in both the composition and degree of fatty acid substitution could be achieved [12-15]. Thus, the idea that a family of lipopolysaccharides can be formed by a microbial production system has now been demonstrated. The bioengineering of the surfactant structure is of fundamental importance since it shows that the acylation of sugar units by the acyltransferase enzyme system can potentially take place with a wide range of acyl donor molecules.

Saturated and unsaturated fatty acids with chain lengths of C11 to C18 were studied as sole carbon sources for *A. calcoaceticus*. Some of the data from these studies are presented in Tables 1-2 [12-15]. Initially, emulsan synthesized either with ethylpropionate, myristic acid (C14:0) or ethanol were studied (Table 1). The study illustrates two important points: (1) the structural complexity of emulsan with a diverse set of fatty acid substituents on the polysaccharide backbone, (2) there is a clear influence of carbon source on the types and amounts of substituents. One clear indication of this effect is the significant level of odd chain length fatty acid groups, ~12 mole percent 17-carbon chain length when ethylpropionate is used, while the use of myristic acid (C14) did not yield odd chain length pendant groups.

In a second series of studies, n-alkanoic fatty acids of chain lengths 11:0, 12:0, 13:0, 14:0, 15:0, 16:0, 17:0 and 18:0 were used as sole carbon sources to explore the impact on emulsan production levels and structural features. In general, chain lengths below 15 carbons had less of an influence on composition than those in the range of 15-17. These results are illustrated in Table 2. In this range a high percentage of the fatty acids provided in the culture medium were directly incorporated onto the backbone. The mole percent of emulsan fatty acid esters having chain lengths equal to that of the C15, C16 and C17 n-alkanoic acid carbon sources were 53.1, 47.2 and 43.9 %, respectively. In contrast, less apparent increases in C11, C12, and C13 emulsan fatty acids above that of a reference emulsan polymer were found in emulsans formed using the corresponding n-alkanoic acids as carbon sources.

To validate that the exogenous fatty acids were incorporated directly, 0.5% (w/v) $^{13}C_1$-palmitic acid and 0.5% acetic acids were used as cosubstrates [13,14]. An emulsan was formed wherein GC-MS analysis of the 74/75 and 87/88 mass ion pairs showed that 80% of emulsan C16 fatty acid esters were incorporated intact from the C16 carbon source. The percent of C14, C15, C16, C17 and C18 fatty acid esters derived from the corresponding n-alkanoic acids that were mono unsaturated were 10.9, 25.6, 33.5, 69.7 and 84.7%, respectively [12-15]. An aerobic desaturation mechanism (Δ9 desaturase activity) was established for emulsan synthesis [12-15]. Unsaturated fatty acids of emulsan were directly synthesized from saturated fatty acids of the carbon source based on $^{13}C_1$-labeling experiments. It was concluded that fatty acid supplements in *A. calcoaceticus* cultivations can be incorporated intact to large extents within chain lengths C15 to C18, and that the percent unsaturation is chain length dependent.

Additional studies were conducted to explore incorporation profiles with a series of 2-hydroxyl fatty acids [C12:0(2-OH), C14:0(2-OH), C16:0(2-OH) and C18:0(2-OH] as sole carbon sources or when co-fed with myristic acid [12-15]. Again, significant levels of modulation in the composition of the fatty acid pendent groups was observed. For example, C12:0(2-OH) increased to over 64% of total fatty acids and 306 nmol/mg emulsan by providing this fatty acid as a carbon source. Significant quantities of 2-OH fatty acids with chain lengths $\geq$C14, up to 96 nmol/mg of emulsan or 15.2 mol% for C16:0(2-OH), were also incorporated into emulsan by providing the corresponding 2-OH fatty acid carbon source. Furthermore, increasing the 2-OH fatty acid content in the medium resulted in large increases in total fatty acid content for emulsans, up to 955 nmol/mg of emulsan or 23 wt%.

The idea of depressing *de novo* fatty acid synthesis to promote increased incorporation was explored [12-15]. Cerulenin, an irreversible inhibitor of β-ketoacyl-ACP synthase I and II activities, was studied at 150 mg/L with ethanol (1% w/v) as the carbon source and palmitoleic acid (16:1, 9-cis) as the fatty acid supplement (0.2%). The increase in the content of palmitoleic acid in emulsan due to the presence of cerulenin was dramatic, 61%. In the control experiment, the content of 16:1 was 19.9 mol% vs. 32.1 mole percent in the presence of cerulinin. The influence of iodoacetamide (IAA, a non-specific inhibitor of fatty acid metabolism) on emulsan fatty acid composition was evaluated using myristic acid as the sole carbon source and cosubstrate mixtures of myristic and acetic acids. At 0.5

mM of IAA the concentration of myristic acid in the product was increased from approximately 10 to about 30 mole percent.

Fatty Acid Content	Ethyl Propionate	Myristic Acid	Ethanol
10:0	0	0.9(4)	2.6(11)
12:0	12(50)	15(57)	7(32)
13:0	2(9)	0	4(17)
12:0, 2-OH	5(21)	9(33)	14(59)
12:0, 3-OH	25(104)	15(56)	13(56)
14:1	0	2(6)	nd
14:0	0.9(4)	12(47)	3(14)
15:1	0	0	1(2)
16:1	8(33)	18(69)	8.2(36)
16:0	13(56)	18(68)	23(99)
17:1	5(23)	0	0.3(1.4)
17:0	7(27)	0	0.4(2)
18:1	15(64)	8(30)	13(56)
18:0	2(10)	1(4)	5(20)
Unidentified	3(11)	2(12)	1(5)
Totals	100(420)	100(386)	100(436)

Table 1. *Fatty acid composition of emulsan produced by A. calcoaceticus grown on ethyl propionate, myristic acid (C14) or ethanol, numbers in columns are mol% with nmol of fatty acids/mg emulsan in parenthesis.*

Emulsan Fatty Acid Composition (nmol/mg emulsan (% wt/wt)[b]

Fatty Acid Source[a]	15:0 + 15:1	16:0 + 16:1	17:0 + 17:1	18:0+18:1	Total Fatty Acids
15:0	217(5.2)	43(1.1)	25(0.7)	29(0.8)	408(9.8)
16:0	---	299(6.8)	---	18(0.5)	651(13.8)
17:0	25(0.6)	14(0.3)	184(4.9)	9(0.3)	418(9.7)
15:0+17:0[c]	119(2.9)	17(0.4)	229(6.2)	10(0.3)	528(12.8)
18:0	---	105(2.7)	---	52(1.5)	555(12.2)

[a]The carbon source concentration in the growth medium was 1% wt/vol.
[b]wt/wt = 100 X (total fatty acid weight/emulsan weight).
[c]A 1:1 mol/mol mixture was used.

Table 2. *Some examples of the influence of fatty acid chain lengths from 15 to 18 carbons as carbon source feed on the fatty acid profiles of emulsans synthesized by A. calcoaceticus.*

Genetic Modifications

Based on the success with modulation of fatty acid content and levels, we decided to explore additional control of incorporation profiles of the fatty acids provided to *A. calcoaceticus* RAG-1 [16]. Transposon mutants of *A. calcoaceticus* were generated using a modified transposon, Tn10 (mini-Tn10PttKm) [17] as described by Leahy *et al.*, [18,19]. Antibiotic selection was used to isolate transformants and mutant selection was based on two phenotypic functions: (a) loss of β-oxidation pathways (to promote direct incorporation of exogenous fatty acids), and (2) loss of fatty acid synthesis capability. The mutants selected for further study to assess fatty acid incorporation profiles contained either one or two insertions of the transposon based on Southern hybridization analysis of restriction enzyme-digested genomic DNA.

Interestingly, unlike *A. calcoaceticus* RAG-1, the transposon mutants can not grow on a defined medium supplemented with ethanol or fatty acids, but have the ability to grow on Luria Bertani broth (LB), a nutrient rich medium. Therefore, in order to make appropriate comparisons between *A. calcoaceticus* RAG-1, the parent strain in these studies, and the transposon mutants, all cultures were grown on LB. To obtain a more complete picture of mutant metabolic behavior, a series of fatty acids including C11 (undecanoic acid), C14 (myristic), C16 (palmitic) and C18 (stearic), were selected as primary carbon sources for *A. calcoaceticus* RAG-1 and the growth of three transposon mutants [two fatty acid oxidation mutants (M1 and M2) and one fatty acid synthesis mutant (M3)].

A.calcoaceticus RAG-1, when grown on ethanol in defined media, produced an emulsan that consisted of a broad range (C10-C18) of fatty acids with approximately 436 nmol of fatty acids per mg of emulsan. In contrast, there were no fatty acids smaller than C16:1 present on the emulsan produced by *A. calcoaceticus* RAG-1 grown on LB. The total fatty acids present per mg of emulsan was only 2.5 nmol/mg in the case of growth on LB. The fatty acid profiles of polymer produced on LB with C14 (Figure 2) and LB with C18 consisted of fatty acids smaller than C16:1 with particularly large mole percentages of C12:0(3-OH) (71 mol% and 66 mol%, respectively). The total fatty acid contents of C14 and C18 (113 and 138 nmol/mg of emulsan, respectively) were considerably higher than the emulsan formed from LB alone (2.5 nmol/mg). In general emulsans produced with LB present in the media did not contain fatty acids of chain lengths less than or equal to C10 (capric acid), nor did they consist of C15 (pentadecanoic acid).

The fatty acid profile for the emulsan formed from one of the transposon mutants (M2) grown on LB shows a large percentage of C12:0(3-OH) (73 mole %) and relatively small mole percentages of the other fatty acids, including less than 1% C14:0 (Figure 3). A comparison between this emulsan and emulsan generated by *A. calcoaceticus* RAG-1 on LB indicated a large difference in the fatty acid profiles. These data support the presence of a mutation in a fatty acid oxidation pathway.

This hypothesis was further strengthened following the analysis of emulsan produced from M2 grown on C14 (Figure 3). As expected, M2 produced an emulsan consisting of a large mole percentage of C14 (70%), indicating direct incorporation of this exogenous fatty acid. However, only 8 nmol of fatty acids/mg of emulsan was present on this C14-derived emulsan.

Similar results were not observed when M2 was grown on C16 or C18 (Figure 4), with only slight increases in the corresponding fed fatty acids. The C16-produced emulsan did show an increase in the mole percent of C16:1 fatty acid; from 2 mol% on LB alone to 10 mol% on LB plus C16. Further, both the C16 and C18-formed emulsans, from M2, consisted of large mole percentages of C12:0 (2-OH), 53 mol% and 43 mol%, respectively.

Explanations for this observed physiological behavior are still under study through the identification of the genes disrupted through the insertion of the transposon. Disruptions in acyl-CoA synthetase, thus altering the feed of fatty acids into the β-oxidation cycle and a role for a Δ-9-desaturase activity, which is responsible for the formation of unsaturated fatty acids from preformed fatty acids, seem logical based on the data observed. The increase in the mole percent of C12:0(2OH) on emulsans produced by longer chain fatty acids (C16, C18) suggests a possible disruption in the *fad*B portion of the β-oxidation pathway [20].

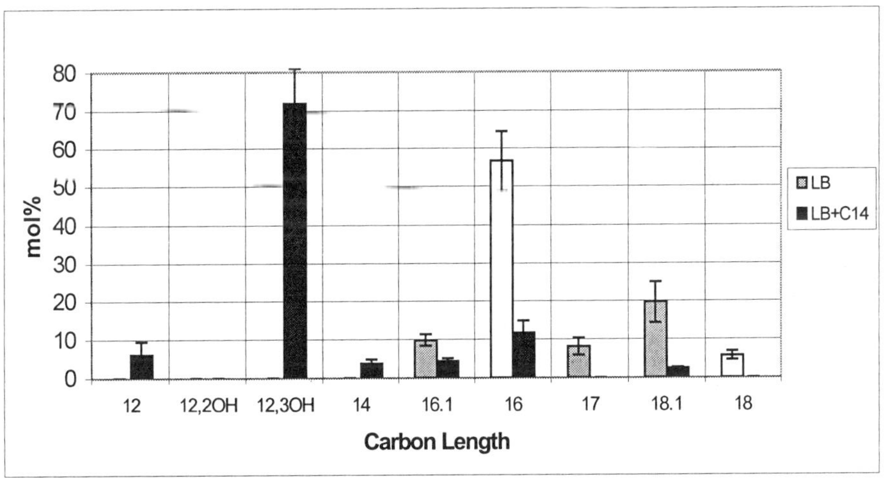

Figure 2. *Fatty acid profiles for emulsans produced by A. calcoaceticus RAG-1 grown on LB and LB with myristic acid (C14) (1% w/v) as carbon sources. Total nmoles of fatty acids per mg of emulsan were 2.5 and 113 for LB and LB with C14, respectively. The x-axis represents the fatty acids (mole percent) present on emulsan. '12,2-OH' represents the fatty acid 2-hydroxylauric acid and '16.1' stands for mono-unsaturated palmitic acid (palmitoleic acid). The molar percentages of the fatty acids are exhibited on the y-axis.*

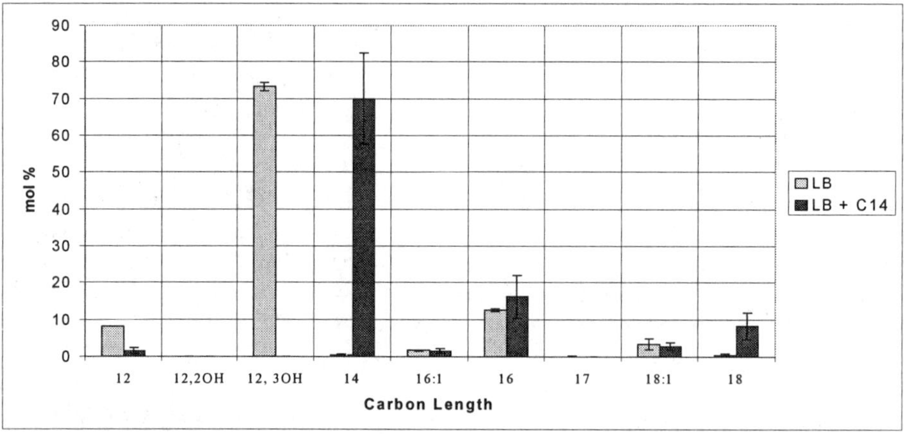

Figure 3. *Fatty acid profiles of emulsan produced by M2 on LB and LB with myristic acid (C14) (1% w/v). Total nmoles of fatty acids per mg of emulsan were 33 and 8 for M2 on LB and LB with C14, respectively.*

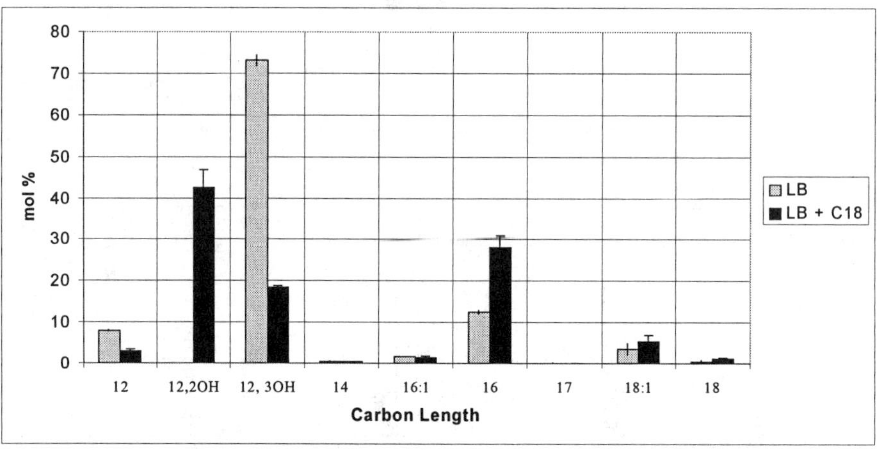

Figure 4. *Fatty acid profiles of emulsan produced by M2 on LB and LB with stearic acid (C18) (1% w/v). Total nmoles of fatty acids per mg of emulsan were 33 and 65 for M2 on LB and LB with C18, respectively.*

The results from the analysis of the fatty acid incorporation profiles by the mutants, only some of which are summarized here, suggest that new options for control of incorporation profiles can be obtained using this genetic approach. These data point the way toward strategies for further structural tailoring of emulsans, particularly once the molecular genetics are clarified. In general, the levels of emulsan secreted from the transposon mutants are significantly lower in comparison to the parent strain, *A. calcoaceticus* RAG-1, while some of the emulsification properties appear to be significantly improved (described in the next section). We anticipate that detailed studies of the molecular genetics, regulation and processing pathways will provide clarification of the physiological and structural data observed, and ultimately lead to novel bioengineering approaches to 'designer bioemulsifiers'.

Emulsification Properties of Emulsan

The combination of hydrophilic anionic sugar main chain repeat units along with the hydrophobic alkyl side chains leads to the amphipathic behavior of emulsan and therefore, its ability to form stable oil in water emulsions. Emulsan causes little reduction in surface tension between an oil and water interface (10 dynes/cm) but binds tightly to oil surfaces protecting oil droplets from coalescence [21]. The polymer has been used successfully in environmental cleanup and degradation studies to disperse polluting oils because of its amphipathic structure [22]. Emulsions formed with emulsan have been shown to be stable for months and can be broken when desired by enzymatic degradation or at elevated temperatures [1,2,3].

We have studied the colloidal properties of the structural variants of emulsans to determine if structural changes in the polymer result in alteration in emulsification behavior and critical micelle concentration (CMC) [12,23]. Surface tension decreased with increasing emulsan concentration, as expected for typical surfactants. The CMCs ranged from 25 to 58 mg/L and surface tension and interfacial tension measurements indicated stability between pH 2 and 10. Structural features influenced emulsifying activity. For example, total fatty acid content and distribution of chain lengths had significant influence. Maximum emulsifying activity was found for emulsans containing around 400 nmol of total fatty acids per mg of emulsan (nmol/mg) and activity towards longer chain length oils increased as the fatty acid side chain lengths increased as well. Thus, substrate-specific interactions between the emulsan fatty acid pendent groups and the dispersed phase are suggested. It is also important to note that the hexadecane-in-water emulsions were prepared with droplet sizes between 9 and 19 μm and many of these emulsions were stable with respect to coalescence for months.

Enhanced incorporation of long side chain groups gave polymers with improved emulsification activity on longer chain n-alkane substrates. The emulsification activities for emulsans produced by *A. calcoaceticus* RAG-1 grown on ethanol, assayed against hexadecane (C16), tetradecane (C14) and dodecane (C12) were 131, 42 and 40 Klett Units (1 KU = 0.002 Absorbance Units), respectively. The corresponding values for emulsan from the same strain grown on LB are shown in Table 3 and indicate; increasing emulsification activity for samples produced on longer-chain fatty acids in the case of hexadecane, and similar general trends against

C14 and C12 substrates. The removal of protein from emulsan (16.4% when the *A. calcoaceticus* RAG-1 is grown on palmitate) by hot phenol extraction or proteinase K treatment, resulted in less than <20% loss in emulsification behavior.

Carbon Source	Emulsification Activity in Klett Units		
	Hexadecane (C16)	Tetradecane (C14)	Dodecane (C12)
LB	5 ± 0	5 ± 0	5 ± 0
LB + C14	21 ± 1	16 ± 2	15 ± 0
LB + C16	30 ± 10	10 ± 2	0 ± 0
LB + C18	133 ±51	93 ± 43	30 ± 13

Table 3. *Emulsification activities from emulsans produced by A. calcoaceticus RAG-1 grown on LB and LB with fatty acids. The averages from three experiments are shown along with the standard deviation.*

The emulsification activities of the emulsans formed by one of the mutants, M2, on C11, C16, and C18 (with LB) decreased as the length of the oil substrate decreased (Table 4). Two interesting results were observed with the emulsans formed from M2: (1) The C18-emulsan emulsified hexadecane extremely well (235 Klett Units), yet consisted of only 65 nmol of total fatty acid per mg of emulsan. This compares with 180 nmol/mg emulsan from *A.calcoaceticus* RAG-1 grown on ethanol. This result demonstrated that the total quantity of fatty acids does not always have an effect on emulsification activity and the nature of the fatty acid distribution may be important in some instances. (2) The emulsan generated on C14 plus LB, which exhibited a high degree of C14 incorporation, did not demonstrate significant emulsification activity. This sample consisted of only 8 nmol/mg emulsan, and suggests that a minimum threshold of fatty acid content is required to initiate emulsification properties. This would be a logical conclusion due to the need for an appropriate hydrophobic/hydrophilic balance in the molecule.

For comparison, similar emulsification tests on hexadecane (C16) with commercially available synthetic surfactants and emulsifiers, including Pluronic F68, sodium dodecyl sulfate, and tween-80 (polyoxyethylenesorbitane), produced activities of 5, 55, and 20 Klett Units, respectively. Furthermore, these commercially produced surfactants are not biodegradable. Interestingly, a new bioemulsifier known as "alasan" (produced by *Acinetobacter radioresistens*) was recently described which showed the ability to emulsify to 1,400 Klett Units [24].

Carbon Source	Emulsification Activity in Klett Units		
	Hexadecane (C16)	Tetradecane (C14)	Dodecane (C12)
LB	5 ± 0	0 ± 0	5 ± 0
LB + C11	20 ± 0	5 ± 0	0 ± 0
LB + C14	0	0	0
LB + C16	75 ± 7	5 ± 1	0
LB + C18	235 ± 25	140 ± 20	52 ± 18

Table 4. *Emulsification activities of emulsans produced by M2 grown on LB and LB with fatty acids. The averages from three experiments are shown along with the standard deviation.*

Biomedical Applications

The structural similarity between emulsans and lipopolysaccharides (LPS) derived from the outer membranes of Gram negative bacteria, suggests important biomedical uses for the emulsan variants. For example, as described earlier, since the emulsans can be 'tailored' biologically to generate a family of structural analogs, this control can be used to explore structure-biological response relationships. Furthermore, aside from the benefit of structural 'tailorability', this type of polymeric compound could be used directly as a delivery system (e.g., for drugs, vaccines) due to its ability to form micelles and to emulsify. This novel combination of structural control with direct delivery would provide important potential biomedical options.

One example of a potential application for these structures, due to their structural similarity to LPS, may be as vaccine adjuvants. Adjuvants by definition are substances used in combination with an antigen to generate enhanced levels of immunity beyond that developed with the antigen alone. The result is a more effective vaccine. In general, the use of adjuvants involves a balance between toxicity and adjuvant effectiveness, based on risk-benefit assessments [25,26]. Currently, human vaccines are limited to aluminum-based adjuvants, while veterinary vaccines use a diverse set of compounds. Even aluminum-based adjuvants have been scrutinized due to concerns related to Alzheimer's disease [27]. The requirements for adjuvants are many (for examples, see 25,26]. The most important issue is safety which has been the major barrier to the successful introduction of new adjuvants since alum appeared ~50 years ago. Safety concerns arise due to the over-stimulation of the immune system or the presence of toxic components in the adjuvant formulation.

A variety of immunostimulatory compounds as candidate adjuvants have been, or are currently, under study in many laboratories (for example, see reviews see 28-31]. A conclusion from current efforts aimed at developing new vaccine adjuvants are that the structural diversity available for a specific class of candidates is relatively limited. This limitation prevents a more exhaustive investigation of structure-activity relationships to modulate toxicity and levels and types of immune responses. In addition, it would appear to be useful to combine the benefits of a particulate-based adjuvant delivery system (e.g., micellar, emulsions) with inherent immunostimulation properties. This type of dual function adjuvant could offer a variety of benefits in expanding delivery options and tailorability to the response level and targets desired. This is the opportunity presented with the emulsan-based systems.

In our preliminary studies to address this concept, we have characterized macrophage activation *in vitro* with one emulsan and a dose-dependent response was observed in terms of cytokine release. In addition, preliminary mouse studies *in vivo* support a sustained and significant immunoactivation at a level comparable to Freunds complete adjuvant, and this activation was antigen-specific.

We characterized emulsan from *A. calcoaceticus* RAG-1 grown on an ethanol feed source with the macrophages. Based on the structural features of emulsan and similarities to traditional microbial lipopolysacharides, we determined that release of the pro-inflammatory cytokine, Tumor Necrosis Factor (TNF), would be an indication of stimulation of macrophages. We examined the effects with and without the bound protein to elucidate differences. Removal of the protein would be advantageous to minimize potential undesirable responses with the cell lines. To determine the immunomodulatory effects of emulsan we first determined whether emulsan or deproteinized emulsan could stimulate macrophages to release TNF. The results suggest that this is the case, for both the proteinated and deproteinated polymers.

Emulsan was tested on both primary macrophages and the macrophage cell line, RAW 264.7. The results (only RAW cell results shown) demonstrate a dose-dependent release of TNF in response to emulsan stimulation from both macrophage sources (Figure 5). The response is approximately 20-fold lower than that of LPS on a per weight basis. Since emulsan is a bacterial product, and because of the possibility of LPS contamination, we conducted the same macrophage activation assays on macrophage cell lines derived from the LPS-responsive and LPS non-responsive mice, C3H/HeJ and C3H/FeJ mice. The cell lines HeNC2 (LPS-responsive) and GG2EE (LPS-non-responsive) were a generous gift of Dr. A. Ding and Dr. C. Nathan (Cornell University Medical College). The results indicate that emulsan induced TNF release in an LPS-independent manner.

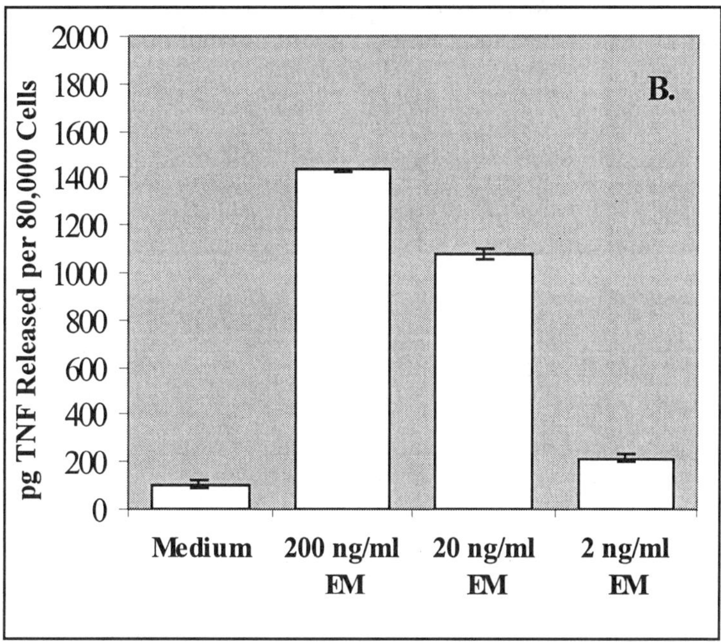

Figure 5: *TNF release by RAW 264.7 cells stimulated with emulsan (EM).*

The *in vivo* mouse studies were conducted with a classical hapten-carrier immunization protocol. These assays were performed with the emulsan from *A. calcoaceticus* RAG-1 in both 'native' proteinated and deproteinated forms. The mice were randomly placed in eight groups and immunized as follows: (1) no emulsan, just 100 μg protein-dinitrophenol (KLH-DNP) [Ag alone], (2) 200 μg crude emulsan [EM alone], (3) 200 μg deproteinized emulsan [Dp-EM Alone], (4) 20 μg crude emulsan and 100 μg KLH-DNP [EM (low) & Ag], (5) 200 μg crude emulsan and 100 μg KLH-DNP [EM (high) & Ag], (6) 20 μg deproteinized emulsan and 100 μg KLH-DNP [Dp-EM (low) & Ag], (7) 200 μg deproteinized emulsan and 100 μg KLH-DNP [Dp-EM (high) & Ag], (8) 100 μl Complete Freund's Adjuvant (CFA)/Incomplete Freund's Adjuvant (IFA) and 100 μg KLH-DNP (CFA was used as adjuvant in primary immunization, and IFA was used at boost) [CFA/IFA & Ag] (Figure 6).

Pre-immune sera were taken 3 days prior to primary immunization. Mice were boosted after 28 days, and sera were taken every 3 days after boost until day 21 post-boost, and then again at 6 weeks and 9 weeks. A time course of total DNP-specific antibody titers is presented in Figure 6. The results indicate that both preparations of emulsan (with and without associated protein) have significant adjuvant activity; on the same order as Freund's adjuvants. The emsulan or antigen alone did not stimulate the same level of immune response. These data suggest that further exploration of emulsan structural analogs may be a useful experimental strategy to explore the mode of action of vaccine adjuvants [32].

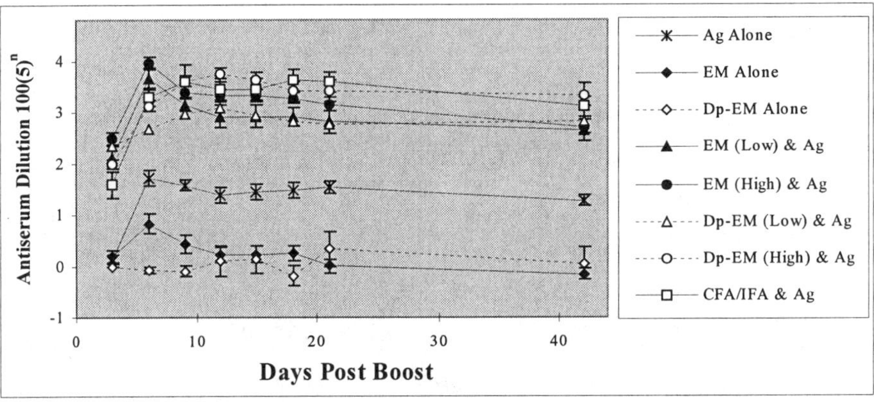

Figure 6. *Profiles of emulsan adjuvant activity.*

Conclusions

The types of studies summarized illustrate the value of bioengineered microbial polymers. Through a variety of strategies we are able to manipulate the biosynthetic machinery of this bacterium to alter the structural features of the biopolymer. These types of structural modifications would be virtually impossible to attain using traditional organic chemistry methods. In turn, through structural manipulation, we are able to illustrate the relationship between structure and function, in this case solution properties for this class of biopolymer. Furthermore, these types of structures have potential utility in biomedical applications, such as those described for vaccine adjuvants.

Acknowledgements

We thank the U. S. Department of Agriculture (CSREES, NRICGP) and the National Science Foundation-Environmental Protection Agency for grant support to DK and RG, respectively, in support of this work

LITERATURE CITED

1. Gutnick D. L.; Y. Shabtai. Exopolysaccharide bioemulsifiers in biosurfactants and biotechnology, In, N. Kosaric, W. L. Cairns, and Neil C. C. Gray (eds), Exopolysaccharide Bioemulsifiers, Marcel Dekker, Inc.: New York, Vol. 25, 211-246, 1987.
2. Gutnick, D. L.; Y. Shabtai. Exopolysaccharide bioemulsifiers in biosurfactants and biotechnology, In, N. Kosaric, W. L. Cairns, and Neil, C. C. Gray (eds), Exopolysaccharide Bioemulsifiers, Marcel Dekker, Inc NY, Vol. 25, 232, 1987.
3. Belsky, I.; D. L. Gutnick; E. Rosenberg. Emulsifier of *Acinetobacter* RAG-1: determination of emulsifier-bound fatty acids. FEBS Letts. 101:175-178, 1979.
4. Hayes, M.E., E. Nestaas, and K. Hrebenar, Microbial Surfactants. Chemtech. April, 239-243, 1987.
5. Gutnick, D. L., E. Rosenberg, and E. Belsky, Apo-beta-emulsans. US patent, 4311830, 1982.
6. Van Dyke, M., S. Gulley, H. Lee, and J. Trevors, Evaluation of microbial surfactants for recovery of hydrophobic pollutants from soil. J. Indust. Microbiol. 11, 163-170, 1993.
7. Shephord, R., J. Rockey, I. Shutherland, and S. Roller., Novel bioemulsifier from microorganisms for use in foods. J. Biotech. 40, 207-217, 1995.
8. Klekner, V., and N. Kosaric., Biosurfactants for cosmetics, in Biosurfactants: production, properties, applications, N. Kosaric, Ed., Marcel Dekker, Inc.: NY. 329-372, 1993.
9. Inoue, S., Y. Kimwa, and M. Kinta, German Patent 2905252. Kao Soap Co., Japan, 1979.
10. Yamane, T., Enzyme technology for the lipid industry: an engineering overview. J. Am. Oil. Chem. Soc. 64, 1657-1662, 1987.
11. Shabtai, Y.; D. L. Gutnick.. Extracellular esterase and emulsan release from the cell surface of *Acinetobacter calcoaceticus*. J. Bacteriol. 161:1176-1181, 1985.
12. Gorkovenko, A.; J. Zhang; R. A. Gross; A. L. Allen, D. Ball, D. L. Kaplan. Biosynthesis of emulsan analogs: direct incorporation of exogenous fatty acids. Proc. Am. Chem. Soc., Div. Polym. Sci. Eng. 72:92-93, 1995.
13. Gorkovenko, A.; J. Zhang; R. A. Gross; A. L. Allen; D. L. Kaplan. Bioengineering of emulsifier structure: emulsan analogs. Can. J. Microbiol. 43:384-390, 1997.
14. Gorkovenko, A.; J. Zhang; R. Gross; D. L. Kaplan; A. L. Allen. Control of unsaturated fatty acid substituents in emulsans. Carbohydrate Polymers,39:70-84, 1999.
15. Zhang, J.; A. Gorkovenko; R. A. Gross; A. L. Allen; D. L. Kaplan. Incorporation of 2-hydroxyl fatty acids by *Acinetobacter calcoaceticus* RAG-1 to tailor emulsan structure. Internatl. J. Biol. Macromol. 20:9-21, 1997.
16. Trapotsis, A. Selection and analysis of bioemulsifiers: emulsan analogs. Masters of Science Thesis, 1999.

17. Herrero, M.; V. d. Lorenzo; K. N. Timmis. Transposon vectors containing non-antibiotic resistance selection markers for cloning and stable chromosomal insertion of foreign genes in gram-negative bacteria. J. Bacteriol. 172:6557-6567, 1990..
18. Leahy, J. G.; J. M. Jones-Meehan; E. L. Pullias; R. R. Colwell. Transposon mutagenesis in *Acinetobacter calcoaceticus* RAG-1. J. Bacteriol. 175:1838-1840, 1993.
19. Leahy, J. G., J. M. Jones-Meehan, R. R. Colwell. Transformation of *Acinetobacter calcoaceticus* RAG-1 by electroporation. Canadian J. Microbiol. 40:233-236, 1993.
20. Neidhardt, F. C. Escherichia coli and Salmonella tyrphimurium – cellular and molecular biology. Am. Soc. Microbiol. 1987.
21. Rosenberg, E.; A. Perry; T. Gibson; D. L. Gutnick. Emulsifier of Arthrobacter RAG-1: specificity of hydrocarbon substrate. Appl. Env. Microbiol. 37:409-413, 1979.
22. Shoham, Y.; E. Rosenberg. Enzymatic depolymerization of emulsan, J. Bacteriol., 156: 161-167, 1983.
23. Zhang, J.; S.-H. Lee; D. Avison; R. A. Gross; A. L. Allen; D. L. Kaplan. Surface activity of emulsan-analogs. J. Colloids and Dispersons. In review, 1999.
24. Navon-Venezia, S., Z. Zosim, A. Gottlieb, S Carmeli, E. Ron, and E. Rosenberg, Alasan, a new bioemulsifier from *Acinetobacter radioresitens.* Appl. Environ. Microbiol., 61(9): 3240-3244, 1995.
25. Gupta, R. K.; G. R. Siber. Adjuvants for human vaccines – current status, problems and future prospects. Vaccine 13/14:1263-1276, 1997.
26. Gupta, R. K.; P. Griffin, Jr.; A.-C. Chang; R. Rivera; R. Anderson; B. Rost; D. Cecchini; M. Nicholson; G. Siber. The role of adjuvants and delivery systems in modulation of immune response to vaccines. In Novel Strategies in Design and Production of vaccines, Eds. S. Cohen; A. Shafferman, Plenum Press, New York, pp.105-113, 1996.
27. Allison, A. C.; N. E. Byars. An adjuvant formulation that selectively elicits the formation of antibodies of protective isotypes and of cell-mediated immunity. J. Immunol. Methods 95:157-168, 1986.
28. Donnelly, J. J. New developments in adjuvants. Mechanisms of Aging and Development 93:171-177, 1997.
29. Usinger, W. R. A comparison of antibody responses to veterinary vaccine antigens potentiated by different adjuvants. Vaccine 15:1903-1907, 1997.
30. Vogel, F. R. 1995. Immunologic adjuvants for modern vaccine formulations. In Combined Vaccines and Simultaneous Administration: Current Issues and Perspectives 754:153-160, 1997.
31. Werner, G. H.; P. Jolles. Immunostimulating agents: what next? A review of their present and potential medical applications. Eur. J. Biochem. 242:1-19, 1996.
32. Fuhrman, J, B. Panilaitis, R. Gross, D. L. Kaplan. Vaccine adjuvants using bioengineered emulsans. Patent Disclosure, 1998.

Biomedical Polymers from Carbohydrates and Amino Acids

Chapter 16

Poly(amino acid)s and Ester–Amide Copolymers: Tailor-Made Biodegradable Polymers

F. Rypáček, M. Dvořák, I. Štefko, L. Machová, V. Škarda, and D. Kubies

Institute of Macromolecular Chemistry, Academy of Sciences of the Czech Republic, Heyrovský sq. 2, 162 06 Prague 6, Czech Republic

Due to their polypeptide backbone, the synthetic poly(amino acid)s have an inherent potential to be degraded in biological environments by enzymes which regularly hydrolyze the peptide bonds, such as proteinases and peptidases. The rate and specificity of enzyme-catalyzed hydrolysis of poly(amino acid)s can be controlled through copolymerization and/or side-chain modifications. A number of structural variations of synthetic poly(amino acid)s, based on the same backbone structure, can be prepared by ring-opening polymerization of respective *N*-carboxyanhydrides. Polymers of trifunctional α-amino acids, such as aspartic and glutamic acids, are of particular interest as they allow for various side chain modifications. Enzymatic degradability and its relation to the structure of hydrophilic, soluble poly(glutamic acid) and poly(aspartic acid) derivatives and crosslinked biodegradable hydrogels was investigated. The controlled living polymerization of NCA makes it possible to prepare telechelic polymers with functional end groups and block copolymers. A new method of preparing ester-amide copolymers by direct copolymerization of α-amino acid *N*-carboxyanhydrides and lactones was developed.

Synthetic biodegradable polymeric materials are often studied for use in biomedical applications, such as sutures, absorbable tissue substitutes and supports, and drug delivery systems. In many biomedical applications, the biodegradability is important not only for toxicological safety reasons but also as a necessary functional requirement. Two important classes of biodegradable polymers currently being most investigated for a variety of medical and pharmaceutical applications are poly(hydroxy acids) and poly(amino acid)s. This chapter focuses on the last class, poly(α-amino acid) derivatives, and points out several new approaches through which these polymers can be tailor-made to achieve specific degradation and functional performance.

Synthetic Poly(α-Amino Acid)s - Polymers with Controlled Biodegradation.

Synthetic poly(amino acid)s are polymers with the backbone formed by peptide bonds. In nature, a combination of twenty different amino acids provides for a vast number of structurally defined polymers, proteins, each of them fulfilling a specific functions in living systems. Although current methods of poly(amino acid) synthesis still remain far behind the specificity of a template-controlled biosynthesis of proteins, their rational use makes it possible to prepare a number of structural variations of synthetic poly(amino acid)s, all being based on the same polypeptide backbone structure. Due to the polypeptide backbone, synthetic poly(amino acid)s have an inherent potential to be degraded in biological environments by the action of enzymes hydrolyzing peptide bonds, such as proteinases and peptidases. In this respect they resemble naturally occurring polypeptides.

On the other hand, the statistical character of common polymerization procedures and unwanted side reactions, as well as the possibility of using unnatural amino acid derivatives as monomers, lead to polymers with molecular and physical properties far from those of proteins, and rather similar to other synthetics. This structural variability applies especially to polymers derived from trifunctional amino acids, such as glutamic and aspartic acids and lysine. The feasibility of modifying side chains of polyGlu or polyAsp by introducing various alkyls as ω - esters or amides, makes it possible to prepare polymers with a broad scale of properties, ranging from linear, hydrophilic, and random coil molecules, either neutral or ionic, to strongly hydrophobic polymers with prevailing helical or β-sheet conformation, soluble in nonpolar organic solvents or forming semicrystalline solids.

Two methods of synthesis are most frequently used for poly(amino acid)s. Direct thermal polycondensation of amino acids is practically applicable only to the synthesis of aspartic acid derivatives (1). On the other hand, this method provides for a low-cost industrial process for poly(aspartic acid) applicable in large scale (2). The ring-opening polymerization of α-amino acid *N*-carboxyanhydrides (NCA) is a generally applicable method that, unlike the thermal polycondensation, can also avoid racemization (3).

Poly(amino acid)s containing carboxylic side chains, such as poly(glutamic acid) and poly(aspartic acid) are promising candidates as water-soluble polymers in industrial applications, such as detergent dispersants, scale inhibitors, and flocculants (4-6). In these applications they could become biodegradable substitutes for polycarboxylates, based on the (meth)acrylate backbone. Both constituents of these polymers, glutamic and aspartic acids, can now be produced in a large scale by low-price biotechnological processes from renewable resources (4).

In the following paragraphs we review our recent work on the relationship between the structure of poly(amino acid) derivatives and the rate and enzyme specificity of their degradation. Examples are given showing how the biodegradation pattern can be controlled by designing the polymer structure and how these specific features can be used to achieve desired properties and function.

Enzymatic Degradation of PHEG Derivatives

The derivatives of polyGlu are typically obtained by polymerization of γ-alkyl glutamate NCAs, thus yielding poly(γ-alkyl glutamates). *N*-Carboxyanhydrides of α-amino acids were usually synthesized by the reaction of triphosgene with respective α-amino acid derivatives, typically in THF (7). By modification of ester groups in side chains, either poly(glutamic acid) or poly(γ-alkyl glutamine)s can be prepared (8,9).

The enzymes involved in biodegradation of poly(amino acid)s are peptidases and proteinases. These classes of enzymes are abundant in living systems in both the intracellular and extracellular compartments. They may cleave the polypeptide chain either by attacking some of the inner peptide bonds within the polymer chain, by endopeptidase mechanisms, or by attacking only the peptide bonds closest to the polymer end and splitting one or two amino acid residues from the chain end, by exopeptidase mechanisms. Some endopeptidases are highly specific as to the amino acid sequence they recognize as a substrate. Hence, the constitution of poly(amino acid)s controls also the rate, at which the polymer can be degraded by a particular enzyme. As different biological compartments may contain enzymes with different specificities, the variation of poly(amino acid) structure affects also the biological compartment in which the degradation will take place (10).

Poly(N^5-hydroxyalkyl L-glutamine)s, in addition to their practical significance as biomaterials (11) can be used as models, suitable for fundamental studies of enzymatic degradation of polymer substrates (12,13). Ring-opening polymerization of γ-benzyl L-glutamate NCA and subsequent total aminolysis of benzyl esters with 2-aminoethanol provide a linear, hydroxylated polymer, PHEG, formed by N^5-(2-hydroxyethyl) L-glutamine units linked through α-peptide bonds(8). PHEG is easily soluble in water, forming a random coil conformation in aqueous solutions (14). Although N^5-2-hydroxyethyl L-glutamine (HEG) is not a natural amino acid, several proteolytic enzymes, such as papain, pepsin, cathepsin B and pronase E, can cleave the peptide bond between two HEG structure units of PHEG by the endopeptidase mechanism (15), with papain being the most active (15,16).

Homopolymers: Random vs. Nonrandom Mechanisms of Cleavage

Enzymatic degradation can be considered as an interaction of two macromolecules, the polymer as a substrate and the enzyme. PHEG as a homopolymer, offering a uniform structure for the interaction with the active site of an endopeptidase enzyme along all the length of the polypeptide chain. In spite of the simplicity of chemical structure, the system displays non-homogeneity in the sense that the cleavable bonds, i.e. the actual substrate, are locally concentrated in relatively small volumes of polymer coils. When analyzing such a system, two viewpoints have to be considered: different accessibility to the enzyme of the bonds located either inside or on the surface of the polymer coil, and the mechanism by which the enzyme attacks individual macromolecules of the substrate. Concerning the latter, two limiting cases have to be considered : (i) a random attack, in which the enzyme after each cleavage is released from the polymer-enzyme complex and attacks, with equal probability, any other substrate molecule in the system, or (ii) a single-chain mechanism, in which the enzyme remains in a complex with the substrate until it has cleaved all degradable bonds in the substrate chain and then it is released to attack another macromolecule. A "multiple attack" would be an intermediate mechanism when the polymer-enzyme complex dissociates after the cleavage of the bond, nevertheless it remains in contact with the polymer coil (or one of its fragments resulting from the cleavage) for a sufficient time so that it significantly increases the probability of the next cleavage occurring in the same coil (17). The extent of the multiple attack would be reflected in the deviation of the molecular-weight distribution (MWD) of the sample during the degradation in comparison with the MWD assuming a random process.

The enzymatic degradation of PHEG and its derivatives was followed by size-exclusion chromatography (18,19). Using kinetic modeling, theoretical MWDs corresponding to states with certain molar fractions of bonds degraded were simulated assuming either a random or various nonrandom degradation models. These simulated MWDs were compared with experimental ones obtained from size-exclusion chromatography (SEC) by applying either enzymatic degradation by papain or chemical degradation by aminoethanol, as an experimental approximation of the random process (18).

These experiments showed that the enzymatic degradation of a polymer molecule cannot be described as a random process even for homopolymers with a monotonous chain structure such as PHEG. The model leading to the best agreement with the experimental data assumed that the subsequent cleavage may take place only in one of the fragments produced by the previous cleavage, while the ratio between the kinetic constants for the process leading to the cleavage in the same fragment and that of preventing the cleavage by diffusion of the enzyme outside the coil was 1/15 (18). Although PHEG is a neutral molecule exhibiting a very weak affinity to the enzyme active site and no apparent sorption of PHEG to protein molecules, these results show a significant influence of a "macromolecular effect" in kinetics of enzymatic degradation. It can be assumed that the effect could be more pronounced for molecules exhibiting certain nonspecific sorption interaction with the enzyme, thus increasing the contact time between the enzyme and polymer molecule.

Copolymer Constitution Effect

The affinity an enzyme to the polymer as a substrate can be changed by modification of the original polymer. Modified derivatives of PHEG can be prepared either by copolymerization with other amino acids, or by a polymer modification reaction. Thus, one copolymer P(HEG-*ran*-Phe) was prepared by copolymerization of γ-benzyl L-glutamate with L-phenylalanine, a natural amino acid, whereas copolymers P(HEG-*ran*-BG) and P(HEG-*ran*-TrnG) were prepared by polymer modification reactions of PBLG. P(HEG-*ran*-BG) was obtained by incomplete aminolysis of PBLG, and P(HEG-*ran*-TrnG) by using both tyramine and 2-aminoethanol in complete aminolysis of PBLG. The resulting PHEG analogues contained about the same molar fractions of aromatic residues in some of their side chains (Figure 1) (20).

Although original PHEG, being a homopolymer, was not degraded by serine proteases, such as chymotrypsin A or elastase, the incorporation of hydrophobic amino acid derivatives created target sites in the polymer molecule, recognizable by these enzymes and, consequently, made the copolymers susceptible to degradation by chymotrypsin. The modification causes a dramatic increase in the rate of degradation by chymotrypsin A, as shown in Figure 2. It is worth noting that besides P(HEG-*ran*-Phe), containing the natural amino acid residues (L-Phe) and being the fastest degraded, also modified glutamine or glutamate residues that mimic natural hydrophobic amino acids significantly increased the susceptibility of the PHEG derivatives to degradation by chymotrypsin.

The introduction of other structure units by copolymerization or by polymer modification creates a definite number of cleavable bonds in the polymer chain. The cleavage of these bonds by a specific enzyme could be very rapid and it produces fragments of the size corresponding to the average size of the polymer segment between the cleavable bonds (21). The possibility of making the polymer degradable by a certain enzyme, or increasing its degradation rate by orders of magnitude through a minor modification becomes a very useful feature in applications in which a rapid lowering of the molecular weight is needed, e.g. excretion through the kidney, or dissolution of a crosslinked system is required to trigger the release of an active compound.

Ultimate Degradation Products

If the purpose of degradation was to convert the polymer to monomer units or metabolizable fragments, we need to identify the ultimate degradation products and the mechanisms leading to them. Endopeptidases, causing rapid decrease in the molecular weight, require that several amino acid residues distant from the ends of polypeptide chain interact with the enzyme active site. As the polymerization degree of fragments decreases, the effect of polymer end groups becomes more pronounced and, at a certain length, the polymer molecule is too short to form a sufficiently stable complex with the enzyme active site and its further degradation stops (13)(18). For PHEG it was found that two amino acid residues from each polymer end are hardly degradable by endopeptidases such as papain or cathepsin B; therefore, ultimate products of endopeptidase degradation

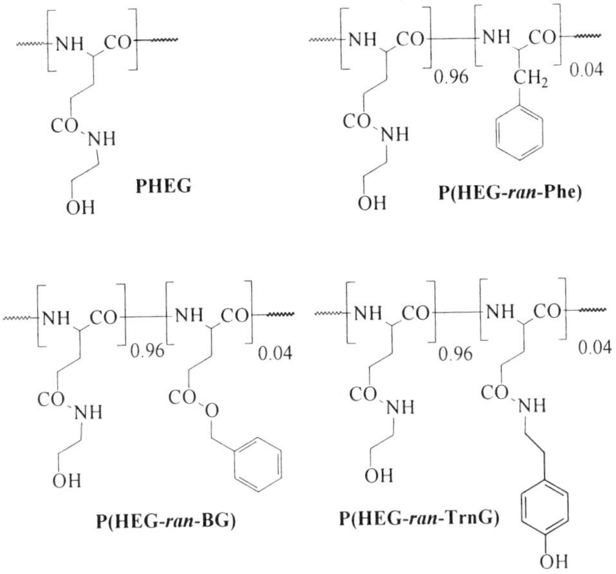

PHEG

P(HEG-*ran*-Phe)

P(HEG-*ran*-BG)

P(HEG-*ran*-TrnG)

Figure 1. Structures of PHEG, and PHEG derivatives.

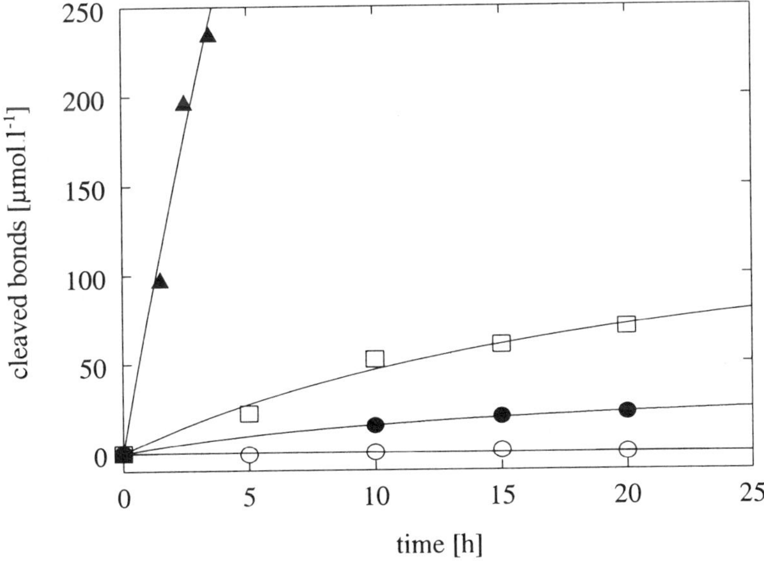

Figure 2. The effect of chemical modification on the rate of degradation of PHEG derivatives by chymotrypsin A. PHEG (○) ; P(HEG-ran-TrnG) 4.0 mol% TrnG (●); P(HEG-ran-BG) 3.9 mol% BG (□); P(HEG-ran-Phe) 3.7 mol% Phe (▲).

are expected to be about the size of tetramers. Polar oligopeptides of that size (MW ≈ 700) would still be too large to pass through the lipid membrane of cell lysosomes, which are the most probable site of accumulation of PHEG if it was used as a drug carrier in medical applications (22).

We studied the enzymatic degradation of N^5-(2-hydroxyethyl)-L-glutamine homopolymer (PHEG) and its random copolymer with L-glutamic acid, P(HEG-*ran*-Glu), by papain, pronase and leucine aminopeptidase (LAP) with the aim to investigate whether the fragments formed from the endopeptidase degradation of PHEG can be further utilized as substrates for exopeptidases and what the structure of the ultimate degradation products is (19). To this end we used a defined *in vitro* system, thus avoiding the complexity of a homogenate (21) or native lysosomes (23) precluding the identification of monomeric fragments.

It follows from the obtained data that the fragments produced by the endopeptidase degradation can be accepted as a substrate by exopeptidases present in mammalian tissues, such as LAP, and that both the HEG oligomers and the HEG-Glu co-oligomers can be cleaved by LAP to respective monomers. No indication was found that aminoethanol needs to be cleaved from HEG to form the glutamic acid residue, before the bonds in the main chain can be cleaved. The same ultimate fragments are produced by treating the polymers with microbial complex of pronase E exhibiting both endopeptidase and exopeptidase activities (19).

The knowledge of relationships between the structure and enzymatic degradation of poly(amino acid) derivatives, can be applied to the design of new biomaterials, a discussion of which follows.

Biodegradable Hydrogel.

Biodegradable hydrogels were prepared from PHEG containing methacroylated side chains (0.02-0.10 mole fraction). Methacryloylated PHEG derivatives were used as "macromonomers" in a crosslinking copolymerization with an additional low-molecular-weight monofunctional acrylic monomer, such as acrylamide. Gels with different crosslinking densities and mechanical properties can be prepared by varying the molecular parameters of starting macromonomers, such as molecular weight, degree of substitution by methacroyl side chains, and the volume fraction of monomers at gel formation. (24). The ratio of PHEG macromonomer to the low-molecular-weight acrylic monomer is an additional adjustable parameter that can affect the properties and the structure of the gel network formed by controlling the ratio of biodegradable poly(amino acid) and nondegradable poly(acrylic) component of the gel. When a monofunctional acrylic monomer is used, the methacroylated PHEG actually functions as a biodegradable macromolecular crosslinker. By action of proteolytic enzymes, such as papain, cathepsin or microbial proteases, the PHEG chains between the crosslinking points were cleaved, resulting in the decreasing crosslinking density, leading to swelling of the gel, eventually, to its dissolution to soluble polymer fragments (24).

Hydrogel Permeability to Macromolecules

The biodegradable hydrogel was investigated as a controlled-delivery system for macromolecular drugs, such as peptides and proteins or polymer-drug conjugates. To this end the permeation of various model compounds through the biodegradable hydrogel was investigated as well as the effect of enzymatic degradation on the gel permeability (25). As model compounds, a low-molecular-weight solute, sodium azide, peptides and proteins varying in molecular weight, trypsin inhibitor (TKI, M_w 6,500), trypsin (M_w 23,400) and bovine serum albumin (M_w 67,000) and a linear polymer, poly-α,β-[*N*-(2-hydroxyethyl)-D,L-aspartamide], (PHEA, $\overline{M}_w$ 22,000), were tested. The aim of the studies was to investigate whether the range of crosslinking densities, feasibly achieved with these types of hydrogels, is suitable to control the diffusion of molecules within a broad molecular-size range.

Diffusion of low-molecular-weight solutes, such as sodium azide, was fast and not significantly influenced by gel network density in the range studied. The diffusion of albumin (M_w 67,000) was slow even in the gel with the lowest network density tested. Albumin did not penetrate through the gels with network densities of 0.183 mol/L and higher. The diffusion of medium-molecular-weight macromolecules can be well controlled by the network density of the hydrogel with changes in the permeability coefficients by up to two orders of magnitude within the range of crosslinking densities studied. The variation of permeability is shown in Figure 3a,b for compounds with different molecular weights in a gel with constant crosslinking density and for trypsin (M_w 23,400), an enzyme which is unable to cleave PHEG chains, for gels differing in crosslinking density.

PHEG can be easily degraded by papain (M_w 20,900) via an endopeptidase mechanism. This cleavage rapidly decreases the gel-crosslinking density and, consequently, increases the permeability of the gel for large molecules (Figure 4).

Triggered Hydrogel Delivery System

A self-degradable system for controlled release of biologically active macromolecules was prepared from a methacryloylated copolymer poly[N^5-(2-hydroxyethyl)-L-glutamine-*stat*-glutamic acid], poly[HEG-*stat*-Glu], containing a proteolytic enzyme, papain, covalently bound to some of Glu residues. The polymer-enzyme conjugate was crosslinked by radical copolymerization with acrylamide. The gel containing the bound papain was stable in buffers void of enzyme activators. Activation of the enzyme can be accomplished with low-molecular-weight compounds, such as dithioerythritol and EDTA, which can freely penetrate into the gel. As a result of degradation by the activated enzyme, the concentration of effective crosslinks decreases, the polymeric network expands and, eventually, the gel dissolves (Figure 5). The systems with "triggered" degradation can be used for a controlled release of macromolecular biologically active agents, entrapped in the gel (26).

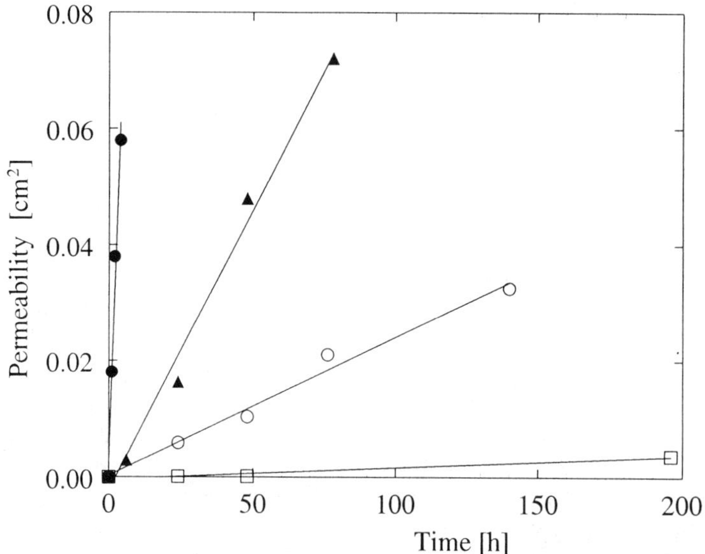

Figure 3a. Permeation of compounds with different molecular weight in the hydrogel with a crosslinking density of 0.040 mol/L: BSA, Mw 67,000 (□); trypsin, Mw 23,400 (○); TKI, Mw 6,500 (▲); sodium azide, Mw 57 (●).

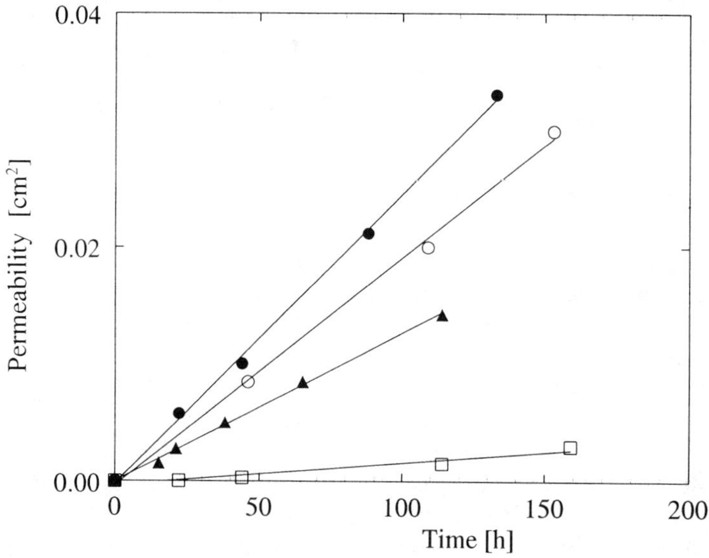

Figure 3b. Permeation of trypsin (Mw 23,400) in the gels with network densities of 0.040 (●), 0.065 (○) 0.183 (▲) and 0.226 (□) mol/L. (Adapted from reference 25. Copyright 1997 Technomic Publishing Co., Inc.)

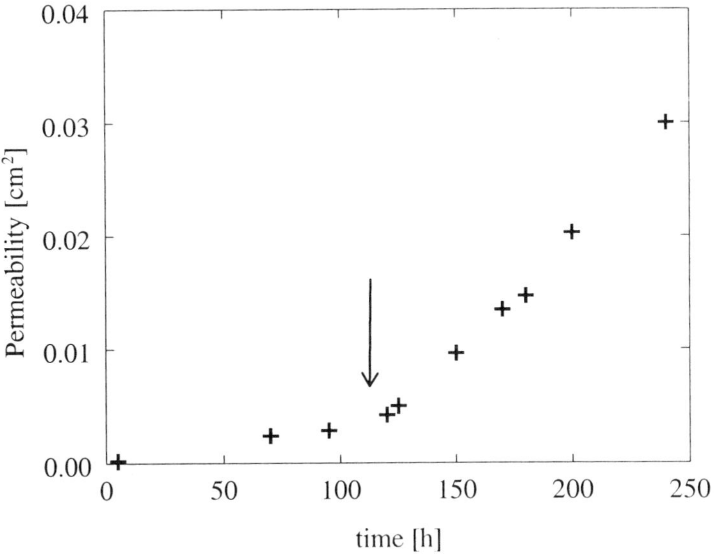

Figure 4. Effect of PHEG-acrylamide gel degradation on the permeation of PHEA ($\overline{M}_w$ 22,000) in the gel with a network density of 0.183 mol/L. Papain was added to both compartments at 120 h. (Adapted from reference 25. Copyright 1997 Technomic Publishing Co. Inc.)

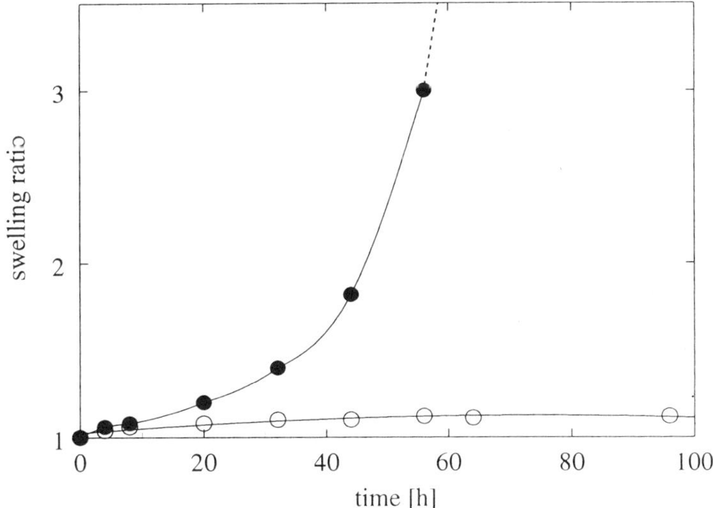

Figure 5. Swelling of the PHEG-acrylamide gel containing a covalently bound papain during its incubation in the presence (●) and absence (○) of enzyme activators. (Adapted from ref. 26. Copyright 1995 Acad. Sci. Czech Rep.)

Ester-Amide Copolymers

Peptide bonds in poly(α-amino acid)s are efficiently degradable by peptidases and proteinases. Their degradation in the organism solely by chemical hydrolysis is too slow to be taken as a significant factor in designing biodegradable medical polymers. In contrast, aliphatic polyesters based on hydroxyalkanoic acids, such as poly(lactic acid), poly(glycolic acid), poly(ε-caprolactone) and related copolymers, are readily degradable by a random hydrolytic mechanism. On the other hand, the absence of functionalized pendant groups in polyesters limits their capacity for a chemical modification. This could be a drawback in applications where either polar groups or functionalized side chains are desirable, e.g. binding of biologically active substances. The functionalized side chains can be provided by L-α-amino acid (AA) residues introduced into polyester molecules. In addition, the presence of peptide bonds in the polymer backbone, can modify degradation patterns of polymers, making them susceptible to peptidases (32).

Copolymers containing α-amino-acid and hydroxy-acid units have been described and prepared by polymerization of cyclodepsipeptides. The first successful copolymerization was performed with p-dioxanone and unsubstituted and alkyl-substituted 2,5-morpholinedione (27). Alternating and random copolymers containing lactic acid or ε-caprolactone in combination with various amino acids, including functionalized ones, were described by Feijen's group (28-30). The effect of the amino acid component in the ester-amide copolymers on the polymer properties and their biodegradability was evaluated (31-33). Here, we present another approach, based on direct ring-opening copolymerization of α-amino acid *N*-carboxyanhydrides (NCA) and lactones.

It is well known that the ring opening polymerization of lactide can be initiated by protic nucleophiles such as alcohols or water. In our previous experiments, the Sn(II) octanoate catalyzed polymerization of lactide was initiated by both alcohols and amines, which could control the molecular weight of the resulting PLA. Under well controlled conditions, the number- average polymerization degree of the resulting PLA corresponded well with the monomer/initiator (nucleophile) ratio and the $\overline{M}_w/\overline{M}_n$ ratio was within the range of 1.1-1.3. The NMR data showed that both types of initiating nucleophiles were incorporated in the polymer chain in a molar ratio well corresponding to the polymerization degree.

Similarly, the ring-opening polymerization of α-amino acid *N*-carboxyanhydrides, when initiated by protic nucleophiles, such as primary amines or alcohols, propagates through a nucleophilic end group (amine or carbamate) (3). Consequently, when both types of monomers, i.e. NCA and lactone, are present, the amine end group of the terminal amino acid residue may attack either another NCA molecule or the lactone monomer, thus giving rise to a peptide or an amide linkage. On the other hand, NCA is reactive enough to react with the nucleophilic hydroxy group of the terminal lactic acid residue as well, thus forming an ester bond while exposing the amine end group of the added amino acid residue for the next propagation step. The objectives of this study were to investigate the feasibility of copolymerization of NCAs and lactones and to evaluate (i) the effect of NCA content in the reaction mixture on properties of the copolymers, (ii) the possibility of controlling the molecular weights of the copolymers, and (iii) the effect of amino acid structure on the copolymerization.

Three types of α-amino acids differing in the structure of side chains were used with emphasis on fuctionalizable amino acids. The side chain of both γ-benzyl L-glutamate and β-benzyl L-aspartate contain an ester bond, while phenylalanine, although having similar steric requirements for the aromatic ring, is without a side-chain ester. All three types of amino acid NCAs formed copolymers with both the lactones studied, i.e., lactide and ε-caprolactone. The composition of copolymers was determined by NMR spectra. The ^{1}H NMR and ^{13}C NMR spectra revealed that both types of structural units and both types of bonds, ester and amide, were present in the copolymers. The combination of NMR results with GPC data revealed that the samples were not merely mixtures of homopolymers and that the amino acid residues were incorporated in the copolymer. If decanol was used as an initiator, its incorporation to one end of the copolymer chain could be detected by NMR spectra. Both types of structures, lactic acid and amino acid, were found as terminal units. The ^{1}H NMR spectrum of the copolymer obtained by copolymerizing lactide and γ-benzyl L-glutamate NCA, using decanol, as an initiator, is presented in Figure 6.

With all amino acids, the increasing the concentration of NCA in the reaction mixture increased proportionally the mole fraction of amino acid residues in the copolymer. In addition, with glutamate and aspartate, the increasing content of NCA in the reaction feed decreased the yield of the copolymerization and the average molecular weight obtained. The decrease in molecular weight was also accompanied by an increase in polydispersity (Table I).

Table I. Copolymerization of L-lactide with BzlGlu-NCA; the effect of NCA content in the feed on molecular parameters of copolymers [a]

Bzl Glu-NCA [b] mol %	*y [c] mol%*	*yield %*	*$\overline{M}_n$ [d] (theor)*	*$\overline{M}_n$ (GPC)*	*$\overline{M}_w/\overline{M}_n$ (GPC)*
0	0	96	7060	6440	1.26
8	8	92	7040	4000	1.37
15.5	15.1	88	6980	3800	1.60

[a] Monomers/decanol = 50 (mol/mol); dioxane, 60 °C, 72 h; [b] Bzl LGlu-NCA in the feed; [c] BzlGlu residues in the copolymer (from NMR); [d] calculated from the feed composition and the copolymer yield..

The effect of the side chain structure of amino acids on the course of copolymerization was evaluated by comparing the copolymerization of aspartate and glutamate derivatives with phenylalanine, an aromatic amino acid with similar steric requirements, however, without the side chain ester bond (Table II).

Table II. The effect of amino acid structure and the monomer-to-initiator ratio (M/I) on the properties of copolymers [a]

Amino acid	*M/I* [b]	*Yield (%)*	*y (mol %)*	$\overline{M}_n$ [c] *(theor)*	$\overline{M}_n$ *(GPC)*	$\overline{M}_w/\overline{M}_n$ *(GPC)*
BzlGlu	50	88	15.1	6990	3140	1.60
	100	86.3	16.2	13600	4670	1.36
BzlAsp	50	86.4	14.2	6780	4510	1.13
	100	85.6	12.7	13300	7230	1.12
Phe	50	85	14.6	6270	7630	1.10
	100	83	12.8	12120	13480	1.11

[a] 16 mol% of NCA in the polymerization mixture ; dioxane, 60 °C, 72 h; [b] mole ratio of monomers to initiator (decanol); [c] calculated from the feed composition and the yield of copolymer.

The lower molecular weight and higher polydispersity of copolymers containing glutamate and/or aspartate esters may indicate possible side-chain reactions involving the pendant ester groups. It is worth noticing that the copolymers with phenylalanine, containing an inert side chain, were obtained with molecular parameters very closely approaching those calculated from the composition of the feed, with the molecular-weight distribution well controlled by the M/I ratio.

Figure 7 illustrates the feasibility of controlling the molecular-weight in copolymers by using decanol as an "initiator" on the example of copolymers of L-lactide and L-phenylalanine NCA. The molecular weight of the copolymers increased with the increasing ratio monomers/initiator (decanol), the highest $\overline{M}_w$ being achieved in the absence of the initiator. In the presence of initiator, copolymers with a narrow molecular-weight distribution were obtained, while in its absence, the polydispersity of resulting copolymers was much higher. The broadening of the molecular-weight distribution indicated an uncontrolled polymerization process, in which new polymer chains are initiated during the course of polymerization by side reaction products.

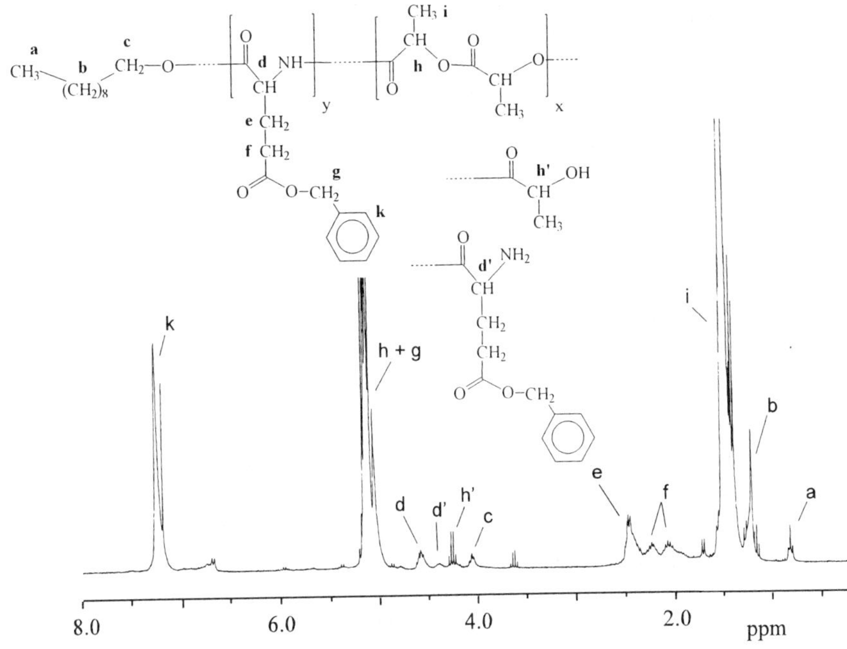

Figure 6. ^1H NMR spectrum of the lactic acid/γ-benzyl L-glutamate copolymer (15 mol % of γ-benzyl L-glutamate residues).

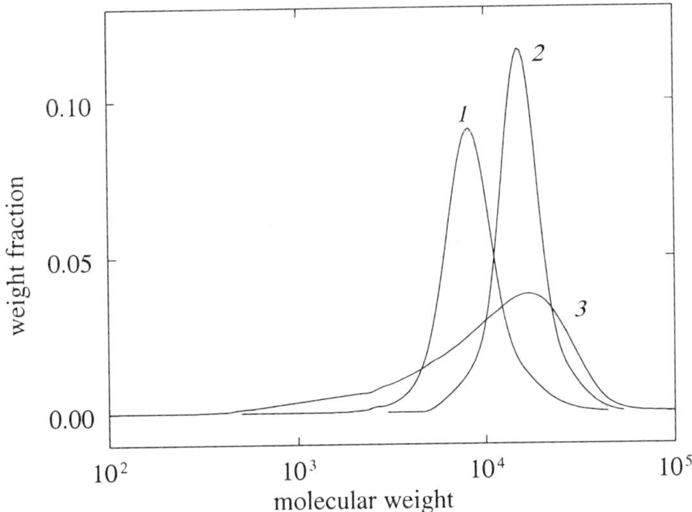

Figure 7. MWD of the lactide/phenylalanine copolymers (15 mol % Phe.NCA in the feed) prepared with different monomer-to-initiator (decanol) ratios: 1 - M/I = 50; 2 - M/I = 100; 3 - no initiator added.

Controlled Living Polymerization of *N*-Carboxyanhydrides

Although the synthesis of NCA and its polymerization, first described by Leuchs already in 1906 (34), is well known and investigated for last four decades, neither of the common polymerization mechanisms produced polymers with narrow molecular-weight distribution (see Ref. (3) for a thorough review on NCA chemistry). An increasing interest in well defined block copolymers and their supramolecular structures boosted recent efforts in developing a controlled process for "living" polymerization of NCA. Two such methods of polymerization of NCAs have been developed recently, reporting on poly(amino acid)s with narrow molecular-weight distribution and making it possible to prepare well defined block copolymers. The recent paper by Deming et al. (35) reports on successful use of transition metal-amine initiators for preparation of well-defined poly(γ -benzyl L-glutamate). Also, through a controlled polymerization in steps, block copolymers can be prepared (36).

In our laboratory we have developed a method for controlled living polymerization of NCAs, that uses *N*-acyl-NCA derivatives as co-initiators with tertiary or secondary amines as base catalysts. The N-acyl-NCA co-initiator provides for a defined concentration of growing centers for polymer chains (37).

In the activated-monomer mechanism, which applies when the NCA polymerization is initiated by a base, that cannot react as a nucleophile, such as a tertiary amine, the growing polymer chain carries *N*-acylated terminal NCA-group. It is well known that, the propagation reaction, i.e. that of *N*-acyl- NCA terminal group with NCA anion (activated monomer) is much faster than the initiation reaction, i.e. that of the NCA-anion with the other monomer molecule (3,38). Consequently, in a typical polymerization, a high-molecular-weight polymer is formed even at low monomer conversions while new chains are still being initiated.

The introduction of *N*-acylated NCA at the start of the reaction provides for a defined concentration of initiating species from which polymer chains start to grow. The growing polymer chains have one acyl-protected end and a "living" end-group formed by the NCA ring N-acylated with the polymer chain (Figure 8). Thus the propagation is limited to only one type of reaction. The polymerization degree could be controlled by the amount of N-acyl-NCA initiator added. We have investigated the feasibility of preparing polyglutamates with a controlled molecular-weight distribution by using this concept.

The persistence of the "living"polymer-chain end, i.e. polymer-acylated NCA, could be confirmed after reacting the polymerization mixture with an excess of *tert*-butylamine and NMR analysis. The living character of polymerization was additionally supported by the dependence of degree of polymerization on the monomer conversion (39). The number- average molecular weight of the polymers obtained from polymerizations carried out to full conversion of NCA corresponds very closely to that calculated from the monomer/co-initiator ratio and the polydispersity ($\overline{M}_w/\overline{M}_n$) is within a range of 1.1 - 1.3. (Table III).

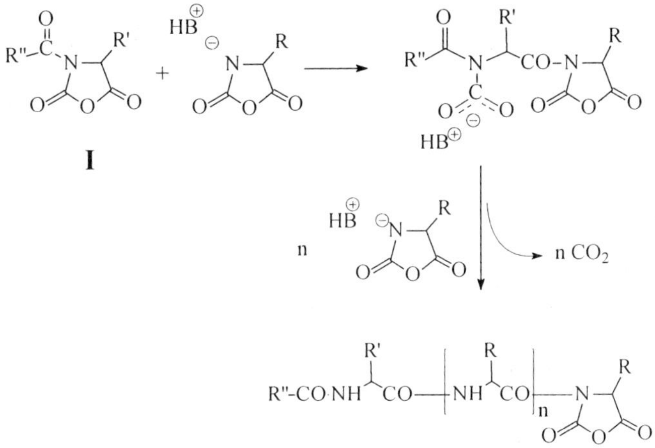

Figure 8. Propagation of the polypeptide chain on the N-acyl-NCA initiator unit.

Table III. Polymerization of BLG-NCA (M) in the presence of *N*-acetylglycine NCA (I) as a co-onitiator

$[I]_0$ (10^{-3}mol/l)	Conv. [a] (%)	$[M]_0/[I]_0$	$\overline{M}_n$ (theor) [b] (g/mol)	$\overline{M}_n$ (exp) [c] (g/mol)	$\overline{M}_w/\overline{M}_n$
1.75	100	20 [d]	4380	4700	1.16
0.7	100	50	10950	10500	1.16
0.52	100	67	14655	15000	1.12
0.35	31	100	6790	8600	1.20
0.35	52	100	11300	11600	1.14
0.35	72	100	15860	17500	1.13
0.35	100	100 [e]	21900	21300	1.23

[a] NCA conversion.; [b] calculated from the M/I ratio for the given monomer conversion; [c] determined by SEC in the isolated polymer sample; [d, e] the ratios of glutamic acid to glycin residues found in the resulting polymer by the amino acid analysis was as follows: [d] Glu/Gly = 25 [e] Glu/Gly = 95.

Conclusions

A rational synthesis of poly(amino acid)s makes it possible to design polymer materials with required pattern of enzymatic biodegradation. Structural variations controlling the biodegradability of poly(amino acid)s can be produced either by copolymerization or via polymer modification reactions.

Poly(amino acid)s can provide for water-soluble polymers, that expose functional side-chains to which compounds of interest can be bond and that can be used as biodegradable drug-carriers.

Biodegradable crosslinked hydrogels can be prepared from hydrophilic poly(amino acid) derivatives.

The copolymers containing ester and peptide bonds in the main chain can be prepared by the copolymerization of lactones with *N*-carboxyanhydrides of α-amino acids. The incorporation of α-amino acid residues provides for linear biodegradable polyesters with functionalized side chains.

Due to their structural versatility and biodegradability, materials based on synthetic poly(amino acid) derivatives are well suited for various medical and pharmaceutical applications that require polymers to be tailor-made for certain particular purposes.

Acknowledgments

The presented work was partly supported by the grants from the Grant Agency CR (No: 203/98/0882) and the Grant Agency of Academy of Sciences CR (No: A4050503).

References

1. Fox, S. W.; Harada, K. *A Laboratory Manual of Analytical Methods of Protein Chemistry, Including Polypeptides*; Pergamon Press: Elmsford, N.Y., 1966.
2. Paik, Y. H.; Swift, G.; Simon, E. S. Polysuccinimide Polymers and their Preparation. **1994,** *EP 578451.*
3. Kricheldorf, H. R. *Alpha-Aminoacid-N-Carboxy-Anhydrides and Related Heterocycles*; Springer Verlag: Berlin, Heidelberg, New York, 1987.
4. Nakato, T.; Yoshitake, M.; Matsubara, K.; Tomida, M.; Kakuchi, T. *Macromolecules* **1998,** *31*, 2107.
5. Roweton, S.; Huang, S. J.; Swift, G. *J. Environ. Polym. Degrad.* **1997,** *5*, 175.
6. Swift, G. *Polym. News* **1994,** *19*, 102.
7. Daly, W. H.; Poché, D. *Tetrahedron Lett.* **1988,** *29* , 5859.
8. Nosková, D.; Kotva, R.; Rypáček, F. *Polymer* **1988,** *29*, 2072.
9. Pytela, J.; Rypáček, F.; Metalová, M.; Drobník, J. *Collect. Czech. Chem. Commun.* **1989,** *54*, 1640.

10. Drobník, J.; Rypáček, F. *Adv. Polym. Sci.* **1984,** *57*, 1.
11. Hayashi, T. *Prog. Polym. Sci.* **1994,** *19*, 663.
12. Hayashi, T.; Tabata, Y.; Nakajima, A. *Prog. Polym. Phys. ,Japan* **1983,** *26*, 591.
13. Yaacobi, Y.; Sideman, S.; Lotan, N. *Life Support Syst.* **1985,** *3*, 313.
14. Antoni, G.; Presentini, R.; Neri, P. *Il Farmaco* **1980,** *35*, 575.
15. Pytela, J.; Saudek, V.; Drobník, J.; Rypáček, F. *J. Controlled Release* **1989,** *10*, 17.
16. Hayashi, T.; Tabata, Y.; Nakajima, A. *Polym. J.* **1985,** *17*, 463.
17. Kondo, H.; Nakatami, H.; Hiromi, K.; Matsuno, R. *J. Biochem.* **1978,** *84*, 403.
18. Pytela, J.; Jakeš, J.; Rypáček, F. *Int. J. Biol. Macromol.* **1994,** *16*, 15.
19. Pytela, J.; Kotva, R.; Rypáček, F. *J. Bioactiv. Compatible Polym.* **1997,** *13*, 198.
20. Rypáček, F. *Polym. Degrad. Stabil.* **1998,** *59*, 345.
21. Rypáček, F.; Pytela, J.; Kotva, R.; Škarda, V. *Macromol. Symp.* **1997,** *123*, 9.
22. Ehrenreich, B. A.; Cohn, Z. A. *J. Exp. Med.* **1969,** *129*, 227.
23. Chiu, H. C.; Kopečková, P.; Deshmane, S. S.; Kopeček, J. *J. Biomed. Mater. Res.* **1997,** *34*, 381.
24. Škarda, V.; Rypáček, F.; Ilavský, M. *J. Bioactiv. Compatible Polym.* **1993,** *8*, 24.
25. Škarda, V.; Rypáček, F. *J. Bioactiv. Compatible Polym.* **1997,** *12*, 186.
26. Rypáček, F.; Škarda, V. *Collect. Czech. Chem. Commun.* **1995,** *60*, 1986.
27. Shakaby, S. W.; Koelmel, D. F. **1983,** *Eur.Pat.Appl. 86,613.*
28. Helder, J.; Kohn, F. E.; Sato, S.; Van den Berg, J. W. A.; Feijen, J. *Makromol. Chem. ,Rapid Commun.* **1985,** *6*, 9.
29. In't Veld, P. J. A.; Dijkstra, P. J.; Feijen, J. *Makromol. Chem.* **1992,** *193*, 2713.
30. In't Veld, P. J. A.; Dijkstra, P. J.; van Lochum, J. H.; Feijen, J. *Makromol. Chem.* **1990,** *191*, 1813.
31. Helder, J.; Dijkstra, P. J.; Feijen, J. *J. Biomed. Mater. Res.* **1990,** *24*, 1005.
32. Ouchi, T.; Nozaki, T.; Ishikawa, A.; Fujimoto, I.; Ohya, Y. *J. Polym. Sci. Part A:Polym. Chem.* : **1997,** *35*, 377.
33. Ouchi, T.; Nozaki, T.; Okamoto, Y., Shiratani, M.; Ohya, Y *Macromol. Chem. Phys.* **1996,** *197*, 1823.
34. Leuchs, H. *Ber. Dtsch. Chem. Ges.* **1906,** *39*, 857.
35. Deming, T. J. *J. Amer. Chem. Soc.* **1997,** *119*, 2759.
36. Deming, T. J. *Nature* **1997,** *390*, 386.
37. Dvořák, M.; Rypáček, F. *Polycondesation `96*: Proceedings, Intl. Symposium on Polycondensation, Paris, Sept. 23-26, **1996,** 532.
38. Bamford, C. H.; Block, H. *J. Chem. Soc.* **1961,** 4989.
39. Dvořák, M.; Dybal, J.; Rypáček, F. *Macromolecules* **1998,** *in press.*

Chapter 17

Surface-Modified Polymers and Polymers with Strong Heparin-Like Activity Based on Vinyl Sugars

Günter Wulff, Susanne Bellmann, Achim Brock, Holger Schmidt, Stefanie Stalberg, and Liming Zhu

Institute of Organic Chemistry and Macromolecular Chemistry, Heinrich-Heine-University, Universitätsstrasse 1, D-40225 Düsseldorf, Germany

The application of polymerizable vinyl sugars for the preparation of speciality polymers was investigated. In one research project hydrophilic surfaces are generated on standard polymers after copolymerization with low amounts of protected vinyl sugars. Copolymerization of several isopropylidene-protected vinyl sugars with styrene, methyl methacrylate, and acrylonitrile were carried out in solution and in bulk. Copolymers with varying composition were obtained. After eliminating the hydrophobic protecting groups from the copolymer surfaces by acid hydrolysis, these surfaces become hydrophilic owing to newly formed hydroxyl groups which could be demonstrated by water contact angle measurement and surface conductivity. These polymers show improved biocompatibility.
For various applications blood-compatible polymers are desirable with the aim to prevent blood clotting. Instead of heparin synthetic vinyl sugar polymers can be used.
Copolymers based on sulfated vinyl sugar monomers are promising candidates for reaching this goal because of their strong anticoagulant activities. In addition due to the stable C-C bonds along the polymer backbone the long term stability is much higher.

Some years ago we have started a systematic investigation on the use of renewable resources for the preparation of speciality polymers. In one of our activities we wanted to modify the surfaces of bulk polymers for biomedical applications in such a manner that the surface becomes biocompatible or is even showing heparin-like activity. Polymers of this type can then be used for the preparation of blood bags, tubes, catheters or prostheses.

We wanted to modify the polymer surfaces by different polyvinyl sugars which can be attached to the surface of the polymer by rather stable covalent bonds and thus the stability of these layers should be considerably increased.

A large variety of polymerizable vinyl sugars has been prepared by different groups. This work has been reviewed in detail (*1,2*). In most vinyl sugar monomers the sugar unit is linked to a polymerizable double bond by ester, amide, ether, or glycoside groups (see Scheme 1). However, these groups are susceptible towards cleavage under certain conditions, e.g. during the presence of enzymes or when polymer-analogous reactions are performed (*3*). Therefore, investigations on new polymers based on natural sugars have been undertaken (*2,3*).

Scheme 1. Vinyl sugars and polymers prepared thereof

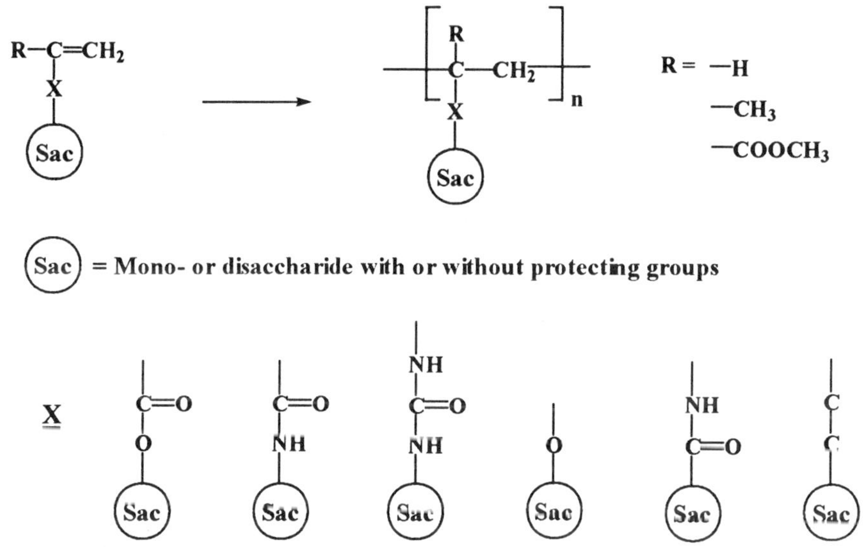

R—C=CH$_2$ (X–Sac) → [C(R)(X–Sac)—CH$_2$]$_n$

R = —H, —CH$_3$, —COOCH$_3$

Sac = Mono- or disaccharide with or without protecting groups

X = C=O / O ; C=O / NH ; NH / C=O / NH ; O ; NH / C=O ; C / C

We have prepared (*3*) a large number of new polymerizable vinyl sugars which can be transformed to polymers with resistance to chemicals, heat, and micro-organisms. Vinyl sugar derivatives carrying hydrophobic protecting groups were usually the starting material. These monomers can be polymerized as such or after deprotection. The new type of polymer should fulfill the prerequisites:

(i) Main chains and side chains should be entirely composed of C-C-connections to avoid any weak bond for chemical and biochemical decomposition.

(ii) Side chains should not be branched, in order to achieve a comb-like arrangement.

(iii) Monomers should be polymerizable by radical initiation and also anionically. Anionic polymerization would enable the generation of polymers with defined tacticity (isotactic, syndiotactic) as well as the preparation of blockcopolymers.

Olefinic double bonds directly connected to sugars by C-C bonds cannot usually be polymerized radically or anionically, but have to be activated in a suitable way. We have worked out procedures to prepare vinyl ketones **1** and 4-vinyl phenyl derivatives **2** from sugar aldehydes (*3-6*).

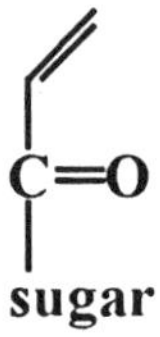

C=O

sugar

sugar

1 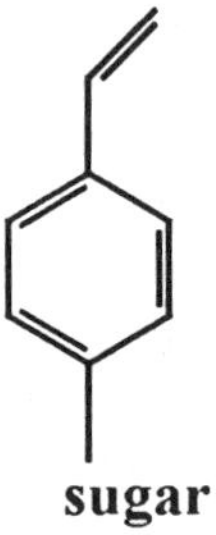 **2**

Another type of vinyl sugars with a C-C linkage based on barbituric acid, was described as well (*7,8*).

The new type of monomers reported in this paper has a higher chemical stability and yields polymers with only C-C bonds in the backbone as well as in the side chain.

Vinyl sugar monomers can easily be polymerized or copolymerized with various comonomers which may achieve defined composition and structure to fulfill the requirements of specific applications (*2,3*). Poly(vinyl saccharide)s, which consist of a chemically stable C-C backbone and hydrophilic sugar side chains, should possess a high potential in surface modification of hydrophobic common polymeric materials. Some of the most widely used synthetic polymers, such as polyethylene (PE), polypropylene (PP), poly(ethylene terephthalate) (PET) and polystyrene (PS), have hydrophobic surface and poor chemical activity due to the absence of polar functional groups. Many problems are caused in commercial applications of these polymers, e.g. the occurance of static electric discharges on the surface, poor dyeability, adhesion and printability. Especially the biocompatibility is rather low.

Considerable work has been carried out by many research groups to modify polymer surfaces by chemical means. For improving the hydrophilicity and dyeability, Manásek et al. used a number of copolymers of acrylate, methacrylate, ethylene, or vinyl ester as modifiers for polypropylene (*9*), Zhang and Rånby modified PET fibers by a photoinitiated graft copolymerization of acrylic acid or acrylamide onto the surface (*10*), and Höcker et al. grafted 2-hydroxyethyl methacrylate (HEMA) on PU (*11*). However, few research work has been done about using vinyl saccharides to improve biocompatibility, hydrophilicity, and the dyeability of conventional polymers. Therefore, saccharides may impart more functions and hydrophilicity to commercial polymers and amend their antistatic property, dyeability, adhesion, printability, and biocompatibility. Nakamae et al. copolymerized 2-(glycosyloxy)ethyl methacrylate with methyl methacrylate or styrene, and investigated the surface properties and protein adsorption of the copolymer films (*12,13*). We have investigated the surface characteristics of copolymers of styrene, methyl methacrylate, or acrylonitrile with the vinyl sugar monomers **3**, **4**, **5**, and **6** by contact angle measurement, XPS, and electrical surface resistance measurement (*14-16*). The use of a hydrophobic sugar monomer carrying isopropylidene protecting groups enhances the compatibility of these units in the bulk of the hydrophobic polymer. Hydrolysis removes the hydrophobic isopropylidene protecting groups from the polymer surface of the copolymers with the result that the surface showed hydrophilic properties and a decrease in electrical surface resistance. Furthermore, to the surface other compounds like dyes or heparin-like substances can be attached by covalent bonds.

Surface Modified Standard Polymers

Surface modified polymers from technically interesting monomers such as styrene, methyl methacrylate, and acrylonitrile were prepared. The vinyl sugar monomers **3**, **4**, **5**, and **6** were used to investigate their potential for modifying polymer surfaces. Copolymers of all of these combinations were prepared. Different copolymerization methods were used for different monomer combinations.

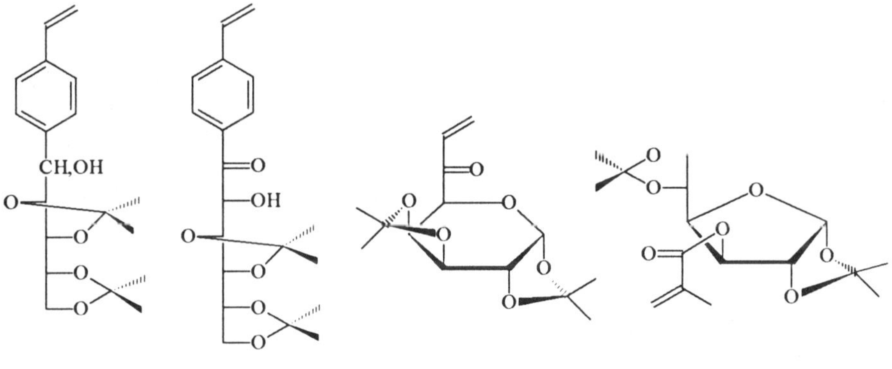

3 (*3*) **4** (*3*) **5** (*3*) **6** (*17*)

Bulk polymerization was used in the case of styrene and vinyl sugar monomers which delivered copolymers in high yield, good purity, and relatively high molecular weight. Because of the better control of the reaction rates copolymers of the vinyl sugar monomers **1** - **4** with MMA or AN were synthesized in toluene or DMSO solution. The copolymers of different composition were investigated in the form of powders, films or thin plates.

The isopropylidene residues from the sugar monomers at the surface of the polymer can be removed with 1 N H_2SO_4 at 80°C (see Fig. 1).

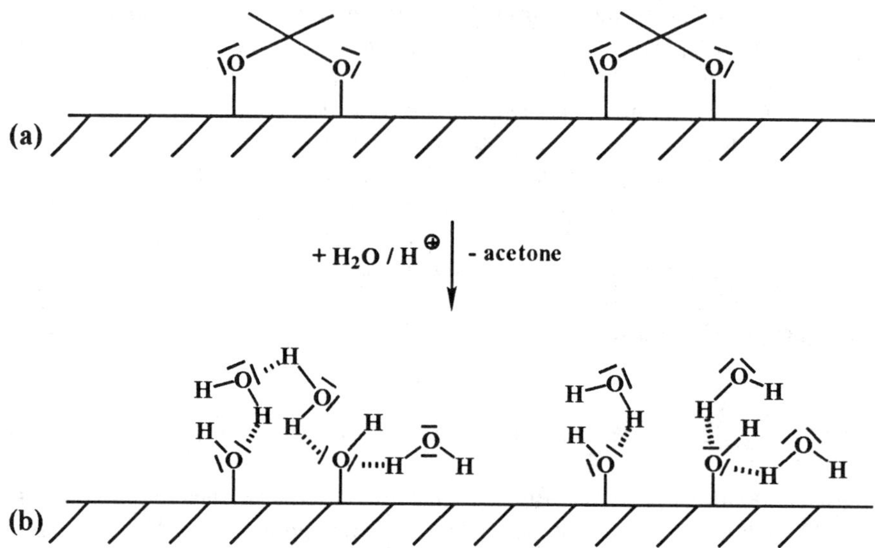

Figure 1 (a) Copolymer with a poly vinylisopropylidene sugar.
(b) The same copolymer after surface hydrolysis.

Fig. 2 shows contact angle measurements at sheets of polystyrene containing 2 mol% of **3** after different times of exposure to acid. After 1-2 min. most of the isopropylidene groups are removed (see Fig. 2) producing surfaces with OH-groups.

Fig. 3 shows styrene polymers with different content of sugar monomer **4**. It is evident that 2 mol% vinyl sugar is sufficient for a strong effect. Contact angle measurements show the increase of polarity of the surface since a water droplet is spreading much flatter on the surface.

Measurements of the electrical surface resistance gave detailed information about the antistatic properties of the modified polymers. Surface resistance is strongly reduced in polymers in which hydroxyl groups have been generated by surface hydrolysis. The behavior can best be explained by a four-phase model (Figure 4) of different layers at the surface of the polymer. The conducting of electric charge is supposed to take place mainly in the adsorbed water layer (B) and in the swelling layer (C). Any factors which build up these two layers are of influence on the surface conductivity.

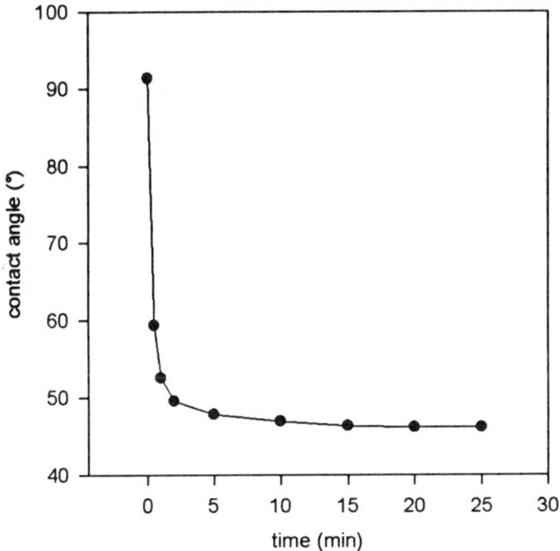

*Figure 2. Change of contact angle with time. Surface hydrolysis of a copolymer of styrene with 2 mol% **3** with 1 N H_2SO_4 at 80°C.*

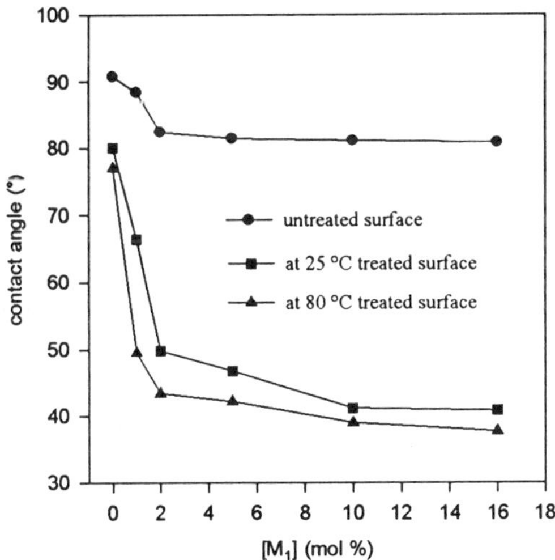

Figure 3. *Contact angle measurement of styrene-**4** copolymers with increasing content of **4**. Surface treatment as in Figure 2.*

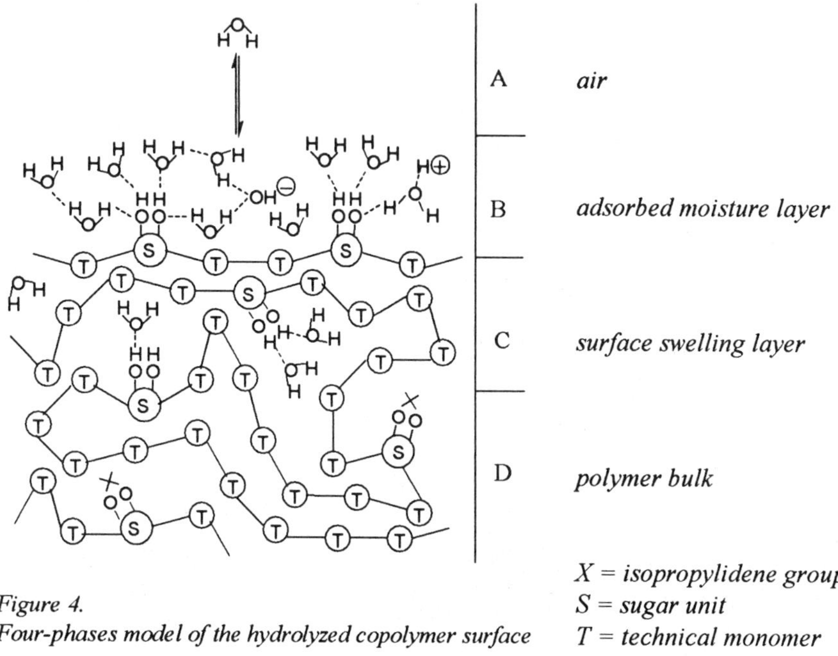

Figure 4.
Four-phases model of the hydrolyzed copolymer surface

X = isopropylidene group
S = sugar unit
T = technical monomer unit

An open-chain sugar has a higher ability to expose hydroxyl groups on the polymer surface. Consequently water adsorption occurs in a higher quantity. For this reason, vinyl sugars **3** and **4** are more efficient than **5**. Figure 5 shows the effect of a 5% vinyl sugar content in copolymers on the surface resistance. For generating an enlarged surface by a swollen layer a cosolvent added during acid treatment is of great importance. The cosolvent must show two properties: it has to be miscible with the aqueous acid, and it has to be a solvent for the polymer. Under these conditions surface penetration by solvent molecules occurs. In our case THF proved to be favorable. It was added at 5% to the hydrolysis mixture. Water, protons, and other ions follow the cosolvent into the swelling layer. A temperature for hydrolysis on the surface sligtly below the boiling point of the cosolvent supports this process.

The exposition of sugar side chains on the surface is forced by the minimization of the surface free energy in contact with a polar medium. In nearly all cases the surface conductivity increased in copolymers with hydrolyzed surfaces. The results of surface conductivity measurements parallel those of the contact angle measurements. According to the literature, the resistance of polystyrene should be around 10^{18} Ω/cm. Owing to limitations in our measurements no higher resistance than $5 \cdot 10^{16}$ Ω/cm could be determined. The same is true for homopolymers of PMMA. Copolymers of styrene with **3** or **4** and with free hydroxyl groups at the surface have a surface resistance of 10^{10} Ω/cm and are thus already approaching the region of antistatic polymers (*18*).

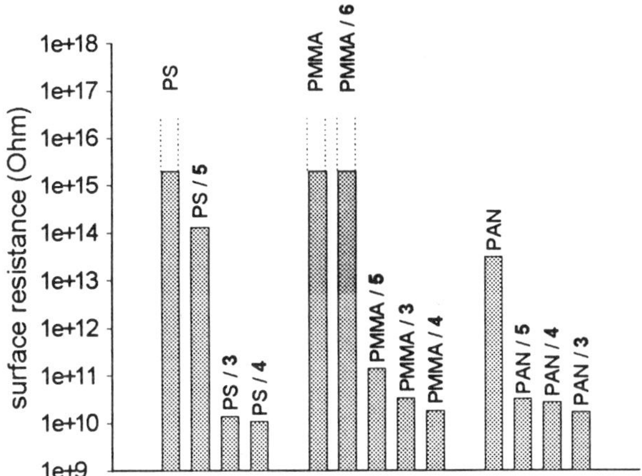

*Figure 5. Surface resistance of copolymers of styrene (PS), methyl methacrylate (PMMA) and acrylonitrile (PAN) with **3**, **4**, **5**, or **6** (5 Mol% of vinyl sugar) after surface hydrolysis. Due to limitations of measurement no higher resistance than $5 \cdot 10^{16}$ Ω/cm could be measured. In case of polystyrene the resistance should be around 10^{18} Ω/cm.*

X-ray photoelectron spectroscopy (XPS) has long been used for characterizing polymer surfaces and the modification of polymer surfaces. Polymers having pendant saccharides were investigated by Nakamae et al. using XPS recently (*12,13*). XPS in our case showed the change of the ratios between typical carbon species and indicated the successful hydrolysis of isopropylidene groups at the surface. A complete removal of the protecting groups was thus proven to be at least in a depth of 10 nm.

If the efficiency of the four comonomers is compared it is evident that monomer **6** gives rather poor results. The reason is that under the condition of hydrolysis of isopropylidene groups at the surface most of the sugar is also hydrolyzed and released from the surface. In contrast, the monomers **3**, **4**, and **5** with C-C connections instead of the ester connection in **6** are very efficient.

The advantage of these new comonomers is that they are also very hydrophobic and similar in copolymerization behavior to the main monomer. Therefore, the properties of the bulk polymer will not be detrimentally influenced. Only at the surface a transformation to a very polar compound is possible. This polar part is covalently attached to the polymer and cannot be removed by diffusion, washing etc..

The reactive hydroxyl groups at the surface of these bulk polymers can be used to attach other compounds. So it is possible to dye these polymers with reactive dyes.

It is also possible by different techniques to attach polysaccharides or hydrophilic polymers to the surface which is of great importance for biomedical applications. In this respect we were very interested to prepare hydrophilic polymers with heparin-like activity and attach these to modified surfaces of bulk polymers.

Vinylsugar Polymers with Heparin-Like Activity

Since the early 1950s several groups have designed blood-compatible polymers and polymer surfaces for use in various medical applications (*19,20*), such as extracorporeal blood-circulating devices, catheters, blood bags and tubes used for blood transfusions, plasma expanders, cardiac valves, aortic bypasses, oxygenators, membranes, hollow fibers, and tubes used for dialysis devices. The development of such materials is very important, because usually blood coagulation is the primary effect which occurs when blood contacts any surface except the natural healthy endothelium wall of the blood vessels. The coagulation cascade is a very complicated biochemical process (*21*).

Under physiological conditions, coagulation of blood is prevented by the action of natural inhibitors, of which the protein antithrombin-III (AT-III) is the most important. AT-III forms inactive stable complexes with the activated coagulation factors. Usually these inhibition reactions are very slow, but in the presence of heparin acceleration occurs. Heparin is a glycosaminoglycuronan consisting of D-glucosamine, D-glucuronic acid, and L-iduronic acid residues that contain *O*- and *N*-sulfate groups to a considerable extent. It is a mixture of polysaccharide chains of variable length, different sulfation pattern, and degree of heterogeneity. It has a strong anticoagulant activity in blood, and it promotes cell adhesion.

In the last fifteen years, great progress has been made in the investigation of the exact mechanism of the catalyzed reaction and the chemical nature of the catalytic sites in natural heparin that bind to AT-III. In the early 1980s a unique pentasaccharide was discovered and characterized as the minimum sequence in heparin which is necessary for antithrombin binding (*22*) (Fig. 6).

Recently, a Dutch research group (*23*) has started to synthesize such pentasaccharides. A methyl α-glycoside of pentasaccharide 7 was selected for clinical development. Today it is possible to produce several hundred grams of this promising derivative in about 55 steps (*24*).

Heparin is widely used to prevent blood clotting during surgical treatments. However, reducing the level of systemic heparinization is desirable because heparin increases postoperative bleeding and requirements for homologous blood transfusion. The use of heparin-coated medical equipment allows systemic heparinization to be significantly decreased.

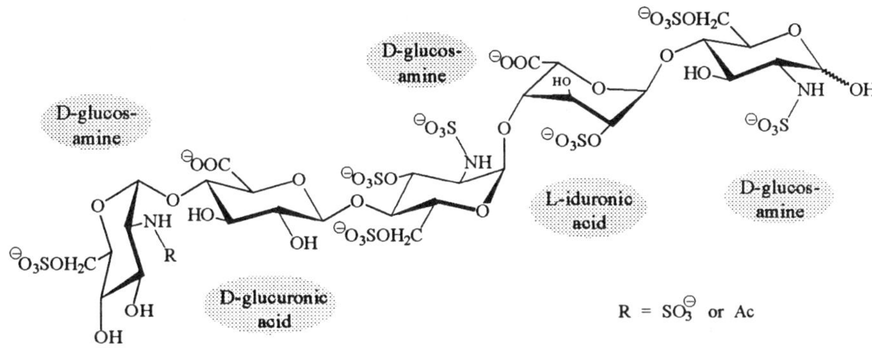

Figure 6. Pentasaccharide 7 fragment with high affinity for antithrombin III

Heparin can be immobilized on polymer surfaces either by electrostatical interactions (anticoagulant-drugs-releasing materials) or by covalent binding (*25*).

The preparation of heparin-coated materials is very expensive because of the cumbersome isolation of heparin from pig intestinal mucosa (100 mg/kg). Besides, the surface-bound heparin is susceptible to degradation by heparinases in the blood stream. For this reason even blood-compatible polymers with covalent bound heparin are unsuitable for long-term applications, such as artificial organs. Several heparin-like materials which may be good alternatives to heparin-coated polymers have been therefore developed since 1951.

Out of these functionalized dextrans with carboxylate groups, benzylamide structures, and carboxymethyl benzylamide sulfonate residues are blood-compatible polymers. Their activities are dependent on the distribution of the different functional groups along the polymer backbones (*26*).

Finally, the selective sulfation of chitin derivatives (*27*) and the preparation of sulfated hyaluronic acids (*28*) are described in the literature as heparin-like substances. The anticoagulant activity increases by increasing the sulfation degree in the case of the hyaluronic acids.

In our group the synthesis of heparin-like substances is approached in a different way. We became interested in polymers with heparin-like activity that are at the same time chemically and mechanically highly stable. In contrast to natural heparin or pentasaccharide 7, both of which are expensive drugs that are administered directly to the human body, our target polymers are only meant to cover the surface of other polymeric materials. For this, we intended to use the vinyl sugars mentioned in the first part. Our investigation aimed at the preparation of a polymer with a very stable C-C backbone and with sugar moieties attached to it also by C-C-bonds. The sugar moieties should roughly have structural similarities to heparin, i.e. contain sugars with *O*-sulfates, *N*-sulfates, and carboxyl groups.

Monomers such as **3**, **4**, and **5** are prepared from isopropylidene-protected sugar derivatives containing an aldehyde group. Reaction with a Grignard reagent, such as 4-vinylphenyl magnesium bromide, then produces, e.g., **3**.

A cheaper possibility for obtaining vinyl sugars makes use of the corresponding amides, whereby the starting materials can be prepared in a one-pot reaction: treatment of D-gluconolactone **8** with diethylamine, removal of excess of amine in vacuo, and trituration of the residue with acetone/sulfuric acid generates the diisopropylidene amides (**9a**, **9b**) in 80% yield. In a second step, **9a**, **9b** is converted to **4** (as regioisomeric mixture) in high yield with 4-vinylphenyl magnesium bromide. Alternatively, acidic work-up provides the unprotected sugar monomer **10** in high yield (see Fig. 7). Compounds like **10** exist as mixture of furanoid and pyranoid ring forms. Unprotected sugar monomers, such as **11**, can also be obtained by hydrolysis of **3** with acidic ion exchangers (*29*).

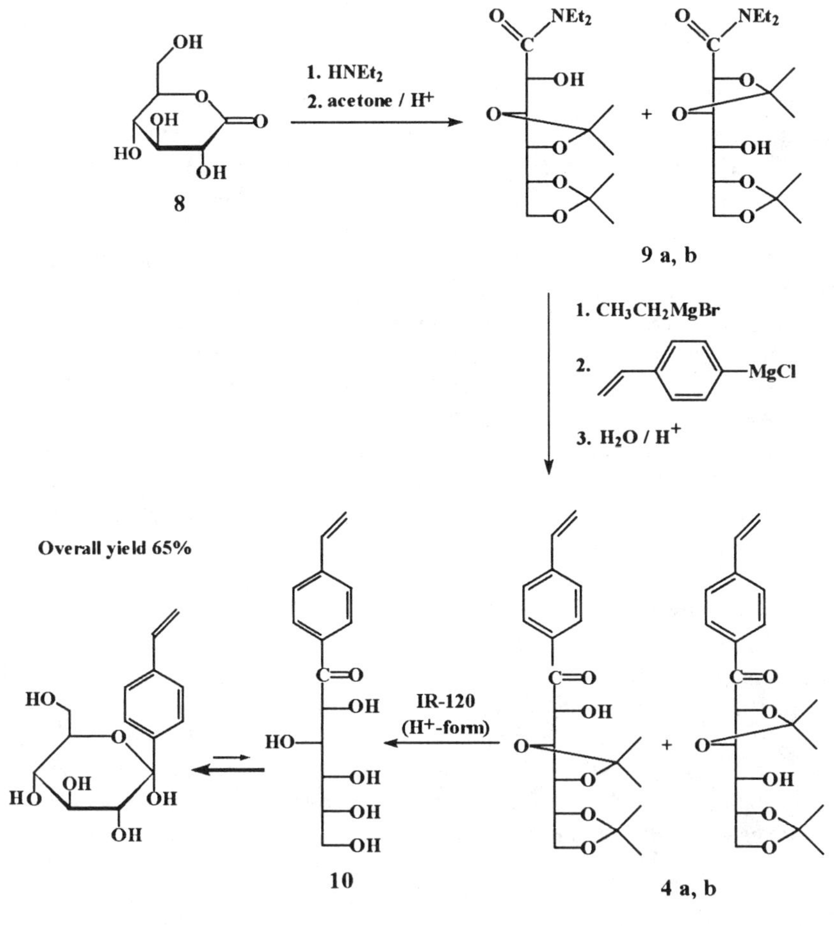

Figure 7. Synthesis of new vinylsugars

Amino sugar monomers are prepared by reductive amination of, e.g., **4** or the oxidized **3**. Again these compounds can be deprotected using ion exchangers to yield the amino sugar monomers **12** or **13**. To obtain *O*-sulfates or *N*-sulfates of different degree of substitution, systematic sulfation with SO_3/pyridine is performed.

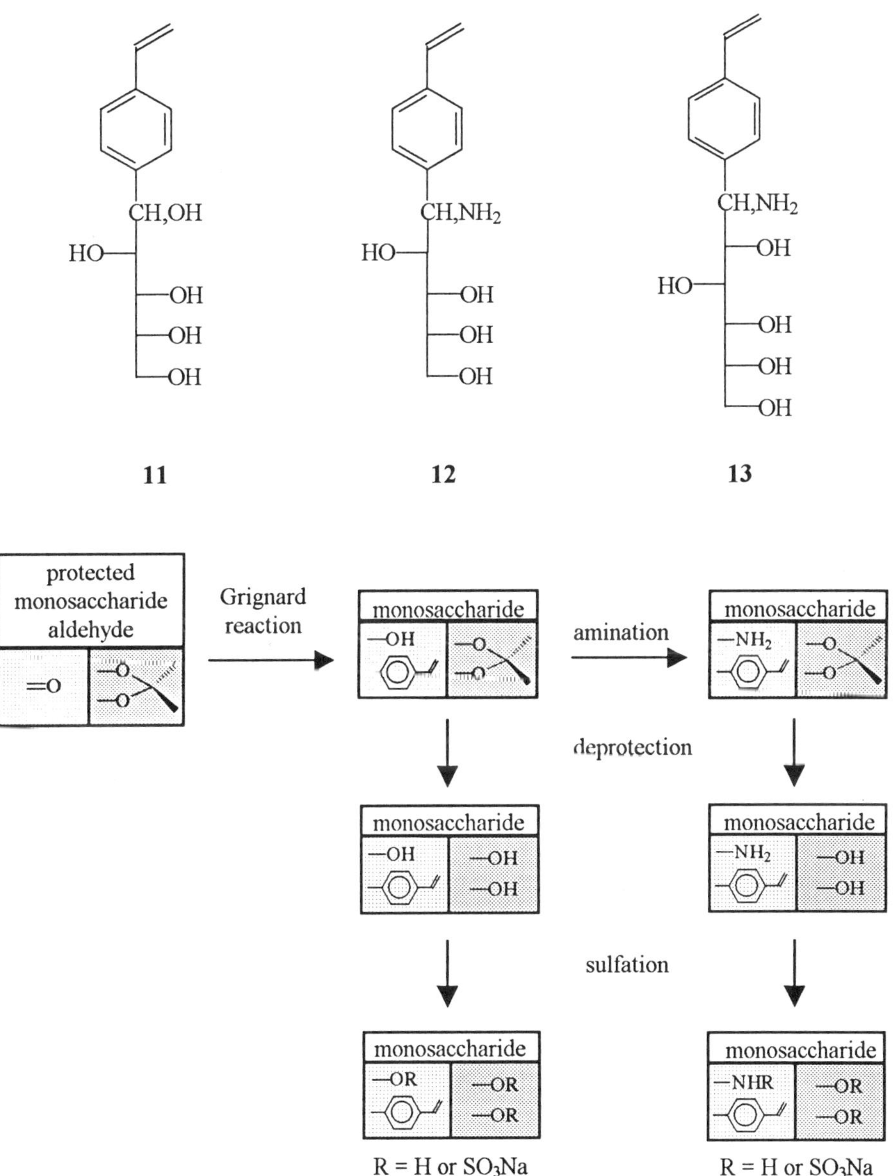

Figure 8. General procedure for the syntheses of sulfated vinyl sugar monomers

Numerous heparin-like copolymers of different vinyl sugar monomers have been prepared by us in a systematic investigation. Like natural heparin, these copolymers should contain normal sugar as well as amino sugar moieties. Furthermore, *N*-sulfates and *O*-sulfates should be introduced in a similar molar ratio as in heparin, together with a certain amount of carboxylic acids to complement heparin analogy. We have recently synthesized a terpolymer that possesses strong heparin-like properties (*30, 31*). It was prepared by radically initiated terpolymerization of sulfated monomer **11**, sulfated monomer **12**, and 4-vinylbenzoic acid. Mole mass was found to be M_n = 85 000. In the monomer the degree of sulfation was DS = 2.9 for **11** and DS = 3.7 for **12**. This means that in **11** 2.9 hydroxyl groups out of 5 are sulfated and in **12** 3.7 hydroxyl groups out of 4 are sulfated. In **12** due to the higher reactivity of amino groups all amino groups are sulfated. The coagulation times of blood (that is partial thromboplastin time, PTT) in the presence of natural heparin or the terpolymer were comparable in both cases (Fig. 9).

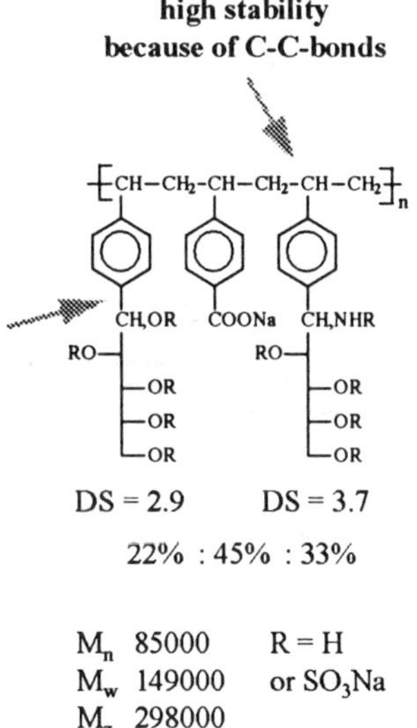

Partial thromboplastin time (PTT) in the presence of terpolymer (TP) or heparin (HP)

concentration of terpolymer or heparin	PTT [sec] (donor 1) HP	TP	PTT [sec] (donor 2) HP	TP
without addition	34.5	33.9	31.9	33.3
0.1	68.5	36.7	68.0	43.3
0.5	248.8	209.3	184.5	128.9
1.0	574.6	>600	387.4	>600
5.0	>600	---	>600	---

Concentration in International Units/mL (6 mg/L).

Figure 9. Comparison of heparin and a terpolymer based on vinyl sugar monomers

This means that a simple and very stable copolymer possesses heparin-like activity compared to heparin itsself. Also cell adhesion is showing a high level. The new type of terpolymer can be attached to polymer surfaces as it is performed with heparin itsself. Another possibility is the attachment of these terpolymers to surface modified polymers as they were described in the first part.

Systematic investigations on structure-property relationships are performed. A large number of different sulfated vinyl sugar monomers are therefore produced. By this method, the influence of chain length, ring form *vs.* open chain, *N*-sulfate *vs.* *O*-sulfate and differences in stereochemistry on the activity of our heparin analogues is investigated (see Fig. 10).

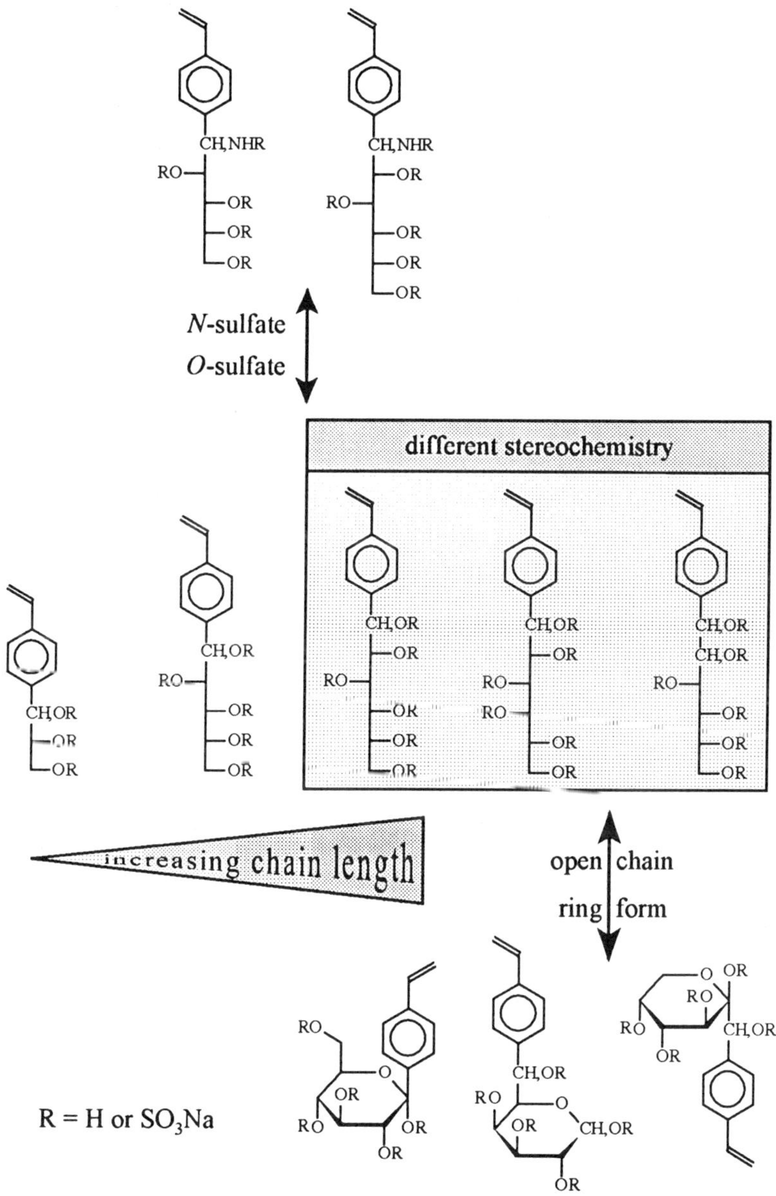

Figure 10. Wide variety of different sulfated vinyl sugar monomers

Acknowledgement

The first part of this work has been supported by *"Fachagentur für nachwachsende Rohstoffe"* (*BML*) (project 93 NR 148-F) and the *"Fonds für Chemische Industrie"*. The second part was supported by *Hüls AG, Marl*.

The biological testing has been performed by *Biomat, Prüflabor BMP, Aachen, Germany*.

References

1. Klein, J. *Makromol. Chem.* **1989**, *190*, 2527.
2. Wulff, G.; Schmid, J.; Venhoff, T. *Macromol. Chem. Phys.* **1996**, *197*, 259.
3. Wulff, G.; Schmid, J.; Venhoff, T. *Macromol. Chem. Phys.* **1996**, *197*, 1285.
4. Wulff, G. in *Nachwachsende Rohstoffe - Perspektiven für die Chemie*, Eggersdorfer, M.; Warvel, S.; Wulff, G., Eds., VCH: Weinheim, 1993, p. 281.
5. Wulff, G.; Schmid, J.; Venhoff, T. in *Carbohydrates as Raw Materials*, Lichtenthaler, F. W. Ed., VCH, Weinheim, 1991, p. 311.
6. Wulff, G.; Schmid, J.; Venhoff, T.; Schwald, D.; Perske, T. in *Nachwachsende Rohstoffe: Polysaccharid-Forschung*, edited in commission of the BMFT, Forschungszentrum Jülich, 1994.
7. Wulff, G.; Clarkson, G. *Carbohydr. Res.* **1994**, *257*, 81.
8. Wulff, G.; Clarkson, G. *Macromol. Chem. Phys.* **1994**, *195*, 2603.
9. Maňásek, Z.; Borsig, E.; Sláma, P.; Romanov, A.; Hrčková, L.; Capuliak, J. *Acta Polym.* **1989,** *40*, 672.
10. Zhang, P. Y.; Rånby, B. *J. Appl. Polym. Sci.* **1990,** *41*, 1459.
11. Anderheiden, D.; Brenner, O.; Klee, D.; Kaufmann, R.; Richter, H. A.; Mittermayer, C.; Höcker, H. *Angew. Makromol. Chem.* **1991**, *185*, 109.
12. Nakamae, K.; Miyata, T.; Ootsuki, N. *Macromol. Chem. Phys.* **1994**, *195*, 2663.
13. Nakamae, K.; Miyata, T.; Ootsuki, N. *Macromol. Chem. Phys.* **1994**, *195*, 1953.
14. Wulff, G.; Zhu, L. M.; Schmidt, H. *Macromolecules* **1997,** *30*, 4533.
15. Wulff, G.; Schmidt, H.; Zhu, L. *Macromol. Chem. Phys.*, in press.
16. Wulff, G.; Bellmann, S.; Schmidt, H.; Zhu, L. *Polymer Preprints* **1998**, *39*, 124.
17. Horton, D.; Tsai, J. H. *Carbohydr. Res.* **1979**, *75*, 151.
18. Deutsche Norm: VDE 0303, *Prüfverfahren für Elektroisolierstoffe*.
19. Hastings, G. W.; Ducheyne, P. *Macromolecular Biomaterials*, CRC Press, Boca Raton, 1984.
20. Szycher, M. (Ed.), *Biocompatible Polymers, Metals and Composites*, Technomic, Lancaster PA, 1983.
21. Voet, D.; Voet, J. G. *Biochemie*, VCH, Weinheim (BRD), 1992, 1100-1110.
22. Choay, J.; Petitou, M.; Lormeau, J. C.; Sinay, P.; Casu, B.; Gatti, G. *Biochem. Biophys. Res. Commun.* **1983**, *116*, 492.

23. Westerduin, P.; van Boeckel, C. A. A.; Basten, J. E. M.; Broekhoven, M. A.; Lucas, H.; Rood, A.; van der Heijden, H.; van Amsterdam, R. G. M.; van Dinther, T. G.; Meuleman, D. G.; Visser, A.; Vogel, G. M. T.; Damm, J. B. L.; Overklift, G. T. *Bioorg. & Med. Chem.* **1994**, *2*, 1267-1280.
24. Basten, J.; Vermaas, D.J.; Westerduin, P.; lecture at the 5th European Training Course on Carbohydrates, 14-19 June 1998, Kerkrade, The Netherlands.
25. Gott, V. L.; Whiffen, J. D.; Dutton, R. C. *Science (Washington D.C.)* **1963**, *142*, 1297.
26. Chaubet, F.; Champion, J.; Maiga, O.; Mauray, S.; Jozefonvicz, J. *Carbohydrate Polymers* **1995**, *28*, 145-152.
27. Tokura, S.; Itoyama, K.; Nishi, N.; Nishimura, S.-I.; Saiki, I.; Azuma, I. *Pure Appl. Chem.* **1994**, *31*, 1701-1718.
28. Barbucci, R.; Magnani, A.; Lamponi, S.; Casolaro, M. *Macromol. Symp.* **1996**, *105*, 1-8.
29. Wulff, G.; Diederichs, H. *Macromol. Chem. Phys.* **1998**, *199*, 141-147.
30. Anders, C.; Wulff, G.; Bellmann, S. (Hüls AG), *German patent* registered 1997.
31. Wulff, G.; Bellmann, S.; Brock, A.; Anders, C. (Hüls AG), *German patent* registered 1998.

Chapter 18

Natural Glucans: Production and Prospects

Donal F. Day and Sun Kyun Yoo

Audubon Sugar Institute, Louisana State University Agricultural Experiment Station, Baton Rouge, LA 70803–7305

An emerging trend is the use of oligosaccharides as functional foods. The current technology for the production of these polymers is limited to acid or enzymatic hydrolysis of polysaccharides or synthesis by transglycosylation reactions. The oligosaccharides which have the greatest applicability are either fructans or branched glucans. Modification of a fermentation synthesizing a specific dextran, by addition of suitable acceptors that limit the enzymatic reaction, yielded large amounts of branched iso-oligosaccharides. These oligomers stimulated Bifidiobacterium growth but were not utilized by pathogenic organisms, indicating their potential for use in intestinal microflora modification.

Homopolymers of glucose are widespread in nature. The linear homoglucans have been broadly investigated and widely utilized. The primary representatives of this class of polysaccharides are cellulose, starch (amylose) and dextran. The branched homopolymers are also common, but less widely known. The best studied is amylopectin. A neglected sub-group that has a wide range of potential applications, branched oligosaccharides, can be derived from dextran. Dextran is the collective term given to that group of bacterial polysaccharides composed of chains of D-glucose units connected by α-1,6 linkages. Their structure (degree of branching), molecular size and other properties are organism specific. For example, the degree of branching can vary from less than 5% to greater than 70% depending on the producing strain. Dextran is produced only from sucrose, by bacteria of the family Lactobacillaceae. The commercial market for linear dextran is several billion kilograms per year.

The term oligosaccharide applies to those polysaccharides, joined by glycosidic bonds, that range in size (degree of polymerization, DP) between two and ten monosaccharide units. Until recently, oligosaccharides were primarily used as sweeteners in foods. With increasing health consciousness of the public, the other physiological functions of oligosaccharides, such as calorie lowering, anti-cariogenicity and Bifidius growth factor have attracted commercial interest (Table 1).

Table 1. Functional Characteristics of Oligosaccharides*

Low calorie
Maltitol, Palatinat, Erythritol
Inclusion complex formation
Cyclodextrins, cycloinulinooligosaccharides
Anti-cariogenicity
Palatinose, Trehalulose, isomaltooligosaccharides, isomaltofructoside, xylosylfructoside
Bifidus factor
Fructooligosaccharides, galactooligosaccharides, isomaltooligosaccharides, xylo-oligosaccharides, lactulose, raffinose

*These oligosaccharides are produced by transglycosylation reactions using microbial enzymes. (data from 1)

Oligosaccharides show great potential for improving intestinal microflora in humans and livestock. Non-digestible oligosaccharides pass through the digestive system to the lower intestine, where they can be used by intestinal microorganisms. Generally, branched oligosaccharides as less readily digested than the linear polymers. Desirable oligosaccharides are those that are utilized primarily by the Bifidiobacteria, a specific class of Gram-positive bacteria (2). Maintenance of stable intestinal populations of Bifidiobacteria lowers the pH of the digestive tract and prevents overgrowth by undesirable species and pathogens such as Escherichia, Salmonella, Clostridia, etc. Both fructooligosaccharides and isomaltooligosaccharides have been used in this context (1,3).

Isomaltooligosaccharides (mixtures of isomaltose, panose , isomaltotriose and branched oligosaccharides DP4 and 5) are commercially produced from starch by a series of enzyme reactions. Starch is first dextrinized using soybean β-amylase and then incubated with α-glucosidase from *Aspergillus niger*. This technology has drawbacks. It is multistep, requiring several enzymes each of which must be produced separately. Significant concentrations of unreacted, digestible sugars, glucose, maltose and maltooligosaccharides remain in the final product limiting the effectiveness of the product for modifying intestinal microflora populations. Isomaltooligosaccharides are degradation products of dextran. They have been produced *in vitro* for cosmetic applications with a dextransucrase isolated from *Leuconostoc mesenteroides* (4). Using a slightly different approach, we have developed a simple modification of the dextran fermentation for the production of a high quality branched isomaltooligosaccharides suitable as Bifidius factors.

Dextran is produced commercially by the fermentation of sucrose by *Leuconostoc mesenteroides* B512F. A large polysaccharide is produced that is 95% linear α 1,6 glucan. Different Leuconostoc strains produce dextrans that vary in quantity, length and arrangement of branch chains. Dextran may have α-1,2, α 1,3 and/or α 1-4 linked glucan branches. *L. mesenteroides* ATCC 13146 produces a highly branched dextran containing as much as 50 % α 1-3 linkages, where the majority of the linkages connect only a single glucose to the main chain (5). This organism appeared to be a good candidate for production of branched isomaltooligosaccharides.

The size of dextran can be restricted either by substrate limitation generated by the presence of large amounts of chain initiating compounds (6) or by restricting chain elongation with chain terminating compounds. Dextransucrase is believed to work by a single chain mechanism (7) where the enzyme normally remains attached to the growing dextran chain. The enzyme hydrolyzes sucrose and transfers the D-glucopyranosyl residue to a growing glucan chain. The termination step of the reaction is not completely understood, but evidence suggests that dextransucrase is released slowly from the polymer (8). Certain low molecular weight compounds can substitute for the glucopyranosyl residue terminating dextran chain growth. These compounds do not act as primers, but rather they terminate polymerization (9). The order of preference for chain terminating acceptors of the dextran reaction is species specific. Maltose and isomaltose have been demonstrated to be the most effective acceptors for dextransucrase (10, 11). Because of its high acceptor efficiency, maltose has been widely used in the cell free synthesis of oligosaccharides by dextransucrase (12).

In vitro the ratio of sucrose to maltose affects the composition of the oligosaccharides produced by dextransucrase reaction. A partially purified dextransucrase from *L. mesenteroides* B512F produced 85% of the theoretical yield of polysaccharide as oligosaccharides with an average DP of 4 when the maltose to sucrose ratio was 2 (13). Cote and Robyt (14) investigated the acceptor reaction of *L. mesenteroides* ATCC 13146 and found that branch formation in this strain, when maltose was the acceptor, was dependent upon reaction conditions. Fermentation of sucrose by *L. mesenteroides* ATCC 13146, under optimum conditions, in the presence of maltose produced 90% of theoretical yield of polymer as oligosaccharides. The fermentation was essentially complete in 24 hours, with oligosaccharide production being linked to growth. The production rate was about 0.9 g/l-hr (Figure 1).

The S/M ratio was found not only to alter the yield of oligosaccharide but also to change the relative proportion of different size oligosaccharides produced by the fermentation (Figure 2). The highest yields of oligosaccharides were obtained when the ratio of sucrose to maltose in the fermentation was 2. This is the same ratio as was reported for optimum oligosaccharide production *in vitro* by the dextransucrase of *L. mesenteroides* B512F (13).

Because individual strains of Leuconostoc do not synthesize identical dextransucrases, differences may be expected in the glucooligosaccharides produced in response to acceptors. Several microbial strains were screened for oligosaccharide size profiles produced in response to maltose. Of the strains tested, *L. mesenteroides* ATCC 13142 produced the most DP 3-5 glucopolysaccharides (Figure 3). Chemical analysis of the products indicated the glucooligomers were primarily pannose and branched DP 4 and 5.

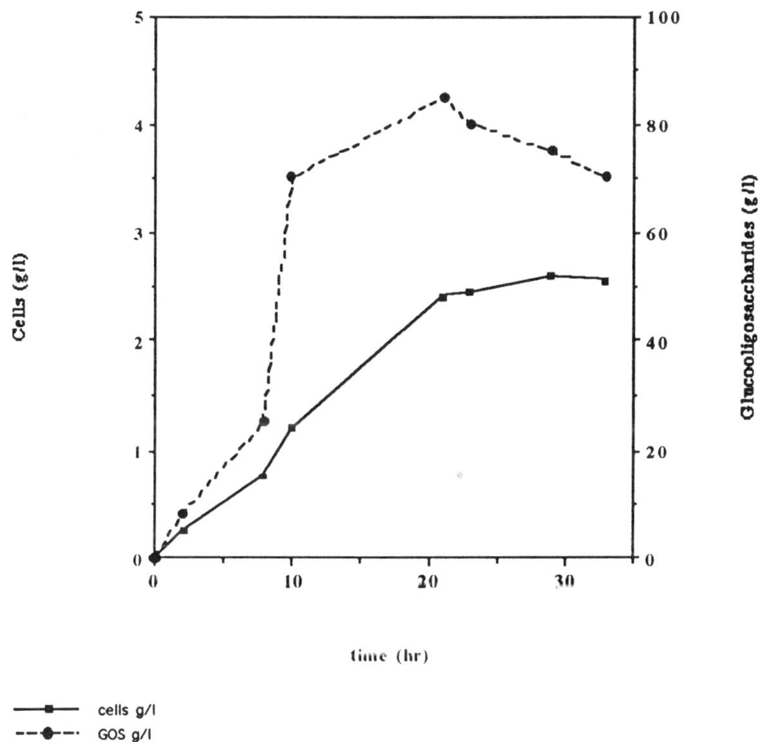

Figure 1. Cell growth (■) and oligosaccharide production by *L. mesenteroides* ATCC 13146 in a media containing 10% sucrose and 5% maltose (data from 15)

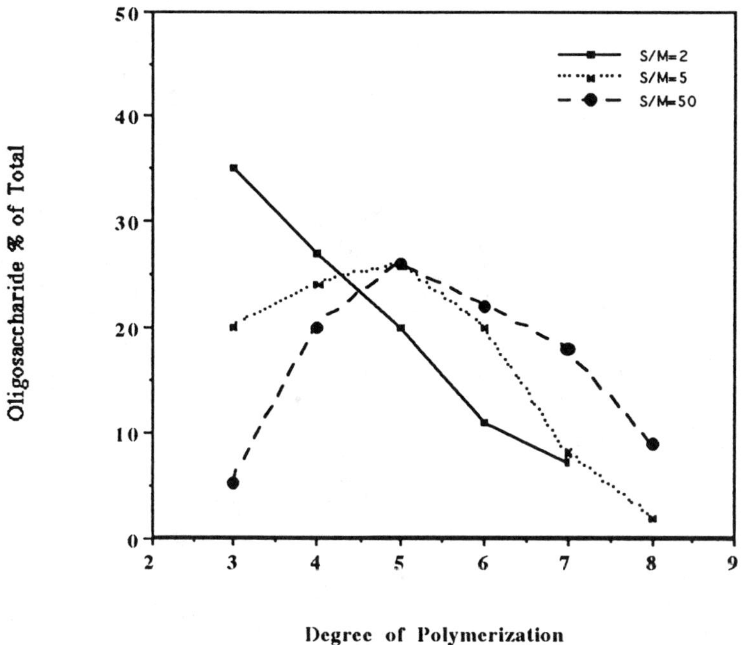

Figure 2. The glucooligosaccharide composition produced in a *Leuconostoc mesenteroides* fermentation as a function of S/M ratio (data from 15).

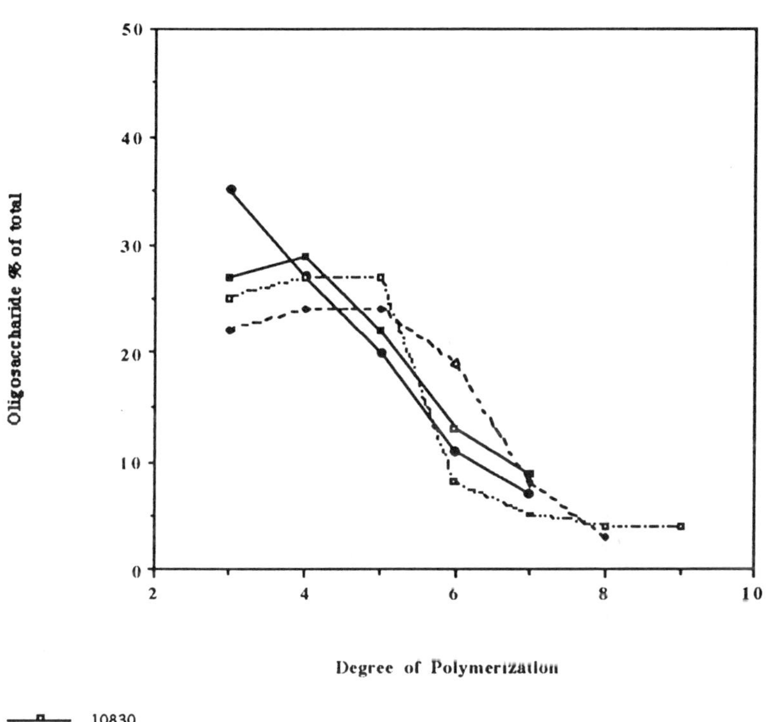

Figure 3. Oligosaccharide profiles produced by 5 different strains of *L. mesenteroides* of sucrose with an S/M ratio of 2 (data from 15).

Oligosaccharides were prepared by drying alcohol precipitated cell free culture broth. Comparative analyses of two commercial preparations and the oligosaccharide preparation from *L. mesenteroides* ATCC 13146 are given below.

Table 2. Composition % of Oligosaccharide Preparations

	Isomalt 500*	Isomalt 900*	ATCC 13146
Glucose	40.4		
Isomaltose	22.9	28.4	
Maltose	4.2	16.8	6.5
Maltotriose	1.1		
Pannose	6.5	14.8	32.7
Isomaltotriose	8.5		
Branched DP4		11.5	30.4
Isomaltotetraose		9.6	
Branched DP5		2.4	21.7
Branched DP6			5.2
Branched DP7			3.6
Other oligosaccharides	16.4	16.5	

*Wako Pure Chemical Co.

This preparation was used to test for microbial growth efficacy. Growth comparisons were made for different species of Staphylococcus, Salmonella, Clostridium and Bifidiobacterium on the ATCC 13146 oligosaccharide mix and a commercial isomaltooligosaccharide mix. Growth on glucose (100%) was used as the standard for comparison.

Table 3.Microbial Growth on Oligosaccharides

Organism	ATCC 13142 Preparation	Isomalt 900
B. bifidium	46	80
C. perfringins	20	
Sal. enteritidis	5	95
Sal. typhimurium	20	100
S. aureus	20	20
S.epidermidis	10	40

Growth of pathogenic microorganisms, relative to growth of Bifidiobacterium was very poor on the ATCC 13146 preparation. The commercial mixture showed lower growth of some, but not all of the test pathogens. These glucooligosacccharides are primarily branched oligomers that are resistant to both gastrointestinal enzymes and utilization by pathogenic microorganisms. The are utilized by the Bifidobacterium and should be expected to bring the concomitant health effects associated with enhanced Bifidobacterium populations. As such they have great potential application in the food industry.

It is obvious that a wide range of potential oligosaccharides and polysaccharides can be produced by the proper selection of traditional microbial fermentation and reaction modifying compounds. In many cases more practically than can be achieved using sequential enzyme reaction sequences. Traditional approaches should not be overlooked in a search for new and useful biopolymers.

Experimental

Oligosaccharide Production by L. mesenteroides in the Presence of Maltose:
Batch fermentations with L. mesenteroides were conducted in a 2 -L Bio Flo II (New Brunswick Scientific Co.) fermentor with a working volume of 1.2 L. The media was a mineral media containing yeast extract (15) with 10% w/v sucrose and 5 % w/v maltose. Fermentations were conducted at pH 6.5, 28°C and 200 rpm. Each fermentor was inoculated from late log phase shake flask cultures at 1.0% the working volume of the fermentor.

Oligosaccharides were recovered from cell-free culture broth by precipitation with ethanol (95%). Ethanol was slowly added at room temperature until it reached 70 % of the volume of the original solution.. The solution was allowed to stand for 2 hours, then the precipitate, which contained the oligosaccharides was removed by centrifugation. The oligosaccharide fraction was re-precipitated with ethanol and then lyophilized prior to use.

Analytical:
Isomaltooligosaccharides were separated by High Pressure Liquid Chromatography using an Aminex HPX-42A (Bio-Rad, Hercules, CA) resin. The detector was a 410 Differential Refractometer (Millipore Corp.). Identification was by comparison with known standards.

Microbial utilization was determined by relative growth of selected bacteria in shake flask culture, grown under the appropriate growth conditions for each test strain. Growth on glucose was used as the positive control for these experiments. All cultures, unless specifically noted, were obtained from the American Type Culture Collection.

References

1. Kitahata, S. *Proc. 94 Korean Food Technology Symposium*, Aug. 29, p47-55, Seoul, Korea, **1994**

2. Fishbein, L; Kaplan, M.; Gough, M. *Vet. Human Toxicol.* **1988**, 30: 104-107.

3. Spiegel J. E.; Rose, R.; Karabell, P.; Frankos, V.; Schmitt, D. *Food Technology* **1994**, Jan., 85-89.

4. Monsan, P.; Paul, F. *FEMS Microbiol. Rev.* **1995,** 16, 187-192.

5. Seymour, F.; Knapp, R.; Chen, C.; Jeanes, A.; Bishop, S. *Carbohydr. Res.* **1979,** 75, 275.

6. Kim D.; Day, D. *Enzyme Microb. Technol.* **1994**, 16, 844-848.

7. Robyt, J. *Proc Symp. Mechanisms of saccharide polymerization and depolymerization- Miami Beach 1978*, Academic Presss: New York, **1980**, 43-54.

8. Ditson, S.; Mayer, R. *Carbohydr. Res.* **1984**, 126, 170-175.

9. Robyt, J.; Walseth, T. *Carbohydr. Res.* **1979**, 68, 95-111.

10. Robyt, J.; Walseth, T. *Carbohydr. Res.* **1978**, 61, 433-445

11. Robyt, J.; Ecklund, S. *Carbohydr Res.* **1983,** 121, 279-286.

12. Smiley, K.; Morey, E.; Boundy, J.; Plattner, R. *Carbohydr. Res.* **1982,** 108, 279-283.

13. Paul F. *Carbohydr. Res.* **1986,** 149, 433-441

14. Cote, G.;Robyt, J. *Carbohydr. Res.* **1983,** 119, 141-156

15. Yoo, Sun K. *The Production of Glucooligosaccharides by Leuconostoc mesenteroides ATCC 13146 and Lipomyces starkeyi ATCC 74054,* Dept. of Food Science, Louisiana State Univ., Baton Rouge, La. **1997**.

Chapter 19

Acyclic Polyacetals from Polysaccharides: Biomimetic Biomedical "Stealth" Polymers

M. I. Papisov

Department of Radiology, Massachusetts General Hospital and Harvard Medical School, Boston MA 02114–2696

Technologically adaptable hydrophilic polymers combining negligible in vivo reactivity with biodegradability would be instrumental in the development of specialized materials for advanced biomedical applications. Such highly biocompatible biodegradable polymers can be obtained via partial emulation of carbohydrate interface structures prevalent in biological systems. These structures are also present in polysaccharides and in some cases can be chemically "carved out" and isolated as acyclic hydrophilic polyacetals.

Novel concepts in pharmacology and bioengineering impose new, more specific and more stringent requirements on biomedical polymers. Ideally, advanced macromolecular materials would combine negligible reactivity in vivo with low toxicity and biodegradability. Polymer structure should support an ample set of technologies for polymer derivatization; for example, conjugation with drugs, cell-specific ligands, or other desirable modifiers. Materials combining all the above features would be useful in the development of macromolecular drugs, drug delivery systems, implants and templates for tissue engineering.

On the chemistry level, developing such materials translates into an intricate problem of developing macromolecules with minimized interactions in vivo, completely biodegradable main chains, and readily and selectively modifiable functional groups. The problem is further aggravated by the fact that both the main chain and the functional groups interact with extremely complex biological milieu, and all their interactions may be amplified via cooperative mechanisms.

Macromolecule interactions *in vivo* are mediated by several components of cells surfaces, extracellular matrix, and biological fluids. For example, both macromolecule internalization by cells and cell adhesion to polymer-coated surfaces can be mediated by several cell surface elements, many of which are functionally specialized (phagocytosis- and endocytosis-associated receptors, adhesion molecules, etc). Macromolecule recognition by cell receptors is often mediated by specialized recognition proteins of plasma, such as immunoglobulins (1), fibronectins (2,3,4), proteins of complement system (5,6,7,8), soluble lectins (9,10,11,12), vitronectin (13) etc. These proteins contain at least one receptor-recognizable site per molecule, and often more than one substrate-binding site. Although other proteins, e.g., albumin, also can bind polymers (via non-specific mechanisms), the distinctive features of recognition proteins relate to their ability to trigger remarkable biological responses. Some recognition proteins, such as C-reactive protein, are acute-phase proteins, i.e., their concentration in plasma increases as a result of inflammation or trauma. Others, such as β2-glycoprotein I (β2GPI), are "reverse" acute phase proteins, i.e., their concentration in blood during the acute phase decreases. Recognition proteins bind a variety of structures; we have reviewed their role in pharmacology of macromolecules and particulates in more detail elsewhere (14,15).

Cooperative binding, often referred to as "non-specific interactions", is another major factor of macromolecule (and surface) reactivity in vivo. Cell interactions with polymers and recognition protein-polymer complexes also have an element of cooperativity (16,17). The very nature of cooperative interactions suggests that *any* large molecule can significantly interact with a complex substrate, for the simple reason that, because the binding energy is additive, the association constant of cooperative binding (K_a) would grow with the number of associations *exponentially* (14). In other words, any polymer of a sufficient length can be expected to interact with at least one of the various components of a biological system. Even if a molecule of certain size shows low interactions in cell cultures and in vivo, a larger molecule of the same type, or a supramolecular assembly, can have a much higher binding activity (18).

The essence of the above is that even if polymer molecules are assembled of domains that do not interact with cell receptors and recognition proteins, such molecules can be capable of cooperative interactions in vivo, i.e., completely inert polymers may not exist at all. However, several biomolecules and biological interfaces do appear to be functionally inert, except their specialized signaling domains. For example, plasma proteins are known to circulate for several weeks without uptake in the reticuloendothelial system (RES), whereas artificial constructs of a similar size have never been reported to have comparable blood half-lives.

Hypothetically, the mutual "inertness" of the natural biomolecules and surfaces may relate to their relatively uniform interface structures, where the potential binding sites are always saturated by naturally occurring counteragents present in abundance. Therefore, emulation of the common interface structures can result in a material that would not actively interact with actually existing binding sites because these sites would be pre-occupied by the natural "prototypes".

Poly- and oligosaccharides are the most abundant interface molecules expressed (as various glycoconjugates) on cell surfaces, plasma proteins, and proteins of the extracellular matrix. Therefore, interface carbohydrates appear to be the best candidates for structural emulation. The main objective of the emulation is to identify and exclude all structural components that can be recognized, even with low affinity, by any biomolecule, especially by cell receptors and recognition proteins.

All interface carbohydrates have common structural domains, which appear to be irrelevant to their biological function. An acetal group and two adjacent carbons are present in all carbohydrates, whereas the receptor specificity of each molecule depends on the structure and configuration of the glycol domains of the carbohydrate rings (Figure 1). We hypothesized that biologically inert ("stealth") polymers could be obtained using substructures that form the acetal side of the carbohydrate ring, i.e., the -O-C-O- group and the adjacent carbons. Although functional groups that are common in naturally occurring glycoconjugates (e.g., OH groups) can be used as substituents, the potentially biorecognizable combinations of these groups, such as rigid structures at C1-C2-C3-C4 (in pyranoses), must be completely excluded. Positioning of the acetal groups within the main chain would ensure polymer degradability via proton-catalyzed hydrolysis.

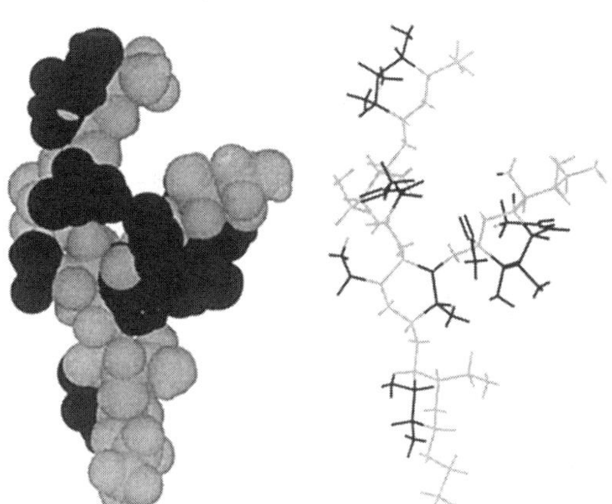

Figure 1. The structure of oligosaccharide interface fragment of glycolipid G_{M1} (space-filled and "stick" models of the same structure)

The signaling domains are shown in black; the biologically inert backbone in gray.

Materials of the suggested general structure (acyclic hydrophilic polyacetals) can be produced using a variety of methods. For example, cleavage of potentially biorecognizable fragments from all carbohydrate residues of a polysaccharide would result in acyclic structures similar to that of interface carbohydrates. We used exhaustive periodate oxidation to transform (1->6)-poly-α-D-glucose into acyclic poly[carbonylethylene carbonylformal] (PCF) with subsequent borohydride reduction resulting in poly[hydroxymethylethylene hydroxymethylformal] (PHF). Both polymers, PCF and PHF, were isolated and characterized in order to evaluate the viability of the concept.

Synthesis

Dextran B512, a product of *Leuconostoc Mesenteroides,* is a linear (1->6)-poly-α-D-glucose with ca. 5% (1->3; β) branching; 95% of the branches are only one or two residues long (19). Periodate oxidation of 1->6 connected polysaccharides has been previously studied (20). In unsubstituted pyranosides the periodate reaction, which is highly specific to 1,2-glycols, starts from breaking either C2-C3 or C3-C4 bond with formation of dialdehydes IIa or IIb. In dextrans, the kinetically controlled

IIa/IIb ratio is approximately 7.5:1 (20). The subsequent, slower stage results in the cleavage of carbon C3, with formation of dialdehyde III (Figure 2).

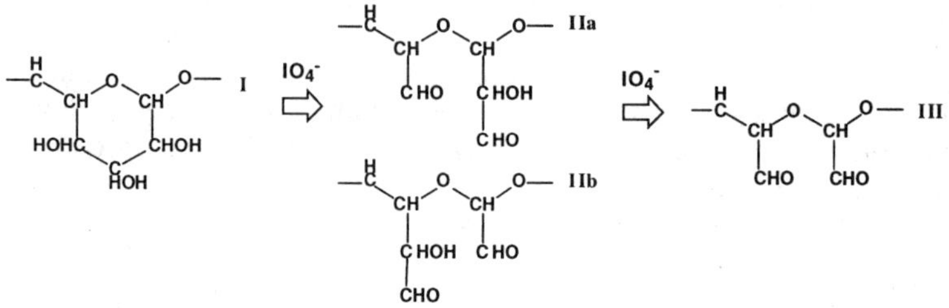

Figure 2. Exhaustive periodate oxidation of an unsubstituted pyranose ring.

Thus, exhaustive oxidation of an entirely 1->6 connected polysaccharide is expected to occur without depolymerization, resulting in macromolecular poly-[carbonylethylene carbonylformal] (PCF). The aldehyde groups can be subsequently reduced with borohydride to obtain a hydroxymethyl-substituted polymer, poly-[hydroxymethylethylene hydroxymethylformal] (PHF, Figure 3).

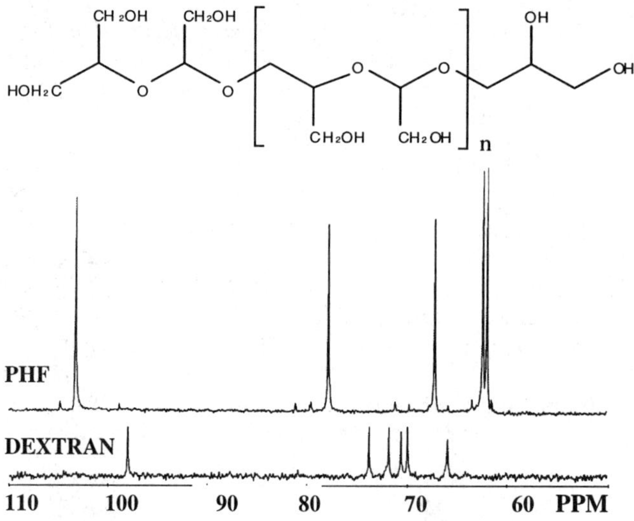

Figure 3. Poly[hydroxymethylethylene hydroxymethylformal] (PHF), structure and ^{13}C NMR spectrum ; 293 K°, 10% solution, 9.4 Tl Brucker system, 100.619 MHz by ^{13}C; proton decoupling, 45° flip angle, recycle delay 1.8 s (Dextran B512 spectrum is given as a reference).

The ^{13}C NMR spectrum of the final product (Figure 3) confirms the expected structure and shows that, unlike some other dextrans, where complete oxidation is blocked (presumably, as a result of formation of intramolecular hemiacetals), Dextran B512 can be completely oxidized with no identifiable residual cyclic structures. The phenol-sulfate analysis (21) also showed only traces (<<0.1%) of the residual carbohydrate.

One of our practical objectives was to develop a technique for large scale polysaccharide processing without significant depolymerization. The major concerns related to (a) possible inclusions of non-1->6 linkages in the poly-(1->6)-α-D-glucose main chain of Dextran B 512, that could be cleaved by periodate oxidation, and (b) relative instability of periodate-oxidized polysaccharides in alkaline media, which could result in depolymerization at the reduction stage (22). Preliminary tests showed that the commonly used versions of the periodate technique (developed for carbohydrate analysis and bioconjugate chemistry) afforded only small amounts of high molecular weight materials. Optimization of both the oxidation and reduction stages for minimal depolymerization resulted in consistently reproducible high yields of polymers with molecular weight distributions similar to the source dextrans (as determined by SEC HPLC) (23,24). Using flow dialysis as a prototype large scale technique for polymer purification and isolation, we obtained PHF with nearly theoretical yields for high molecular weight dextrans (MW>100 kDa). Low molecular weight polymers (MW=20-50 kDa) showed lower yields. The latter were attributed to inadequate polymer retention by low molecular weight cutoff filters, mainly at the final stage of PHF purification (PCF is reversibly associated in aqueous media, especially at 5<pH<7, which facilitates polymer retention by flow dialysis filters). Low molecular weight preparations of PHF were obtained with high yields via alternative procedures: (a) polymer purification by size exclusion chromatography, and (b) partial hydrolysis of 150-200 kDa polymers.

Properties

Both polymers, the intermediate PCF (Figure 2, III) and PHF (Figure 3), were obtained in >99% pure form (by SEC HPLC) as colorless solid compounds.

PCF was found to be stable in aqueous media below pH≈9. Depending on the pH, PCF undergoes transitions that appear to be similar to the previously described for partially oxidized dextrans (25). At pH=4÷5, most aldehyde groups seem to exist in a gem-diol form. At lower pH the aldehyde absorption peak (267 nm) becomes apparent, and above pH 5 both enol and enolate forms are present (240 and 290 nm). Formation of the enol form appeared to correlate with significant intermolecular association at pH=5÷7. PCF was found to be soluble in water, dimethylsulfoxide (DMSO), dimethylformamide (DMFA), pyridine and water-alcohol mixtures, and insoluble in acetone, acetonitrile, dioxane, methanol, ethanol, glycerol, methylenechloride, toluene and triethylamine. Solubilization of dehydrated (lyophilized) preparations in water was slow, except at pH>7.

The reduced (polyalcohol) form, PHF, was found to be highly hygroscopic. Samples exposed to humid air were viscoelastic at ambient temperature. The apparent melting range of lyophilized PHF (MW=50-200 kDa) was within 100±20°C, depending on the molecular weight, and dramatically decreased after exposure to the ambient (humid) air. High molecular weight PHF is readily soluble in water, DMSO, DMFA and pyridine; slowly soluble in glacial acetic acid and ethyleneglycol, and insoluble in acetone, acetonitrile, dioxane, methanol, ethanol, glycerol, methylenechloride, toluene and triethylamine. Preparations with MW<5 kDa were soluble in methanol.

As expected, the stability of the PHF main chain was pH-dependent. While incubation at the neutral and high pH over several days did not change SEC elution profile, incubation at pH<7 showed significant fragmentation (Figure 4). In the presence of 50 mM sodium phosphate buffer, the hydrolysis rate at pH=3 was almost twice higher. Solubilization of crosslinked PHF gels in aqueous media showed an analogous pattern. At pH=7.5, both soluble and crosslinked PHF were resistant to a one hour incubation at 100°C.

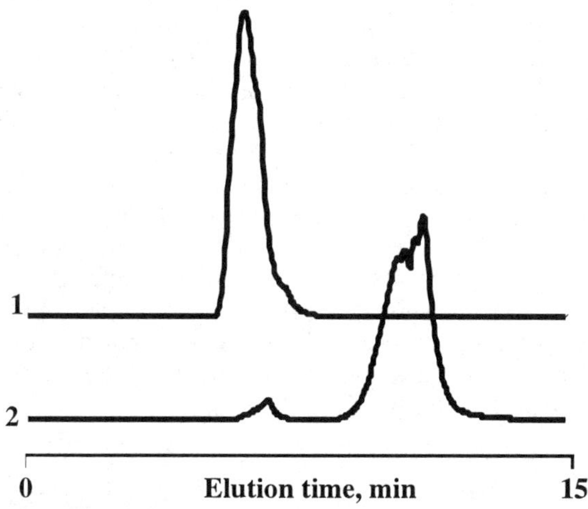

Figure 4. Size exclusion HPLC profile of 200 kDa PHF before (1) and after (2) 4 days incubation at pH=3, 37°C.

This pH dependence of main chain stability is valuable in several biomedical applications, where polymer-based products should be stable and functional in biological milieu (pH=7÷7.5) but undergo depolymerization after internalization by cells. Degradation of the cell-internalized polymer is important to avoid adverse effects associated with long-term polymer deposition in cells, in the first place in the glomerular mesangium and reticuloendothelial system (26,27).

Acidic conditions (pH≈5) are characteristic for the intracellular lysosomal compartment where polymers are transferred after internalization by cells. Therefore, cellular uptake of PHF-based preparations can be expected to result in non-enzymatic main chain hydrolysis at a moderate rate. This appears to be a significant advantage, as compared to several synthetic polymers, e.g., polyethyleneglycol, polyacrylates and vinyl polymers, which are hydrolysis-resistant. The final products of the PHF hydrolysis, glycerol and glycol aldehyde, have low toxicity; both are metabolized via major metabolic pathways. This may be one of the underlying reasons for the observed extremely low toxicity of PHF (see below).

Derivatives

Modification of either polymer did not present significant difficulties. Due to the availability of well-developed methods for alcohol and aldehyde group modification, the reaction conditions can be selected such as to ensure the integrity of the polyacetal main chain (e.g., at 4<pH<9 in aqueous media). Although neither polymer is soluble in most organic solvents, several desirable lipophilic derivatives, e.g., PHF conjugates with lipids, can be successfully synthesized in suitable solvent mixtures (e.g., pyridine-DMSO or pyridine-methanol).

To investigate the technological flexibility of PCF/PHF system and to characterize PHF-based preparations, several model linear and branched forms of derivatized PHF, model gels and bioconjugates were successfully synthesized and studied in vivo. The examples are given below.

PHF derivatization

Direct derivatization of PHF through primary alcohol groups. The alcohol groups of PHF can be acylated or alkylated in DMSO, DMFA or in water. Acylation with diethylenetriaminepentaacetic acid monocycloanhydride in DMSO was utilized to obtain PHF modified with diethylenetriaminepentaacetic acid (DTPA), a chelating group suitable for polymer labeling with metal ions such as ^{111}In (radioactive γ-emitter). Indium-111 labeled preparations were used in biokinetics and imaging studies. Alkylation with epybromohydrine in water was utilized to produce model epybromohydrine-crosslinked gels (that were used to investigate the resistance of PHF-based matrix to hydrolysis).

Derivatization through terminal 1,2-glycol group was used for producing terminus-activated PHF. The 1,2 glycol is formed at the former reducing end of the polysaccharide chain (whereas at the former non-reducing end a 1,3 glycol is present), see Figure 3. The 1,2-glycol is readily transformed into active aldehyde group via periodate oxidation. For example, a terminus-activated polymer with apparent molecular weight of 3.6±0.4 kDa per aldehyde group (I_2 titration) was produced and subsequently conjugated with lipids (in pyridine-methanol media) and proteins (in water) (28).

Derivatization through non-terminal glycol groups. Non-terminal 1,2-glycol groups were introduced into PHF structure via modification of the polysaccharide oxidation technique. Oxidation of the original dextran was ca. 10% incomplete (all carbohydrate rings were open but 10% of the C3 were not eliminated), so the product of subsequent reduction (PHF) contained 1 glycol per 20 functional groups. The glycol groups were further oxidized with periodate resulting in PHF comprising active aldehyde groups along the main chain. The latter were conjugated with several model reagents via aldehyde condensation with amino-, hydrazido-, aminooxy- and other groups (see below).

Partial fragmentation of the PHF backbone with simultaneous incorporation of new functional groups was used to produce PHF with activated terminal groups. Treatment with mercaptopropionic acid in DMFA (mercaptolysis) resulted in fragments containing terminal carboxyls. The fragmented polymers were fractionated by precipitation (DMSO/chloroform or DMFA/acetone) and further subfractionated by HPLC. The terminal carboxylic groups were activated in DMSO with N-hydroxysuccinimide in the presence of dicyclohexylcarbodiimide. The resultant polymer containing terminal N-oxysuccinimide ester group was precipitated and washed with chloroform and lyophilized. Terminal N-oxysuccinimido-PHF was used to produce soluble terminal graft copolymers (comb copolymers) with polyamines, e.g., with poly-L-lysine via direct reaction in water (24), and conjugates with lipids (distearoylphosphatidylethanolamine, DSPE) via condensation in DMSO/pyridine mixture. The DSPE-PHF conjugates were used for liposome stabilization (28).

PHF derivatives via modification of aldehyde groups of PCF

Modification of aldehyde groups of PCF (or PHF comprising aldehyde groups generated via glycol oxidation as described above) presents a set of synthetic approaches for producing a vast variety of PHF derivatives in mild conditions. For example, aldehyde groups can be conjugated in aqueous media with amines via formation of enamines with subsequent cyanoborohydride or borohydride reduction; this approach is widely used in protein immobilization on polymers (29,30).

Whenever conjugation through amines is not desirable, e.g., the reagent to be coupled with PHF has a biologically functional aminogroup, a variety of aldehyde group reactions with hydrazides, hydrazines, O-substituted hydroxylamines and 2-mercaptoamines (e.g., N-terminal cysteine) can be utilized. These reactions can be carried out in conditions where enamines are not formed (for example, in aqueous media at pH=4÷6).

Selectivity of aldehyde-mediated reactions opens the way to fast synthesis of complex functional conjugates, for example graft copolymers carrying multiple labels on the backbone (31) and several cell-specific ligand groups (of one or more types) on the side chains. Aldehyde-mediated reactions can also be used for assembling complex PHF-based functional matrices e.g., for tissue engineering. Examples of PHF derivatization via aldehyde reactions are given below.

Partial derivatization of PCF was used to produce linear functionalized PHF derivatives and random-point PHF graft copolymers.

Linear PHF conjugate carrying fluorescein, DTPA and formyl-Met-Leu-Phe-Lys (f-MLFK, a chemotactic peptide) was synthesized via PHF condensation with cystamine (H_2N-C_2H_4-SS-C_2H_4-NH_2) and f-MLFK, with subsequent cystamine reduction and modification of the formed mercaptogroups with fluorescein maleimide (fluorescent label) and DTPA (chelating group for ^{111}In). This preparation was used as a model cooperative vector for targeting formylpeptide receptors of white blood cells (37,38).

Random-point graft copolymers of PHF and DTPA-modified poly-L-lysine (backbone) were prepared, using previously developed technique (31), via DTPA-Polylysine condensation with an excess of PCF, with subsequent reduction and separation of the unbound PHF. A dextran-polylysine graft copolymer was prepared analogously as a control for animal studies. The hydrodynamic size of both products, as determined by photon correlation light scattering, was 16±4 nm. Graft content was

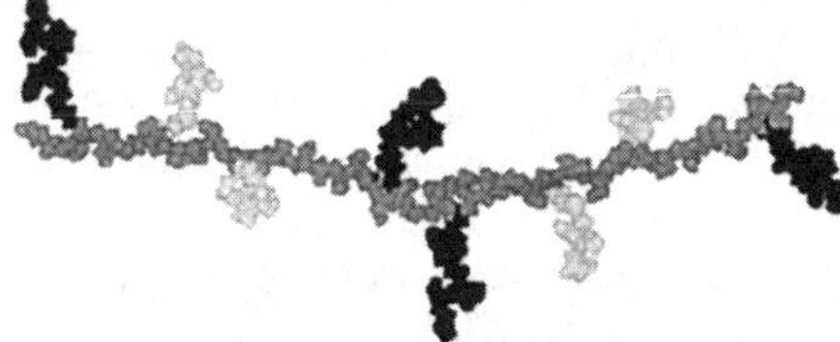

Figure 5. The structure of fMLFK-DTPA-PHF conjugate. The PHF backbone (ca. 1 kDa chain fragment shown) is modified by fMLFK (black) and DTPA (light gray) at random positions. Reproduced from reference (37).

20-25 molecules per backbone. Both copolymers were labeled with Indium-111 for animal studies.

In vivo studies

Because the central practical objective of this study was to develop a polymer with minimized interactions in vivo, we studied biokinetics of PHF and various PHF derivatives and attempted to identify the dose level at which toxic effects of PHF would become noticeable. Biokinetics provide valuable data on polymer interactions in vivo because, for particles and large macromolecules circulating in blood, blood half life is a mathematically exact measure of the overall polymer interactions with the biological milieu (14). Biologically inert ("stealth") polymers are expected to have insignificant accumulation in RES and other tissues. Low rates of tissue binding and uptake by cells result in a long blood half-life, except relatively small molecules (generally, MW<50 kDa) which can be cleared from blood via renal filtration.

Acute toxicity in mice. PHF of the highest molecular weight available at the time of the experiment (approximately 0.5-1 MDa) was used to minimize renal excretion that would mask the potential toxic effects. Although the injected dose reached 2 g/kg, all animals survived. After 32 days, all animals were alive, and their weights did not significantly differ from the control group (24±3 g vs. 25±2 g). None of the animals showed any noticeable symptoms of toxicity, including anaphylactoid reactions (e.g., paw edema) that develop in rodents in response to administration of many polymers, including dextran B 512. Absence of adverse reactions indicated that PHF interactions with immunocompetent cells and recognition proteins were biologically insignificant, which is in agreement with the underlying hypothesis. Administration of large doses of PHF-based preparations in rats and rabbits also did not cause any signs of toxicity nor anaphylactoid reactions.

Circulation of PHF was studied in normal anesthetized rats. Radiolabeled preparations were administered via tail vein. The initial biokinetics were studied by dynamic γ-scintigraphy (32). Blood half-life of the low molecular weight [^{111}In]DTPA-PHF (50 kDa fraction) in rat was found to be 45 min (clearance via renal filtration). The polymer was cleared by 24 post injection, with very little accumulation in tissues (<0.05% dose/g in any tissue). The highest label accumulation (0.16% dose/g) was found in kidneys. The high molecular weight [^{111}In]DTPA-PHF (500 kDa fraction) demonstrated significantly longer circulation (blood half-life ca. 26 hr.), with almost even distribution among tissues. Accumulation in RES was only twice as high as in other tissues, and thus was related, most likely, to a higher rate of spontaneous endocytosis in RES, rather than to PHF recognition by RES phagocytes.

Biokinetics of graft copolymers. Biokinetics of graft copolymers depend (at high graft densities) on the structure of the graft, whereas the effect of sterically hindered main chain is minimal. The graft copolymer model is sensitive to cooperative interactions because several graft chains can interact with a substrate (e.g., functional components of cell surface) simultaneously. For example, multiple chains of dextran B-512 in dextran/polylysine graft copolymers (and dextran-coated nanoparticles) are readily recognized by lymph nodes and spleen phagocytes, whereas single dextran molecules are not (18,33).

Biokinetics of graft copolymers were studied in normal outbred rats as described above. A series of graft copolymers of PHF with different graft densities showed the following results. Terminal (comb) copolymers with graft densities of two, seven, and ten PHF chains per backbone showed blood half-lives of 5.4±0.3, 7.2±1.2, and 9.8±1.5 hours, respectively. The long blood half-lives at higher graft densities, where copolymer molecule interactions are mediated essentially by the side chains, indicated low overall level of cooperative interactions of PHF in vivo.

In the subsequent comparative study, random-point graft copolymer of dextran showed blood half-life of ca. 1.5 hr. and a highly characteristic uptake in lymph nodes and spleen, with somewhat lower accumulation in liver and kidneys. Graft copolymer of PHF with analogous structure showed a much longer 25.3±2.5 hr. blood half-life, and a dramatically lower uptake in RES (Table 1).

Thus, the results of in vivo studies showed that neither linear nor highly branched PHF derivatives were efficiently recognized by RES, unlike the original Dextran B512. In studies with partially oxidized dextran (23), loss of recognition correlated with elimination of the rigid stereospecific structures of the carbohydrate molecule.

Table 1. Biodistribution of Dextran and PHF graft copolymers in rat (% dose/g tissue), 24 hr. after intravenous administration of 1 mg/kg body weight (adapted from reference 15).

Tissue	Graft	
	Dextran B-512	PHF
Blood	0.3	3.7
Lymph nodes, paraaortic	58.9	0.9
Lymph nodes, mesenteric	81.8	0.8
Spleen	19.9	1.3
Liver	9.0	2.1
Kidney	2.7	3.7
Muscle	0.1	0.4
Heart	0.3	0.9
Lung	0.2	1.2

Biokinetics of PHF modified with chemotactic peptide was studied to evaluate PHF as a biodegradable "stealth" backbone polymer for targeted macromolecular drugs.

The model chemotactic peptide, f-MLFK, binds formylpeptide receptors of white blood cells. As a result, administration of labeled f-MLFK preparations results in label accumulation in the areas of white blood cell invasions, such as acute inflammations (34). Peptide conjugation with macromolecules hypothetically can open the way to dramatic improvements in pharmacokinetics by means of (1) regulating the blood clearance via decreasing the rate of renal and, possibly, RES clearance and (2) increasing the agent-leukocyte association constant via cooperative binding effect of multiple peptide molecules exposed on the carrier. The cooperative character of agent-leukocyte interaction suggested an additional opportunity to explore a (3) hypothetical thermodynamic discriminatory effect that is expected to result in a more selective agent association with leukocytes and suppression of non-specific interactions with other tissues. The improvements in biokinetics, however, would be diminished if the backbone polymer interactions prevailed in the overall conjugate interactions in vivo.

Biokinetics of [^{111}In]DTPA-mercaptoethylamino-PHF-fMLFK, 15 and 70 kDa (Figure 5), was studied in a rabbits. Animals were normal or bearing focal bacterial inflammation induced by inoculation of E.Coli (clinical isolate) in thigh muscle. ^{111}In-labeled PHF-DTPA and monomeric DTPA-fMLFK were used as control preparations. Images were acquired over a 20 hr. period, followed by a biodistribution study.

The blood clearance rate of the 15 kDa preparation was fast; approximately 80% of activity was cleared from blood during the first 15 minutes through kidneys; the rest was cleared with a half-life of 45 min. The 70 kDa preparation showed half-life of 2 hr. with no initial fast phase. Both preparations significantly accumulated in the infection site. Scintigraphic images of the final biodistributions are shown in Figure 6.

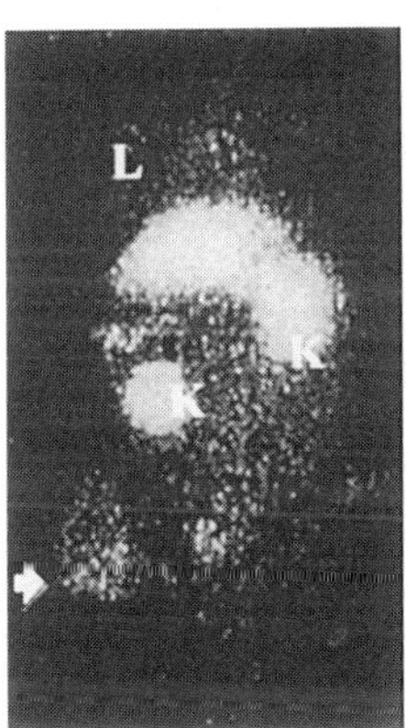

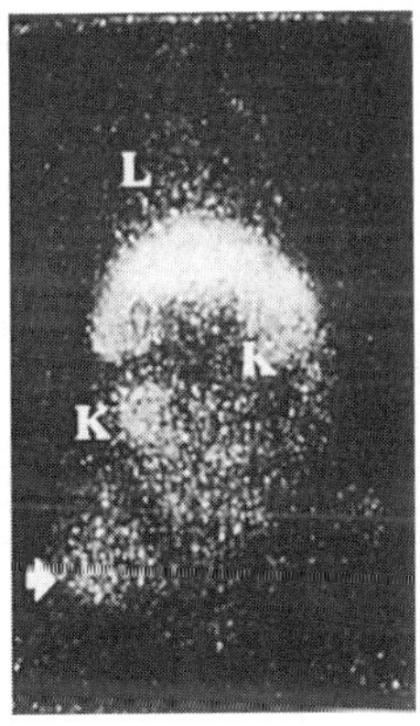

Figure 6. Whole body scintigraphic images of rabbit (inflammation model).

Anterior view, 20 hr. after administration of radiolabeled f-MLFK (left, control) and f-MLFK-PHF conjugate (right). K: kidneys; L: liver.

Note accumulation of both preparations in the inflammation (arrow), and significantly lower out-of-target accumulation of the f-MLFK-PHF conjugate, especially in kidneys.

The biodistribution data showed that immobilization of multiple f-MLFK molecules on PHF did not increase label accumulation in RES as compared to monomolecular f-MLFK, and decreased accumulation in kidneys by 80% (37). This study showed feasibility of PHF (from both technological and biological points of view) as a backbone polymer in targeted bioconjugates.

Discussion

The goal of this study was to determine whether a polymer emulating common acyclic structures of bilogical interface carbohydrates (hydrophilic polyacetal) would have a combination of properties close to an "idealized" biomedical material, such as: "inertness" in vivo, biodegradability of the main chain, low toxicity, and technological flexibility.

The model hydrophilic polyacetal, PHF, was produced via complete elimination of carbon 3 from carbohydrate residues of poly-(1->6)- α-D-glucose main chain of Dextran B 512. The blood clearance rates of PHF and PHF-protected macromolecules (graft copolymers) were close to that of similarly structured derivatives of polyethyleneglycol, (35) which is currently the "gold standard" of biological inertness, and significantly longer than of analogously structured derivatives of dextran B-512 (36).

The potential advantages of hydrophilic polyacetals, as compared with polyethyleneglycol, are biodegradability and availability of readily modifiable groups along the main chain, which opens the way to producing various functional conjugates (37,38).

Advantages of polyacetals as compared to polysaccharides relate to both biological functionality and safety. For example, Dextran B512, (a.k.a. “pharmaceutical dextran”), which is known as one of the least biologically active polysaccharides (39), is a product of a microorganism (*Leuconostoc Mesenteroides*). Dextran is known to produce anaphylactoid reactions that are mediated by immunoglobulins specific to isomalto-oligosaccharides (40). The origin of the immunity is unknown; however, it has been shown recently that *Streptococcus Sanguinus*, an oral streptococcus prevalent in dental plaques (41), produces isomalto-oligosaccharide containing lipoteichoic acid (42,43). The latter was shown to bind recognition proteins of plasma (44) and stimulate immunocompetent cells (45,46). Therefore, *S. Sanguinus* can potentially induce production of oligoisomaltose-reactive antibodies, and the associated sensitivity to dextran-containing preparations, practically in any individual. Obviously, biomaterials lacking receptor-recognizable domains and antigenic determinants of wide-spread bacterial species would convey a much lower risk of anaphylactoid reactions.

Conclusion

The experimentally determined properties of the synthesized model acyclic hydrophilic polyacetal (PHF) were in a good agreement with the hypothesis that polymers obtained via partial emulation of polysaccharides may have an excellent combination of useful features. Properties of PHF suggest the potential utility of polymers of this type in pharmacology and bioengineering, for example as structural or protective components in macromolecular drugs, drug delivery systems, and templates for tissue engineering. Development of carbohydrate-derived and fully synthetic hydrophilic polyacetals may become a promising direction in the development of new biomedical materials.

Acknowledgments

The described work was supported by grants from The Whitaker Foundation, Genetic Therapy Inc. (A Novartis Company), and US Army. Author is grateful to R. Wilkinson and S. Hillier for assistance in animal experiments. Thanks are due to Drs. A.J. Fischman, J-P. Dotto, T.J. Brady and R. Gross for valuable discussions.

Literature cited

1 Atkinson, J.P., and Frank, M.M.. J. Clin. Investigation 1974, 54, 339-348.
2 Walton, K.W., Almond, T.J., Robinson, M., Scott, D.L. Br. J. Exp. Path. 1984, 65, 191-200.
3 Benecky, M.J., Kolvenbach, C.G., Wine, R.W., DiOrio, J.P., Mosesson, M.W. Biochemistry 1990, 29, 3082-3091.
4 Khan, M.Y., Medow, M.S., Newman, S.A. Biochem. J. 1990, 270, 33-38.
5 Kinoshita T. Immunology today 1991, 12, 291-301.
6 Law, S.K., Lichtenberg, N.A., Levine, R.P. Proc. Natl. Acad. Sci. USA 1980, 77, 7194-7198.
7 Law, S.K., Minich, T.M., and Levine, R.P. Biochemistry 1984, 23, 3267-3272.
8 Sim, RB. and Reid, K.B.M. Immunology today 1991 12, 307-311.
9 Young, N.M., and Leon, M.A. BBRC 1987 143, 645-651.
10 Ezekowitz, R.A.B., Day, L.E., and Herman, G.A. J. Exp.Med. 1988, 167, 1034-1047.
11 Wright, S.D., Tobias, P.S., Ulevitch, R.J., Ramos, R.A. J. Exp. Medicine 1989, 170, 1231-41
12 Tobias, P.S., Soldau, K., Kline, L., Lee, J.D., Kato, K., Martin, T.P., Ulevitch, R.J. J. of Immunology 1993 150, 3011-3021.
13 Savill, J., Dransfield, I., Hogg, N., Haslett, C. Nature 1990. 343, 170-173.
14 Papisov M.I. Adv. Drug Delivery Rev. 1995, 16, 127-137.
15 Papisov M.I. Adv. Drug Delivery Rev. 1998, 32, 119-138.
16 Ehlenberger, A.G., and Nussenzweig, V. J. Exp. Medicine 1977 145, 357-371.
17 Ross, G.D., Cain, J.A., Lachmann, P.J. The J. of Immunology 1985 134, 3307-3315
18 Papisov M.I., Weissleder R. Crit. Rev. in Therap. Drug Carrier Systems 1996, 13, 57-84.
19 Jeanes A. Molecular Immunology 1986, 23, 999-1028
20 Ishak M.F., Painter T.J. Carbohyd. Res. 1978, 64, 189-197
21 Dubois M., Gilles K.A., Hamilton J.K., Rebers P.A., Smith F. Anal. Chem. 1956, 28, 350-356.
22 Richards E.L. Aust. J. Chem. 1970, 23, 1033-37
23 Papisov M.I. US Patent # 5,811,510, 09/22/1998.
24 Papisov M.I. et al. The biologically inert backbone of Dextran B 512, 2000, under review.
25 Drobchenko S.N., Isaeva-Ivanova L.S., Kleiner A.R., Lomakin A.V., Kolker A.R., Noskin V.A. Carbohyd. Res. 1993, 241, 189-199.
26 Miyasaki K. Virchows Arch. A Path. Anat. And Histol.1975, 365, 351-365
27 Sterzel B., Eizenbach G., Seiler W., Hoyer J. Am. J. Pathol. 1983, 111, 247-257.
28 Papisov M.I. et al. 1997-99, unpublished data.
29 Mosbach K., Editor. Immobilized enzymes. Meth. Enzymol. 1976, 44.
30 Torchilin V.P., Editor. Immobilized Enzymes in Medicine. Springer-Verlag, Berlin, 1991.

31 Papisov M.I., Weissleder R., Brady T.J. In: Targeted delivery of imaging agents; Torchilin V.P., Editor, CRC Press, Boca Raton, 1995; 385-402.
32 Papisov M.I., Savelyev V.Y., Sergienko V.B., Torchilin V.P. Int. J. Pharm. 1987, 40, 201-206.
33 Papisov M.I., Bogdanov A.A., Schaffer B.S., Nossiff N., Shen T., Weissleder R., Brady T.J. J of Magnetism & Magnetic Materials 1993, 122, 383-386.
34 Babich J.W., Graham W., Barrow S.A., Dragotakes S.C., Tompkins R.G., Rubin R.H., Fischman A.J. J. Nucl. Med. 1993, 34, 2176-2181.
35 Bogdanov A.A., Weissleder R., Frank H.W., Bogdanova A.V., Nossiff N., Schaeffer B.K., Tsai E., Papisov M.I., Brady T.J. Radiology 1993, 187, 701-706.
36 Papisov M.I., Weissleder R., Schaffer B.K., Tsai E., Bogdanov A.A., Jr, Nossiff N., Bogdanova A.V., Brady T.J. In: Proc. Soc. Mag. Res. in Medicine, Twelfth Annual Scientific meeting, New York, NY; SMRM, Berkley, CA, 1993, 499.
37 Papisov M.I., Babich J.W., Dotto P., Barzana M., Hillier S., Graham-Coco W., Fischman A.J. In: 25th Int. Symp. Controlled Release of Bioactive Materials, 1998, Las Vegas, Nevada, USA; Controlled Release Society, Deerfield, IL, 1998, 170-171.
38 Papisov M.I., Babich J.W., Barzana M., Hillier S., Graham-Coco W., Fischman AJ. J. Nuc. Med. 1998, 39 (5, Suppl.), 223P.
39 Larsen C. Adv. Drug Delivery Rev. 1989 3, 103-154
40 Ljungstrom, K.G., Willman, B. and Hedin, H. Ann. Fr. Anesth. Reanim. 1993, 12, 219-222.
41 HardieJ.M., Marsh P.D. in: Streptococci; Editors: Skinner F.A., Quesnel L.B., Academic Press, London, 1978, 157-206.
42 Kochanowski B., Fischer W., Iida-Tanaka N., Ishizuka I. Eur. J. Biochem. 1993, 214, 747-755.
43 Kochanowski B., Leopold K., Fischer W. Eur. J. Biochem. 1993, 214, 757-761.
44 BradeL., BradeH., Fisher W. Microb. Pathog. 1990, 9, 355-362.
45 Bkhakdi S., Klonisch T., Nuber P., Fisher W. Infect. Immun. 1991, 59, 4614-4620
46 Keller R., Fisher W., Keist R., Basctti S. Infect. Immun. 1992, 60, 3664-3672

Biodegradable Polymers: Status and Testing

Chapter 20

Biodegradable Polymers in the Marine Environment: A Tiered Approach to Assessing Microbial Degradability

Jo Ann Ratto[1], Joshua Russo[1], Alfred Allen[1], Jean Herbert[1], and Carl Wirsen[2]

[1]U. S. Army Soldier System Command, RD&E Center, Kansas Street, SSCNG-VM, Natick, MA 01760–5020
[2]Biology Department, Woods Hole Oceanographic Institution, Woods Hole, MA 02543

With the oceans comprising more than 99% of the world's biosphere by volume, it is essential to keep it clean for future generations. The assessment of the biodegradability and impact of polymeric materials disposed in the marine environment is an important consideration in achieving this goal. Due to the variation in conditions that can control microbial mineralization rates in the marine environment (e.g. hydrostatic pressure, temperature, and nutrient limitations), approaches need to be examined that are reasonable in both cost and time, which give the best estimate of biodegradability under actual conditions. Utilizing a variety of biodegradable polymers, biodegradation studies with a tiered test approach have been conducted. Carbon dioxide evolution experiments (Tier I), closed (static) and open (dynamic) incubations, laboratory hydrostatic pressure and radiolabled polymer experiments (Tier II) as well as coastal and deep sea in situ mooring incubations (Tier III) can all be used to evaluate the level of biodegradation.

Introduction

"International Year of the Ocean" was declared by the United Nations to be celebrated in 1998.[1] This year recognizes the importance of our oceans, the marine environment and its resources for sustainable development at regional and global levels. A major goal is to diminish pollutants and waste entering the oceans. Plastic debris degrades coastal areas and it is a threat for injuring marine animals by entanglement or ingestion situations.[2,3,4]

In 1975, the National Academy of Science estimated that there were 14 billion tons of garbage dumped in the ocean every year with the United States being responsible for about one third of this pollution.[4] There are now many important programs that have been initiated to help control this pollution problem. For example, the Center for Marine Conservation coordinates coastal cleanups while the Ocean Pollution Research Center develops strategies and methodologies to deal with ocean pollution.[5] The Marine Plastic Pollution Research Control Act (MARPOL) Annex V was initiated in 1988 and currently this international treaty has been adopted by 69 countries to protect the marine environment from various types of garbage.[5] Under this treaty, the discharge of all plastic is prohibited anywhere in the ocean.

The U.S. Navy has conducted several studies to characterize the solid waste generated on its vessels and found on average that each person generated 0.57 gallons of plastic waste per day.[6,7] Retention of waste on board ship is one method to comply with the MARPOL treaty but this has a significant impact on the consumption of manpower and the attitude and behavior of the Navy personnel.[7] The waste-induced odors on board ship can reach unacceptable levels and become hazardous to one's health. The U.S. Navy is currently considering biodegradable polymers for on board ship packaging items. If the polymer is fully biochemically degradable, the plastic can potentially be disposed of in the ocean as long as the rate addition of plastic does not exceed the rate of biodegradation.

However, the biodegradation rates of biodegradable polymers are generally slower in the marine environment than in other ecological compartments (compost, soil and/or wastewater). Microbial activities in the ocean are often limited by a variety of parameters: high hydrostatic pressure, cold temperatures, low oxygen concentration, and sparse nutrients. Studies at the Woods Hole Oceanographic Institution have been done on the microbial degradation activity of both soluble and solid organic substrates and some of these methods are used with biodegradable polymers.[8,9,10] The U.S. Army Soldier System Command at Natick also has pursued the development of biodegradation methods and is currently working jointly with Woods Hole.[11-17]

Other laboratories have also studied some polymer systems and microbial activity with various types of tests in the marine environment.[18-26] Several groups have studied oligosaccharides and polysaccharides such as chitin, chitosan and cellulose in the marine environment.[10,18-21] Poly (3 hyroxybutyrate-co-valerate (PHBV) polymers[22-23] and starch polymer blends[24-28] have also been studied in the marine environment. Clamon-Dicriaud et al. have a recent review on the variety of methods that exist globally for biodegradation.[29] There are marine methods for water soluble materials, but there is clearly a need for global biodegradation methods for water insoluble
polymers.[29-32]

Based on the current methods and biodegradation studies of polymers [33-34] in compost, a similar tiered approach was adopted as shown in Table I.

Table I. Tier Test Summary for Marine Environment

Tier I (*Screening Level*)	*Tier II* (*Confirmatory Level I*)	*Tier III* (*Confirmatory Level II*)
CO_2 Evolution Method Respirometry Modified Sturm (ASTM 5209)	**Incubation Methods** Static (closed constant laboratory conditions) Dynamic (open flowing seawater aquarium) **Pressure Vessel Methods** **Radiolabeled Polymer Methods**	**Incubation Methods** Coastal Studies Deep Sea Moorings

The three different levels are as follows: Tier I is a screening level to indicate inherent degradability of the polymer and has been used in screening biodegradability of polymers in compost and sewage sludge environments.[35-36] The experiments are done under controlled conditions to ensure reproducibility and precision. For the marine environment, a modified Sturm test and respirometry experiments are used. Tier II has been referred to as Confirmation Level I testing that encompasses laboratory and/or pilot scale tests where practical degradability is evaluated. These are experiments under controlled conditions but the design of the experiment resembles the actual environment. For this type of experiment, laboratory and aquarium incubations under both static and dynamic conditions are employed with natural microorganisms from sea water. Pressure and temperature are incorporated as a variables into the experiment. Also, ^{14}C-labeled polymers, if available, can be used to generate confirmatory data to the Tier I results. This Tier II approach was used to measure the respiration of [1,2-^{14}C] vinyl alcohol/ethylene copolymer (EVOH),[37] a compound used in many starch based biopolymers. These studies showed that biodegradation of this material in the marine environment ranged from 10 to 500 years depending on the chemical and physical conditions of incubation. If one just measures the biodegradation of the complete polymer there is no way of assessing the actual biodegradation of the specific component of the plastic. By using specific radiolabeled components such as this or others that are of potential interest in polymer production the actual question of biodegradation can be addressed in a precise and quantitative fashion.

Tier III, known as the Confirmation Level II test, covers the *in situ* degradation of a polymer in a full scale or natural field environment. In this scenario, examples of the different types of testing can be used to assess the degree of biodegradability with a variety of polymers and some of the advantages and disadvantages inherent for these methods will be discussed. The polymers presented in this study were chosen to demonstrate the validity and trends of the Tier testing approach. However, a variety of other polymers also have been studied using these same methods.[38] Previous studies have focused on specific polymers and how the polymer's composition affects biodegradation rates.[11-15] For example, the copolymer poly(3-hydroxybutyrate-co-valerate) with different valerate contents has been systematically studied.[12] In the present study, data is presented for the PHBV polymer, in manufactured forms (i.e. film vs. paper cup vs. powder) to see how it behaves with the various methods keeping in mind the realistic fate of these materials.

Experimental Methods

Tiered I Tests

Respirometry

Materials

Difco® Marine Agar and Difco® Marine Broth were used for culturing microorganisms in these experiments. The mineral marine salt solution composition is found in Table 2. The Instant Ocean is a commercial synthetic seawater from Aquarium System, Inc. and has the following ionic composition (wt%) Cl^-, 55.007: Na^+, 30.473; SO_4^{-2} 7.632; Mg^{2+} , 3.741; Ca^{+2}, 1.6262; K^+, 1.136; HCO_3^-, 0.570; BO_3^{-2}, 0.158; and Sr^{+2}, 0.030. All of the components were prepared using distilled water and sterilized by autoclaving.

Table II. Composition of the Marine Solution for Respirometry Experiement

Substance	*Formula*	*MW g/mol*	*Concentration g/L*
Ammonium Chloride	NH_4Cl	53.49	2.0
Instant Ocean®	-	-	17.5
Magnesium Sulfate, 7-Hydrate	$MgSO_4 7H_20$	246.48	2.0
Potassium Nitrate	KNO_3	101.1	0.5
Potassium Phosphate	$K_2HPO_4 3H_20$	228.2	0.1

The inoculum consisted of thirteen marine microbes collected from the northeast Atlantic coast and one from a tropical location in the Pacific (Hawaii) . The microorganisms were identified by using the Biolog system and various bacterial identification tests (i.e. Gram stains). Their identification to the genus and further characterization to the species level are as follows: *Alteromonas haloplanktis, Xanthomonas campestri, Vibrio alginolyticus, Vibrio proteolyticus, Actinomycete sp., Bacillus megaterium, Bacillus sp., Zooster sp. and Pseudomonas sp.* More than one microorganism occurs at some genus levels.

Test polymer substrates of known weight, 10- 40 mg, had sufficient carbon content to yield carbon dioxide volumes that could be measured accurately using the respirometer. The carbon content was determined by calculation or elemental analysis. Test polymers were in the form of powders, films, fragments, formed articles, or aqueous solutions. All samples were either cryogenically milled to obtain powders or cut into small pieces. Sucrose and/or dextrose serve as the positive control and the inoculum without added substrate was the negative control.

Methods

Sterile flasks containing 40 mL of marine broth that contains nutrients essential for cultivating marine bacteria were inoculated with 250μl stock cultures of each of the 12 microorganisms. Each culture was grown in a separate flask and allowed to grow overnight at 30°C. The optical density (O.D.) was approximately 2.0.

Each culture was harvested by centrifugation at approximately 27000 (g) for eight minutes. Cell pellets were rinsed twice with minimal marine medium and each organism was resuspended in 4mL of minimal marine medium. Each cell suspension was combined together in a 50 mL sterile tube and vortexed to disperse any clusters.

One hundred microliters of the combined cell suspension was used to inoculate 75 mL of the marine medium in 250 mL respirometry flasks. A micro-oxymax version respirometer (Columbus Instruments) was used to quantify the oxygen consumption and carbon dioxide evolution. The system is capable of sampling 20 chambers in sequence during a given experiment. A controlled-environment shaker maintained the temperature of the digestion flasks at 30 ± 2 °C with agitation at 175 rpm. The polymer test samples and positive/negative controls were studied in triplicate. The flasks were sampled at regularly timed intervals and experiments were terminated once the CO_2 production rate approached zero. Typical respirometer experiments ran from 250 to 500 hours.

Using a respirometer to measure the total carbon dioxide (CO_2) produced as a function of time; biodegradability was assessed by determining the proportion of polymer-carbon converted to carbon dioxide. The sole carbon source in these experiments was the polymeric substrates, thus the quantity of carbon dioxide evolved determines the minimum extent of biodegradation, since some carbon from the substrate was used to build biomass. The percent of actual carbon content of the CO_2 produced, expressed as a fraction of the measured or theoretical carbon content of the test material, was graphed as a function of time.

Modified ASTM 5209 (Sturm Test)

The modified Sturm test measures carbon dioxide evolved as a function of time by titration methods. The procedure followed is detailed in the ASTM methods 5209 (Determining the Aerobic Biodegradation of Plastic Materials in the Presence of Municipal Sewage Sludge)[39] and only significant procedural changes are mentioned in this text. The major modification was that natural seawater containing approximately 1.44 x 10^6 cells/mL of microorganisms was used instead of activated sludge as the inoculum. The inoculum and samples were run with and without nitrogen (0.5 g/L NH_4Cl) and phosphorous (0.1g/LKH_2PO_4) additions. The temperature was constant at 20 $\pm$ 2 °C and pH was recorded at the start and end of the experiment. Other modifications to this method include: a decrease in experimental volume to 700 mL volume from 2,470 mL, a decrease to 50 mL titrated aliquots with a correction calculation rather than 100 mL of barium hydroxide.

Tiered II Tests

Static Incubation

Time course (1 year) static laboratory incubations of pre-dried and weighed samples were carried out in compartmentalized plastic containers. The compartments (250 mL volume) contained either aerated natural seawater, or natural seawater overlaying 2 cm of sediment. The system remained aerobic other than the bottom side of the polymer materials in contact with sediment. The size of sample materials added was approximately 3 x 8 cm in the case of films or smaller depending on the thickness and amount of material on hand. Triplicate compartments were set up for each incubation condition so that time course sampling was possible. Each set of triplicates was run under two nutrient conditions a) unsupplemented and, b) supplemented with 5.0 mg/L nitrogen (as NH_4Cl) and 1.0 mg/L phosphorous (as KH_2PO_4). Incubations were carried out at 3°C and 23°C.

Dynamic Incubations

Time course (1 year) dynamic aquarium incubations were conducted in tanks of continuously flowing seawater. Containers and sample size were the same as in the static incubations. There were three sets of exposure conditions in these aquarium incubations which included a) flowing seawater, b) contact with sediment and flowing seawater, and c) contact with *Spartina altiniflora* (eelgrass) enriched sediment and flowing seawater. Nylon mesh with 5mm square openings was placed over the flowing seawater containers to prevent test materials from floating away. The eelgrass additions were introduced to the systems as a long term supply of slowly releasable nitrogen. The addition of soluble nitrogen or phosphate salts to these open systems was not feasible due to the flow of water through the tanks. Aquarium incubations allow for open system conditions (flushing, removal of metabolic by- products, resupply of oxygen and nutrients etc.) and seasonal effects such as temperature and nutrient supply variations. To demonstrate the seasonal effects over the course of a year, the temperature and nitrogen concentration (as ammonia and nitrate) were monitored.

Hydrostatic Pressure Vessels

Time course laboratory static incubations aimed at duplicating deep sea conditions (350 atm hydrostatic pressure and 3°C), were conducted using steel pressure vessels. The polymers tested were significantly smaller in sample size (approx. 1.5 cm^2) than in the above incubations and were weighed on a Mettler micro balance. To ensure aerobic conditions over the lengthy 1.5 year incubation, the samples were individually heat sealed in oxygen permeable non-biodegradable plastic pouches containing natural seawater collected from 3500m depth in the Atlantic Ocean. By suspending these pouches in the large pressure vessel, oxygen could diffuse from the surrounding water into the samples if the level was reduced due to biological activity. A duplicate set of pouches for each material was prepared which contained nitrogen and phosphorus nutrient supplementation (as noted above) and was incubated in the same pressurized fashion.

Sample Processing

At various time points, polymer samples were harvested for determination of dry weights, photography and/or scanning electron microscopy. For dry weight analysis, the samples were carefully rinsed in fresh water to remove salts and deposited sediment/silt and then dried to constant weight before weighing. Sterile controls were used for the purpose of determining weight loss due to solubilization of plasticizers, non-microbial hydrolysis or other ingredients. This was accomplished by incubating samples of the different test polymers in sterilized seawater for 48 to 72 hours and determining dry weight loss. Certain materials lost a notable percentage of initial dry weight by this sterile incubation (3-10%) while others demonstrated weight loss due to soaking of ~ 1% or less. The reported values of actual dry weight loss in the tables are corrected for this effect.

Radioisotope Experiments

Isotopically labeled samples were used to determine biodegradation potential in the marine environment. Samples of a ^{14}C-radiolabeled polymer were weighed out in a hood under low exhaust velocity using a Mettler micro balance. The specific activity (μCi/mg) of the polymer was determined gravimetrically. In general, the weighed samples were placed in sterile 125 mL serum bottles to which 50 mL of an appropriate medium was added. In the case of natural population experiments, 50 mL of collected seawater, which in some cases was supplemented with approximately 10 g of sediment as an additional inoculum, was added to the bottles. Pure culture experiments were studied in a similar fashion. The bottles were crimp sealed using teflon faced rubber serum closures and shaken at 100 rpm. The 75 cc headspace of air ensured that aerobic conditions prevailed over the course of the incubation. Non-radioactive controls containing the same amount of polymer carbon were monitored by oxygen electrode to ensure that oxygen limitation was not the controlling factor in the level of measured respiration. At desired time points, microbial activity in individual bottles was terminated and the radiolabeled carbon dioxide ($^{14}CO_2$) produced by metabolism of the copolymer was collected. For this purpose, the original serum closure was replaced quickly with another rubber serum stopper which holds a plastic center well (Kontes 882320-0000) containing a fluted piece of filter paper (Whatman #1) saturated with 0.3 mL of Hyamine Hydroxide (methyl benzethonium hydroxide). This was followed by injection of 1 N H_2SO_4 through the stopper into the medium to acidify to pH 3.0. The necessary amount of acid to achieve the desired pH was predetermined on a non-radioactive aliquot. Following acidification, the bottles were shaken slowly (100 rpm) for 6 hours during which time CO_2 was trapped in the hyamine hydroxide. After this time period the whole center well cup containing the filter paper was snipped off and placed in a scintillation vial and counted on a liquid scintillation counter. The amount of polymer (or polymer carbon) that was respired to CO_2 was determined knowing the weight (or carbon content) of substrate added, the specific activity, and the amount of activity recovered as CO_2. In determining this respiration data, corresponding sterile natural seawater controls were run to account for volatilization and varied pH effects during trapping and the experimental DPM values adjusted accordingly. A variation in the procedure was made when incubating samples at elevated hydrostatic pressure.[44, 45]

Tiered III Tests

Coastal incubations

Coastal field incubations were conducted in two ways. Off shore shallow water moorings were set at a site 19 miles south west of Woods Hole with a water depth of 41 m. A plexiglass box containing several compartments each covered on two sides by Nitex® screening of different mesh size was attached to the mooring cable and deployed on the bottom for incubations lasting up to an 8 months. Inshore harbor incubations were set and retrieved using SCUBA. In this case, polymer samples were pre-placed in covered mud boxes or in mesh enclosed float arrays and transported to the bottom of Woods Hole harbor at a depth of 22 m. The mud boxes were opened after placement on the bottom. Samples of polymer were incubated in the water column, in contact with the surface of the sediment, or 2 to 3 cm below the sediment surface. Samples were incubated in the presence and absence of marsh grass amendments (*Spartina alterniflora*), as a long term source of potentially available nitrogen to the microbial communities involved in polymer biodegradation.

Deep sea *in situ* incubations

These types of incubations for certain materials were also conducted by attaching nylon mesh (3 mm openings) packages containing polymer samples to deep sea moorings at a depths of 4000 to 4500 m in the Atlantic Ocean south of Bermuda. Following retrieval, these samples were processed as mentioned previously for Tier II incubations.

Results and Discussion

Tier I respirometry tests were used to determine the degree and rate of aerobic biodegradation of plastic materials, including formulation additives, exposed to aerobic microorganisms. Carbon that was consumed by microorganisms was converted into microbial biomass and CO_2, the ratio of which may vary with the particular polymer. Respirometry measures CO_2 evolution within the chamber and does not account for the carbon which has been converted into biomass and is therefore representative of "minimum" evidence of biodegradation. The test was designed to yield reproducible results under controlled laboratory conditions. Figure 1 shows the actual cumulative CO_2 evolved as a function of time during PHBV biodegradation. A powder sample was used and the experiments were performed in triplicate. There is some deviation in the data which may be due to the sample geometry or possible inoculum inhomogeniety in each chamber. An average was taken for the three trials and the net mineralization was calculated to be 65% .

Figure 2 illustrates the net mineralization (%) of a sucrose positive control, inoculum negative control and wheat gluten samples unsupplemented and supplemented with ammonia. The wheat gluten samples follow the same trend for biodegradation, yet the supplemented ammonia wheat gluten sample had slightly higher percent mineralization. One would expect this higher value because nitrogen as well as carbon is necessary for microbial growth.

Currently the Modified Sturm Test (ASTM D 5209) is used to determine the aerobic degradation of plastic like materials using a sewage sludge inoculum. Further modifiying this test to use a natural seawater inoculum, with nutrient supplementation in some cases, respirometry data for glucose, kraft paper

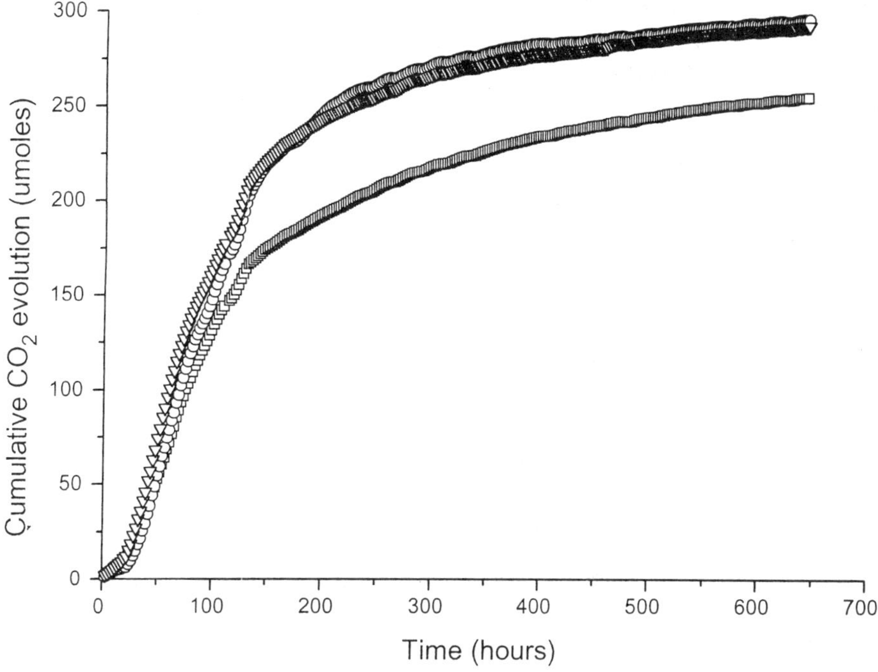

Figure 1. Tier I respirometry data: cumulative carbon dioxide versus time for PHBV (0 = trial 1, ∇ = trial 2, □ = trial 3).

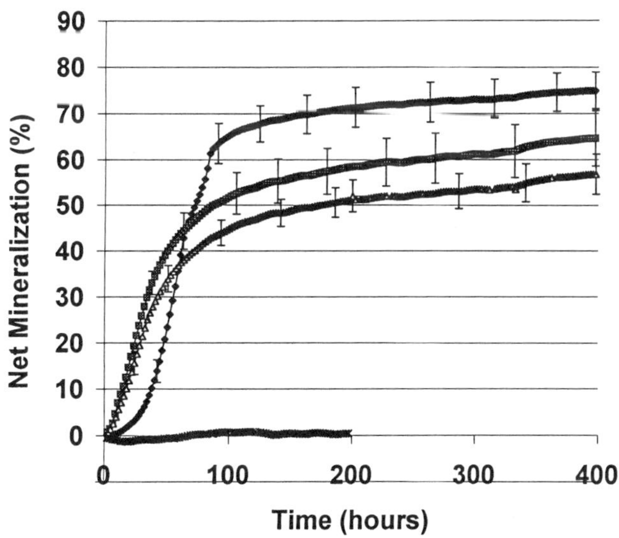

Figure 2. Tier I respirometry data: net mineralization (%) versus time for wheat gluten resins. (* = inoculum, Δ = wheat gluten, □ wheat gluten with ammonia, ◊ = sucrose)

and PHBV were obtained (Figure 3). Note that that materials were run with nitrogen and phosphate nutrients. Without added nutrients, minimal CO_2 evolution takes place for all materials tests, and this is clearly seen in Figure 3 for glucose. After 340 hours, the unsupplemented glucose shows 3% mineralization while the supplemented glucose nutrients has plateaued in 200 hours at approximately 56% mineralization. Both kraft paper and PHBV with nutrients continue to produce carbon dioxide after 340 hours.

This modified (for the marine environment) Sturm test was also based on the evolution of CO_2 from the polymers which can be trapped from a continuous flowing air stream. This air must be scrubbed free of carbon dioxide prior to entering the test flasks[39,40] This approach works well in composting for very metabolically active systems as described for the Sturm test. However in natural seawater and in particular the deep sea, microbial activity is generally reduced in comparison to other environments and microbes would usually be nitrogen and perhaps phosphorous limited in relation to the high content of carbon present in a "biodegradable polymer." The Sturm test has some procedural limitations such as the volume sizes used are excessive for the slowly degrading polymers. The NaOH absorbing potential must be calculated accurately as to how much CO_2 it can absorb from the air before being changed but very necessary so as not to bubble "foreign" CO_2 through the system. For these experiments, a large scale compressor was necessary to keep the air flow steady. Also, backpressure due to $BaCO_3$ forming in the $BaOH_2$ solutions on the frittered glass can cause variations in airflow thus requiring periodic cleaning in acid during a long test period. Overall, the test is cumbersome, somewhat costly, takes up considerable space and requires significant attention from the scientist.

The degree and rate of aerobic respiration of a polymer under the conditions of these Tier I tests may be used to estimate the potential for biodegradation persistence of that compound in biologically active marine environments; e.g. seashore and open-ocean. However, it should be recognized that predicting long-term environmental fate and effects from the results of short-term exposure to a simulated marine environment is difficult. Thus, caution should be exercised when extrapolating the results obtained from this or any other controlled-environment test to biodegradation in the natural environment.

Tier II and III

Radiolabeled Polymer Experiments

To truly assess if a polymeric material is biodegraded by natural microbes, it must be demonstrated that the carbon from the polymer is transformed into cell biomass and respired CO_2. A very precise way to do this and eliminate the need for

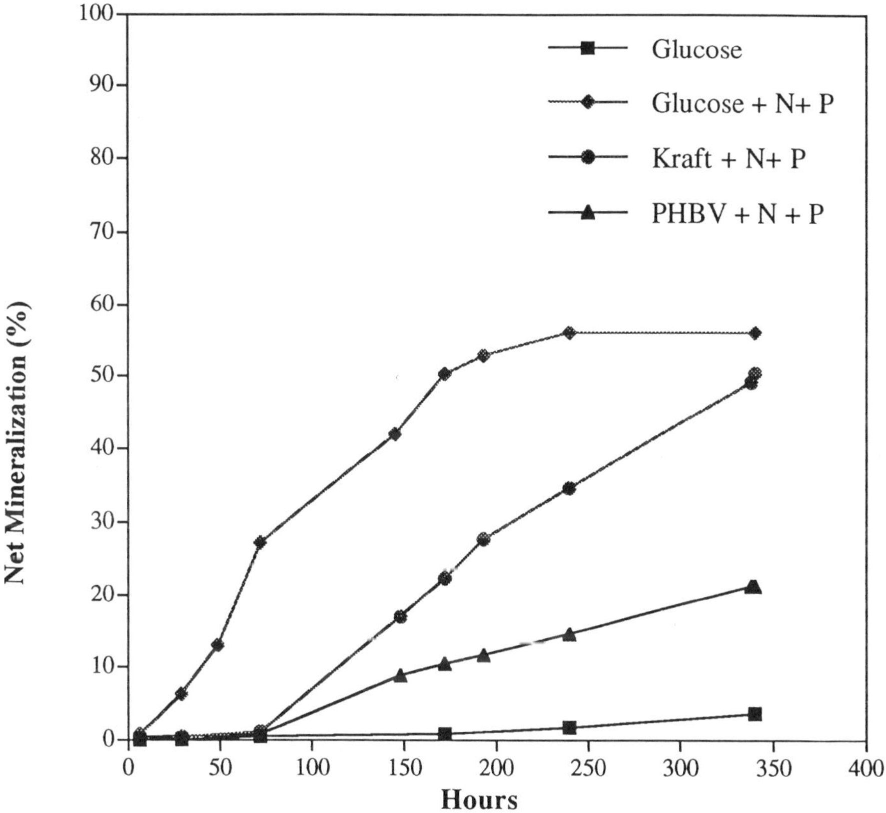

Figure 3. Tier I sturm test data: net mineralization (%) versus time for PHBV, glucose and kraft paper. (N + P designates the addition of nutrients)

continuous airflow and possibility of non-polymer carbon (i.e. carbon dioxide from the air) being trapped as in the modified Sturm test, is to use small amounts of radiolabeled polymers. The ^{14}C isotopically labeled polymer can be microbially degraded producing $^{14}CO_2$ which can be trapped, as in the Sturm test, from samples over a time course experiment (see methods). This approach has been used for many years in addressing the microbial activity of cellular incorporation and respiration of a host of organic compounds in the marine environment under many varied incubation conditions.[41, 42, 43] Most recently, experiments were carried out using radiolabeled polyhydroxybutarate (PHB) with an activity of 0.152 μCi/mg. Approximately 6.0 mg of poly hydroxybutarate (PHB) was used in each 100 mL seawater sample and time course collections of $^{14}CO_2$ were made. The results are shown in Figure 4. The PHB in sea water was amended with nitrogen and phosphate while the PHB in sea water / sediment did not have these nutrients added. There is considerable difference in the results with the nutrients strongly enhancing the rate and level of respired carbon dioxide. The percent carbon dioxide respired for the PHB water sample increased steadily after 50 hours and reached a plateau at 70% after 550 hours while the sample in sediment with no nutrients reached a plateau after 300 hours at only 10%. The sample in contact with the sediment should have a supply of nutrients greater than the sea water and one would expect a higher net mineralization. However, there is not enough sediment in this closed experimental system to allow for extensive mineralization of the polymer. Further experiments with supplementation of nutrients with sediment will determine any additional benefit or effects from the nutrients and sediment.

Incubations

Incubation experiments are based on weight loss though this is not always an indication of the biodegradation (e.g. mechanical disintegration may occur) or biodegradation rate. That is why the Tier I tests were essential for an initial screening. Tables III and IV show the static and dynamic incubation results for PHBV coated cup material; a product that is of interest for potential use by the Navy.

Table III. Tier II weight loss data for the PHBV coated cup in the static incubation. (N + P designates the addition of nutrients)

	Weight Loss (%)			
Time (months)	(3°C) N + P	(3°C) N + P Sediment	(23°C) N + P	(23°C) N + P Sediment
2	3.9	2.6	17.7	76.9
6	4.6	2.8	23.7	86.0
12	7.1	37.0	53.5	100

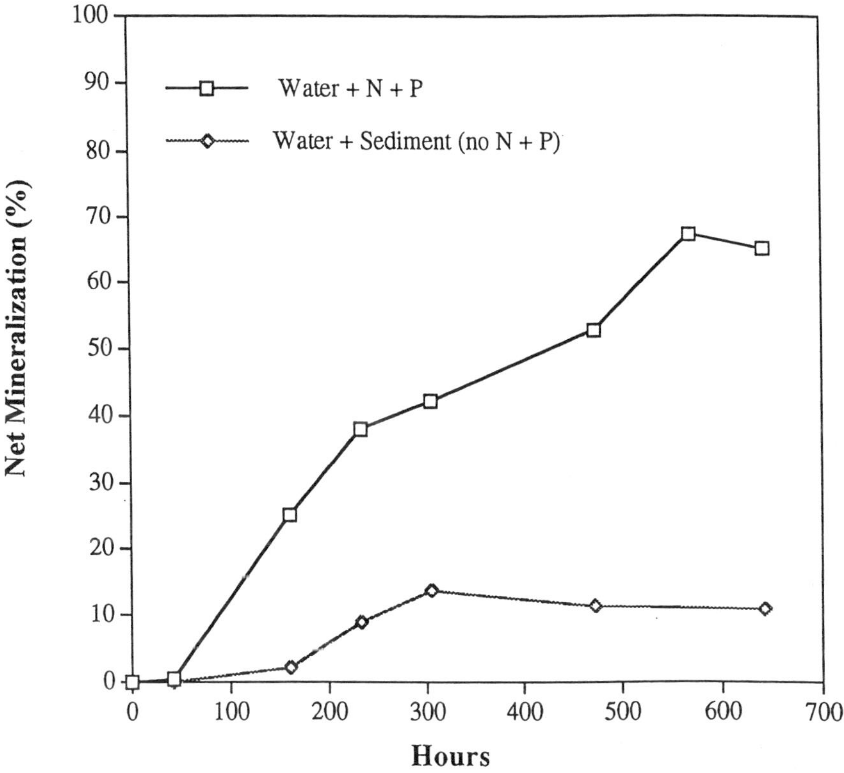

Figure 4. Tier II radiolabeled polymer test data for PHB: net mineralization (%) versus time. (N + P designates the addition of nutrients)

For static experiments, the results are highly dependent on both temperature and nutrient availability. Incubations carried out in water without added nitrogen and phosphorous produced no weight loss after a 12 month test exposure (data not presented). Conversely, there was significant weight loss from the samples incubated in the nutrient supplemented waters (Table III). At 23°C, with the samples in contact with sediment and nutrients there was a 100% weight loss compared to a significantly lower percentage of 37% at 3°C.

Table IV. Tier II weight loss data for the PHBV coated cup in the dynamic incubation.

	Weight Loss (%)		
Time (months)	Flowing Seawater	Flowing Seawater and Sediment	Flowing Seawater, Sediment and Spartina
2	0	22.0	28.5
5	61.5	96.0	99.7
6	100	100	100

Under dynamic aquarium conditions (Table IV), there was 100% biodegradation after 6 months incubation independent of nutrients, sediment or pure sea water. For this example and for other polymer systems, the dynamic incubation method consistently yielded higher weight losses most likely due to the open system associated with seasonal temperatures. The seasonal temperatures typically range from 3 to 24°C.

Incubation data of a pure PHBV film in both the dynamic aquarium (Tier II) and a coastal harbor incubation (Tier III) are shown in Figure 5a and 5b. In both cases the temperature of the water follows seasonal effects (range 3-23°C). There was no significant weight loss until after 6 months. It is interesting to note that during a 12 month test exposure in the water column weight loss from the samples amounted to only 40-50% in the water, whereas the samples completely degraded (i.e. 100% weight loss) in the sediment exposures. Thus, both the Tier II and Tier III tests achieve 100% weight loss at 12 months when the samples were in the sediment. Anaerobic processes may be important during sediment exposures.

Pressure vessel incubations in a closed system containing the PHBV at low temperature (3°C) and high pressure (350 atm) resulted in a complete inhibition of any biodegradation activity by the indigenous bacteria in the deep-water samples over one year incubation. Similar results also occurred for a variety of other biodegradable polymers.

This inhibition effect exists for the open system *in situ* mooring incubations (Tier III). A weight loss of 5.5 % after 8 months was determined for silk samples used for a sonar detector parachute. This material had 100% weight loss in a dynamic incubation test at 5 months and 70% mineralization in the respirometry test.

The same PHBV films as presented in Figure 5 were also studied by the *in situ* incubation. At 6 and 14 months, there were 39% and 53% weight losses respectively in the ocean incubated at a depth of 4000 m which are lower than what was observed in the coastal and dynamic incubation. An *in situ* incubation of an actual product, a PHBV coated cup produced for the U.S. Navy, was performed at 4500 m with the following results: 57% weight loss after 5 months and 82% after 12 months, 73% weight loss after 12 months with *spartina*. These higher values are due to some mechanical washing action and biodegradation of the paper component of the cup, not total biodegradation.

Overall, the deep sea *in situ* (Tier III) and laboratory pressure vessels (Tier II) experiments pose an extreme environment in which biodegradation of polymers may occur because pressure and temperature are significant factors in reducing microbial activity.

Summary

Polymeric materials will biodegrade in the marine environment at varying rates. Contact with sediment often will enhance biodegradation and thus anaerobic process should be considered important in overall biodegradation potential. Nutrients or lack thereof and temperature variation play major roles in the rate of biodegradation under these experimental conditions. If it were possible to bioengineer polymers which contain nitrogen and phosphorous in a compatible polymer matrix, the resulting biodegradation in the environment might be greatly enhanced. The conditions of the deep sea e.g. high pressure, low temperature and low nutrient concentrations all act to reduce biodegradation of polymers to perhaps environmentally unacceptably slow rates.

It is necessary to emphasize the Tier I and II experiments as the costs and logistics of field experimentation (Tier III), particularly for deep sea studies, prove to be cost prohibitive. We have presented our suggestions (Figure 6) on how to proceed with experiements in the marine environment. These include a Tier I test, a Tier II test of using the radiolabled polymer and the dynamic incubation. A Tier III test will only be carried out if the results from the Tier II experiments are favorable. Continued studies in the actual marine environment will be necessary as new materials are produced and postulated to be biodegradable under specific laboratory conditions.

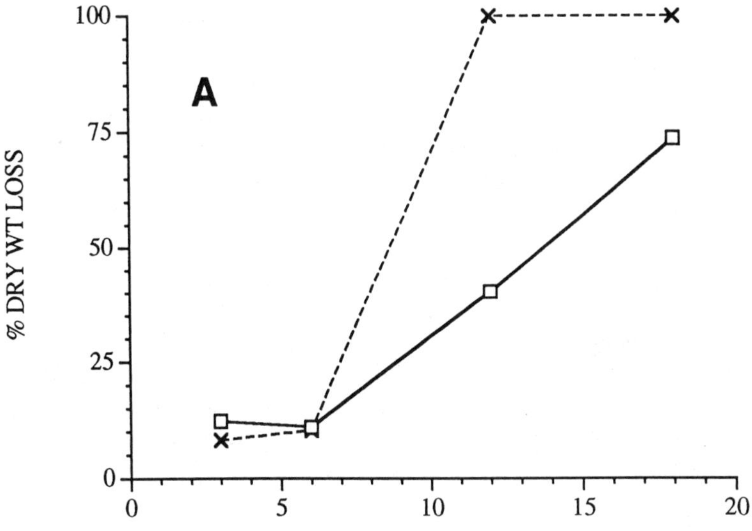

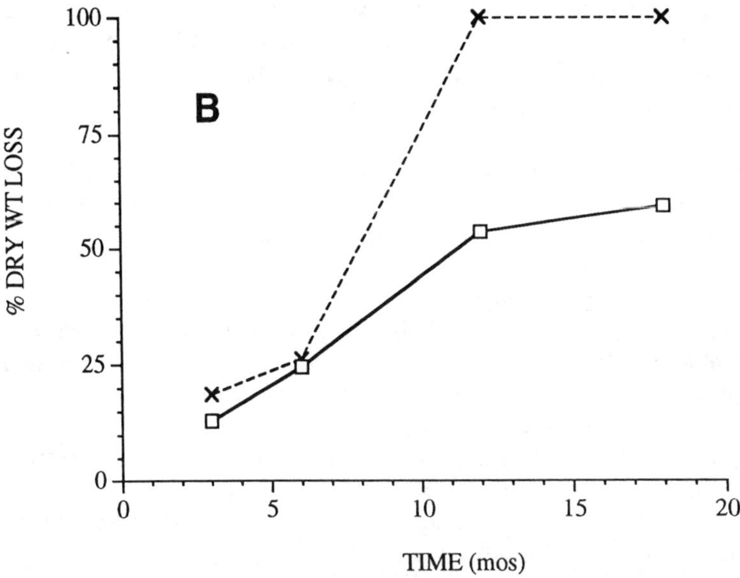

Figure 5. Weight loss data for PHBV film A.) coastal study (Tier III). B.) dynamic incubation (Tier II). (□ = in water – no *spartina*; ◊ = in sediment – no *spartina*)

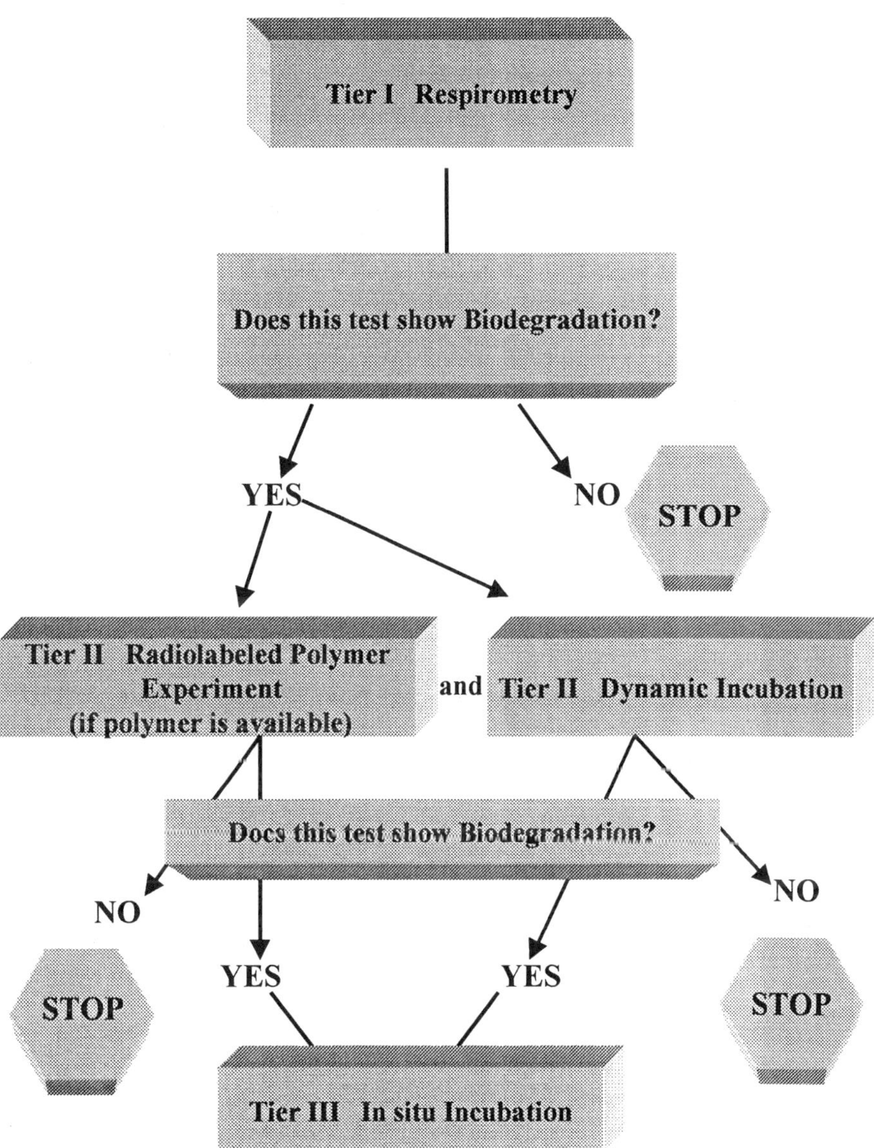

Figure 6. Scheme for performing biodegradation tests in the marine environment.

Future Research and Development

Based on some ongoing work, our laboratories will try to continue using radiolabeled polymers, as they become available with perhaps some synthesis of incorporating nitrogen and phosphorous moieties into the polymer.
It is important to conduct some controlled laboratory degradation studies on select materials of promise under both fully aerobic and fully anaerobic conditions with added nutrients in order to get the maximum degradation in the minimum time. Also, we should continue the dynamic aquarium approach as it is the most applicable to real life degradation without the costs of a Tier III test.

From the laboratory incubations, if certain polymers show fairly active biodegradation, some experiments that involve enrichment and isolation of pure cultures of marine microbes that degrade these polymers could be performed.

On obtaining such isolates, their activity on the polymers could be tested in pure culture by using respirometry to perhaps make some more general statements about the potential fate of the polymer(s) in the marine environment. Overall, by developing standard methods for testing in the marine environment, we hope to more uniformly characterize, in the most efficient time and cost efficient manner, what may be defined as biodegradable in the marine environment.

Acknowledgements

The authors acknowledge the U.S. Navy especially Vickie Edgar, the Program Manager for the Plastic Removal in the Marine Environment (PRIME) and Waste Reductions Afloat Protects the Sea. (WRAPS). The authors are also grateful to Robert Stote of the U. S. Army and Kristen Dixon currently at Millenium Pharmaceutical who helped with the respirometry tests. We thank Professor Steve Goodwin of the University of Massachusetts Amherst for the PHB sample, Key Consulting and Mid West Grains for the wheat gluten samples, ICI for the PHBV polymer, and Peter Stenhouse of the U.S. Army for the silk samples. The authors also acknowledge Deb Cobban of the U.S. Army for her graphics assistance.

References

1. *International Year of the Ocean,* http://www.marine.ie/age/ **1998**.
2. Weisskopf, M. Smithsonian 1988, March.
3. O'Hara, K., Iudicello, S., Bierce, R., Center of Marine Conservation, **1988**.
4. Wilber, R. J., *Oceanus*, **1987**.
5. Krzenski, J.W.; *Maritime Envrionmental Regulations* **1997**, December 15, 1-2.

6. Willson, L. *Characterization of Solid Waste from U.S. Navy Surface Combatants.* Office of the Chief of Naval Operations: Arlington, VA; **1996**, October 31.
7. Willson, L. *Solid Waste Retention Aboard US Navy Ships: A Study on the Limits to and Effects of Holding Trash on Board Sea while at Sea* Office of the Chief of Naval Operations: Arlington, VA; **1996**, July 4.
8. Jannasch, H. W., Wirsen, C. O., Farmanfarmaian, A. *Science*, **1971**, 171, 672-675.
9. Wirsen, C. O. and Jannasch, H. W. *Environmental Science and Biology*, **1976**, 10:880-886.
10. Wirsen, C.O., Jannasch, H. W. *Fundamentals of Biodegrad. Polym. and Packag.* **1993** . Editor Ching, C., Kaplan, D., Thomas, E. Technomic, Lancaster, PA 297-310.
11. Mayer, J., Kaplan, D. L.; Stote, R. E.; Dixon, K. L., *ACS Symposium Series*, **1996**, 627, 159-171.
12. Allen, A. L., Mayer, J., Stote, R., Kaplan, D. L.; *J. of Environ. Polym. Degrad.* **1994**, 2(4), 237-244.
13. Kaplan, D. L., Mayer, J.M., Greenberger, M., Gross, R., McCarthy, S. *Polymer Degradation and Stability*, **1994**, 45 (2) 165-172.
14. Mayer, J. M.; Allen, a. L.; Dell, P. A.; McCassie, J. E.; Shupe, A. E.; Stenhouse, P.J., Welsh, E. A.; Kaplan, D. L. *205th ACS Meeting*, **1993**, 401.
15. McCassie, J., Mayer, J., Stote, R., Shupe, A. E., Stenhouse, P., Dell, P. A., Kaplan, D.L. *Proceedings of ACS Division of Polymeric Materials and Engineering*, **1992**, 353-354.
16. Mayer, J. M. ; Greenberger, M.; Stote, R. E.; Wright, J. R., Reslewic, P. A.; Kaplan, D. L.; *ACS Symposium*, **1991**, (1-2), 201.
17. Mayer, J., Greenberger, M., Kaplan, D.L.; Gross, R.; McCarthy, S.; *Polymer Preprints*, ACS, **1990**, 31(1), 434.
18. Kirchner, M.; *Helgolaender Meeresuntersuchungen*, **1995**, 49(1-4), 201-202.
19. Poulicek, M.; Jeuniauz, C. *Biochem. Syst.. Ecol.* **1991**, 19 (5), 385-394.
20. Andrady, A. L., Pegram, J. E., Olson, T. M. March **1992**, 49. Report No. DTRC/SME-CR-19-90. Research Triangle Institute, Research Triangle Park, NC.
21. Arnosti, C.; Repeta, D.J.; *Limnol. Oceanogr.* **1994**, 39(8), 1865-1877.
22. Mergaert, J.; Wouters, A.; Anderson, C.; Swings, *J.; Can. J. Microbiol. Rev. Can. Microbiol.*, **1995**, 41(1), 154-159.
23. Leschine, S.B.; *Annual Review of Microbiology* **1995**, 49, 399. 1516.
24. Poulicek, M. *Bull.Soc. r. Sci. Liege*; **1996**, 64(6), 327-43.
25. Poulicek, M. *Bull.Soc. r. Sci. Liege*; **1994**, 63(5), 369-82.materbi
26. Sullivan, B. K., Oviatt, C. A., Klein-MacPhee, C. *Biodegrad. Polym. Packag.* **1993** 281-286. Editor Ching, C., Kaplan, D., Thomas, E. Technomic, Lancaster, PA

27. Doering, P. H., Sullivan, B. K.; Jeon, H. *J. Environ. Polym. Degrad.*, **1994**, 2(4), 271-5)
28. Spence, K. E., Allen, A. L., Jane, J. *ACS Symposium Series*; **1996**, 627, 149-158.
29. Calmon-Decriaud, H.ellon-Maurel, V. ,Silvestre, F. Fr. *Adv. Polym. Sci.* **1998**. 207-226.
30. Ogi, P. *Ronza* **1998**, 7, 24-33.
31. OECD Guidelines, chapter 36, **1993**.
32. DeWilde B. OWS, personal conversation
33. American Society for Testing and Materials (ASTM). *ISR Degradable Polymeric Materials Program,* West Conshoshocken, PA; **1996**.
34. Pettigrew, C. ASTM Standardization News **1996** 32-35.
35. Pagga, U., Beimborn, D. B., Boelens, DeWilde, B. *Chemosphere* **1995,** 31:4475-4487.
36. Itavaara, M., Vikman, M. *Chemosphere*, **1995**, 31:4349-4373.
37. Wirsen, C. O. and Jannasch H. W. Environ. Sci. Tech. **1976**, 10:880-886.
38. Ratto, J. A., Herbert, J., Wirsen, C. *7th Annual BEDPS meeting*, **1998**.
39. American Society for Testing and Materials. *Annual Book of ASTM Standards.* ASTM: West Conshoshocken, PA; **1997**, Vol. 08-03; Standard D 5209.
40. Itavaara, M., Vikman M.. *Chemosphere*. **1995**, 31:4359-4373.
41. Pagga, U., D. B. Beimborn, J. Boelens and B. Dewilde. *Chemosphere*. **1995**, 31:4475-4487.
42. Bazylinski, D. A., Wirsen C. O. and Jannasch H. W. *Appl. Environ. Microbiol.* **1989,** 55:2832-2836.
43. Jannasch, H. W. and Wirsen C. O.. *Appl. Environ. Microbiol.* **1982**, 43:1116-1124.
44. Wirsen, C. O. Scientific and Technical Final Report- SFRC DAAK60-92-K-0008 submitted in **1995** to U. S. Army Natick.
45. Wirsen, C. O. Wood Hole Oceanographic Institution, internal reports **1997.**

Chapter 21

Biodegradable Bags Comparative Performance Study: A Multi-Tiered Approach to Evaluating the Compostability of Plastic Materials

R. E. Farrell[1,2], T. J. Adamczyk[1,2], D. C. Broe[1,2], J. S. Lee[1,2], B. L. Briggs[1,2], R. A. Gross[3], S. P. McCarthy[1,4], and S. Goodwin[1,5]

[1]NSF-Biodegradable Polymer Research Center, Departments of [2]Biological Sciences, [4]Plastics Engineering, and [5]Microbiology, University of Massachusetts at Amherst, Amherst, MA 01003
[3]Department of Polymer Chemistry, Polytechnic University, Brooklyn, NY 11201

The compostability of nine commercial and developmental lawn & leaf bags made from "environmentally degradable" plastics was assessed using a three-tiered approach involving both standard lab-scale tests and field studies. Results obtained for the test plastics were compared to those obtained for both positive (Kraft paper) and negative (LLDPE) controls. Statistical analyses were used to rank the test materials in terms of their overall degradation; the degree of agreement in the relative rankings was then assessed using Kendall's coefficient of concordance (*W*). The calculated coefficient of concordance was 0.724** (significant at the 1% level of probability), indicating that the comparative biodegradability of the test/reference materials was relatively unaffected by variations in the test conditions. Five of the test plastics exhibited significantly more degradation than the positive control, with plastics whose degradation pathway included a significant abiotic contribution having an apparent advantage.

Introduction

Faced with concerns surrounding the safe, efficient, and cost-effective disposal of ever-increasing amounts of municipal solid waste, many communities are now taking an integrated approach to waste management. Such an approach involves a multi-tiered waste management system in which the solid waste stream is processed according to the unique characteristics of the various waste fractions (*1*), which are then directed to an appropriate treatment facility. At the community level, the current focus is on recycling – with as much as 25 to 30% of the municipal solid waste stream being recovered as recyclables (e.g., glass, cans, paper, and the so-called commodity plastics). Yard trimmings (primarily leaves, grass clippings, and brush) make up another 16 to 20% of the municipal solid waste stream, and the recycling of these materials via aerobic composting represents an attractive, and cost-effective, alternative to their disposal in landfills. Moreover, composting (the biologically mediated oxidative decomposition of organic matter) produces highly stable, humus-like material that is suitable for use as a soil amendment or organic fertilizer – thus, not only recycling the waste but giving it added value.

In a recent survey of solid waste management programs, Glenn (*2, 3*) reported that an increasing number of municipal and state governments are banning the disposal of yard trimmings in landfills – with 22 states having at least a partial ban in 1997. Not surprisingly, therefore, Glenn also reported that there are now nearly 3,500 yard trimmings composting facilities operating in the United States. As a consequence of these developments, there has been a resurgence in the development and marketing of "biodegradable" plastic lawn & leaf bags. From an environmental perspective, biodegradable plastics possess a number of attractive features; for example, many are derived from renewable (agricultural) resources and, because they can be recycled via composting, will eventually re-enter normal biogeochemical cycles. At the same time, however, there are a number of concerns regarding how (if at all) the composting of these plastics will affect (i) the decomposition of materials enclosed in the plastic, (ii) the final composition and appearance of the composted material, and (iii) the commercial value of the resultant compost.

"First generation" biodegradable plastics (primarily polyethylene-starch composites) did not perform well in actual compost environments and, though degradable, were not considered *bio*degradable (i.e., there was little or no microbial conversion of the polyethylene to CO_2, water, and biomass). As a result of their continuing legacy, claims of biodegradability (compostability) associated with the latest generation of plastic lawn & leaf bags are generally met with an air of skepticism by both the general public and regulatory agencies. Consequently, as this new generation of biodegradable products enters the marketplace, questions regarding the long-term fate of these materials in the environment and their effects on the environment are being asked by government regulators, environmental groups, compost operators, and consumers: e.g., *Are these materials truly biodegradable?*, *How well do they perform relative to the traditional paper leaf bag?*, and *Are the results of lab-scale "screening" tests, which form the basis for most biodegradation testing programs, relevant to what happens in the "real world"?* In the fall of 1996, a study to try and answer some of these questions was initiated by the NSF-

Biodegradable Polymer Research Center, the Massachusetts Department of Environmental Protection, and the Chelsea Center for Recycling and Economic Development. Here, we report the results of this study.

Materials & Methods

The compostability of a number of biodegradable leaf bags was assessed using a three-tiered testing scheme involving both laboratory and field components. Wherever possible, these tests were conducted in accordance with the ISR *Standard Guide to Assess the Compostability of Environmentally Degradable Plastics* (since adopted as ASTM Standard Guide D 6002). Laboratory-scale reactors were used to simulate bioactive disposal sites in which the biodegradability of the test plastics could be evaluated under controlled aerobic composting conditions. Standard (ASTM or ISO equivalent) laboratory test methods were used to assess the degradation/disintegration (measured as weight loss; Tier I test) and mineralization (conversion of polymer-C into CO_2; Tier II test) of the plastic bags. Field trials (Tier III tests) were designed to take advantage of innovations developed during tests conducted by the Institute for Standards Research, Degradable Polymeric Materials Program (*1*).

Test Materials

The materials tested (Table I) included six commercially available and three developmental plastic bags, a POSITIVE REFERENCE material (i.e., a material know to be biodegradable - Kraft paper) and a NEGATIVE REFERENCE material (i.e., a non-biodegradable material – linear low density polyethylene). One sample, the ECO Bag, arrived after the field trials were initiated and, therefore, was included in only the Tier I (lab-scale weight loss) and Tier II (mineralization) tests.

All test materials were used as provided by the manufacturer, except that prior to biodegradation testing, appropriate sub-samples of each material were dried to a constant weight in a convection oven (50°C for 12 to 18 h), cooled to room temperature in a desiccator, and stored in polyethylene ziplock bags at 4°C until needed. The carbon content of the test and reference materials was determined using an Exeter Analytical 240 Elemental Analyzer. Carbon, hydrogen, and nitrogen analyses were obtained simultaneously by combusting the samples (ca. 3000 μg; weighed to the nearest micro-gram) at 1000°C in a stream of oxygen to form CO_2, H_2O, and N_xO_y. The nitrogen oxides formed were then reduced to dinitrogen and the three gaseous products of interest (CO_2, H_2O, and N_2) quantified in a He stream using individually calibrated and blanked thermal conductivity detectors (*4*).

Tier I – Rapid Screening Test: Weight loss from plastic materials exposed to a simulated municipal solid-waste (MSW) aerobic compost environment (*5*)

The Tier I test is used to determine the biodegradability of plastic materials, relative to that of a standard biodegradable material, in a controlled composting

Table I. "Biodegradable" bag materials evaluated

Sample No.	*Trade Name*[a]	*Manufacturer*	*Key Components*	*Thickness (mil)*
6	BAK 1095	Bayer Chemical	polyester amide	1.1 ± 0.1
4	Bionolle®	Showa High Polymer	aliphatic polyester (Bionolle 3000)	1.5 ± 0.1
10	Bio-Solo™	Indaco Bio-Solo	recycled & virgin LLDPE + activators	1.4 ± 0.1
3	Eastar Bio™	Eastman Chemical Co.	copolyester 14766	1.6 ± 0.1
11	ECO Bag	Novon International	polyethylene & starch + activators	1.2 ± 0.1
2	EcoPLA®	Cargill Dow Polymers	polylactic acid	1.8 ± 0.1
8	Garbax™	International Paper	2-ply, 50# wet strength kraft paper[b]	5.1 ± 0.2
9	LLDPE	Berry	linear low density polyethylene[c]	1.9 ± 0.1
1	PCL-starch-PE	Epic Enterprises, Inc.	polycaprolactone + cornstarch + LDPE	1.7 ± 0.0
5	UML1	UMass–Lowell	80:20 Polylactic acid:Bionolle (3001) blend	2.0 ± 0.5
7	UML2	UMass–Lowell	50:50 Polylactic acid:Bionolle (3001) blend	1.4 ± 0.3

[a] Note: this study was carried out as a series of "blind" tests; i.e., the source and exact nature of the test materials was withheld from the investigators until all phases of the study were completed.

[b] Positive reference.

[c] Negative reference.

environment. In addition, the degree of abiotic degradation (primarily hydrolytic) that occurs is assessed by measuring weight loss from test specimens exposed to a "poisoned" control compost. Testing is carried out under carefully controlled laboratory conditions to ensure reproducibility and precision and facilitate inter-laboratory comparisons. However, because test conditions are optimized, the observed biodegradability of the test material may not correlate well with that observed in an actual composting environment. Consequently, the rapid screening test is only intended to provide information regarding the *inherent* degradability of the test material.

Test materials were cut into 2-cm × 3-cm pieces, assigned an identification number, and (to facilitate recovery) placed in 5-cm square litter bags made from nylon-coated fiberglass screen (1.5-mm square mesh). Each sample bag was supplemented with one-half teaspoon of matured compost (L96-0506; screened to pass a 2-mm sieve; see Table II) on each side of the film; the sample bags were then placed in 4-L bioreactors containing approximately 300-g of fresh compost (see Table II for a more detailed description of the compost matrix), and incubated at 52 ± 1°C and a water content of 55 ± 5% (w/w). A schematic of the bioreactor system is presented in Figure 1.

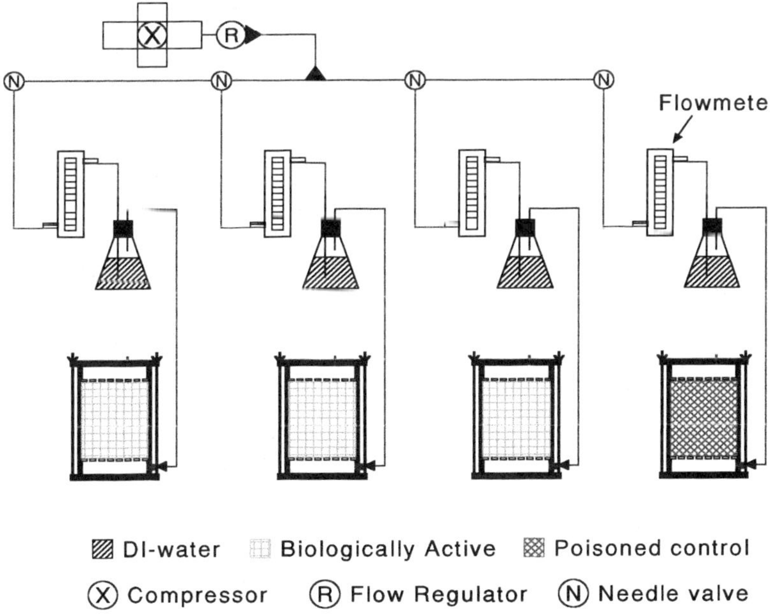

Figure 1. Schematic of the film weight-loss bioreactor system.

Table II. Characteristics of the compost matrices used in the various lab- and field-scale biodegradation tests

Test: Tier	Test: Scale	Test: Parameter	Matrix composition	Compost Characteristics[a]
I	lab	weight loss	*Simulated municipal solid waste compost* (sMSW): 940 g shredded leaves (1:1, w/w, mix of oak & maple); 340 g shredded paper (1:1, w/w, mix of newspaper & computer paper); 140 g mixed, frozen vegetables; 120 g meat waste (added as a 1:1, w/w, mix of dried dog food & dried cat food); 360 g dehydrated cow manure; 40 g sawdust; 40 g urea; and 20 g commercial compost seed. (*ca* 300 g sMSW per bioreactor)	WHC = 1.00 Water content = 60% WHC Air-filled pore space = 45% Ash content = 6%; Volatile solids = 94% pH = 6.7 (initial), 8.1 (final) C:N = 14:1
II	lab	CO_2 evolution	L96-0506: mature (finished) compost from Laughton's Garden Center (Westford, MA); leaves & grass clippings; horticultural waste; shredded brush; and soil. The compost was screened to pass a 2-mm sieve prior to use.	WHC = 2.00; pH = 6.6 OM = 56.7%; TN = 0.94% TOC = 16.4%; C:N = 17:1
III	field (LGC)	weight loss	Leaves & grass clippings; shredded brush; soil; and paper sludge.	Water Content = 50-60% Ash content = 2%; Volatile solids = 98% pH = 7.1; C:N = 23:1
III	field (UMA)	weight loss	Food waste (both pre- and post-consumer) from the university cafeteria system (separated by the cafeteria staff), manure and bedding materials from the university farm system, leaves & brush from the university grounds (when available), and wood chips (added as a bulking agent). Food waste-to-amendment ratio = 3:2.	--- *Pre IVC-composting* --- Water content = 56% Ash content = 4%; Volatile solids = 96% pH = 4.4; C:N = 33:1 --- *Post IVC-composting* --- Water content = 63% Ash content = 8%; Volatile solids = 92% pH = 7.3; C:N = 16:1

[a] WHC = water holding capacity (g H_2O g^{-1} dry compost); OM = total organic matter content; TOC = total organic carbon; TN = total nitrogen.

The compost was aerated by passing a continuous stream (100 to 200 mL min^{-1}) of water-saturated air through the bioreactors; the water content of the compost was checked twice weekly. If the water content dropped below 50%, deionized water was added to the compost; if the water content exceeded 60%, dry air was passed through the compost until the correct water content was re-established. The test exposure was considered valid only if dry-weight loss from the compost matrix itself exceeded 40% (w/w) in the active bioreactors and was <10% in the poisoned-control reactor.

Triplicate samples of the test materials were recovered at 7-day intervals for a total of 6 weeks. Note: on each sampling date, one replicate of each material was selected at random and removed from each of the three active reactor vessels (see Figure 1). Triplicate samples of each test material also were recovered from the poisoned-controls on DAYS 21 and 42. Each sample was carefully removed from its litter bag and cleaned by (i) brushing any loose compost from the film surface with a camels hair brush, (ii) washing the film in a mild, phosphate-free detergent solution, (iii) cleaning gently with a cotton swap, and (iv) rinsing the sample with deionized water and gently blotting the film dry with a lint-free tissue. The residual plastic was then dried to a constant weight in a convection oven (50°C for 12 to 18 h), cooled to room temperature in a desiccator, and weighed.

In some instances, ultrasonic cleaning was used to remove compost particles adhering strongly to the test material. Nevertheless, samples that were highly fragmented could not be hand-cleaned and, unlike studies where an inability to separate residual polymer from the compost matrix itself is equated with complete (i.e., 100%) degradation, we chose a solvent extraction procedure to separate the residual plastic from the compost and yield a sample that could be accurately weighed. Briefly, the solvent extraction procedure was as follows: (i) after removal from the litter bag, the compost was screened through a 2-mm sieve to facilitate the recovery of any large pieces of plastic, and all material passing through the sieve (compost + residual plastic) transferred to a 25 × 100 mm screw cap test tube; (ii) 5 mL of an appropriate solvent (chloroform or THF) was added to the test tube and the contents gently heated using a hot air gun until the plastic was dissolved; (iii) the dissolved plastic was then recovered from the compost by filtering the suspension through Whatman No. 42 filter paper – rinsing the test tube with three 5-mL aliquots of solvent – and quantitatively transferring the solvent to a pre-weighed aluminum weighing dish; (iv) the solvent was then evaporated under a hood at room temperature and the resultant plastic films dried in a convection oven at 50°C overnight; (v) the residual plastic was then cooled to room temperature in a desiccator and weighed. Preliminary studies demonstrated that the recovery of test plastics from the different compost matrices ranged from 96% to 106%. Note: this procedure was used only in those cases where small particles of plastic were clearly intermingled with the compost matrix itself.

Tier II – Polymer Mineralization Test: Aerobic biodegradation of polymeric materials under controlled composting conditions (*6–8*)

Unlike weight loss, which reflects structural changes in the plastic (i.e., deterioration or material erosion), CO_2 evolution provides an indication of the ultimate biodegradability (i.e., mineralization) of the plastic. Polymer mineralization studies were conducted using a *static compost biometer system* incorporating elements of both the soil biometer system of Bartha & Pramer (*9*) and ASTM standard D 5988 (*7*) as well as those of the forced-air composting system (ASTM D 5338) (*6*). In general, the test consists of: (i) collecting and characterizing a matured compost inoculum from a full-scale commercial composting facility (i.e., the yard waste composting facility operated by *Laughton's Garden Center* in Westford, MA); (ii) exposing representative samples of the plastic bags to the compost matrix under controlled isothermal and aerobic conditions; (iii) measuring the amounts of CO_2 produced as a function of time; and (iv) assessing the degree of biodegradability of the plastics by comparing the net amount of CO_2 produced from the test materials to that produced from the POSITIVE CONTROL (i.e., Kraft paper). Note: CO_2 production is expressed as a fraction of the measured carbon content of the test material.

Film samples weighing ca. 500 mg – and yielding a substrate loading of 5 ± 1 mg polymer-C g^{-1} compost – were cut into 10-mm × 10-mm pieces; buried in test reactors (2-L Mason jars) containing 50 g of a matured compost (L96-0506; screened to pass a 2-mm sieve) at a water content of 60% water-holding capacity (WHC); and incubated in a controlled environment chamber at 52 ± 1°C. Carbon dioxide produced during the biodegradation process was trapped in 20 mL of 1.0 *M* KOH added to a glass well at the bottom of the bioreactor. The CO_2 traps were changed at 12 to 72 h intervals and a 3.0-mL aliquot of the KOH from each trap was titrated with standardized 0.1000 *M* HCl. Daily and cumulative CO_2 production (total and net) and percent mineralization were calculated relative to a control flask containing only compost. The water content of the compost was monitored by weighing the bioreactors every third time the CO_2 traps were changed and, when necessary, adding an appropriate amount of deionized water. The systems were aerated by allowing the atmosphere in the bioreactors to exchange and equilibrate with atmospheric air for 10 to 15 min; at the same time, the test matrix was hand-mixed to ensure that anaerobic microenvironments did not develop in the compost.

Tier III – Field Testing

Field-scale biodegradation studies are desirable and ultimately may be necessary; however, they are generally expensive, difficult to control and replicate, and are generally unsuitable for establishing biodegradation pathways (*8*). Nevertheless, if we are to ensure that laboratory-scale tests do an adequate job of predicting a plastic's performance in "real world" disposal environments, it is imperative that we first determine the biodegradability of these materials under conditions representative of various commercial/municipal and backyard composting systems.

The field studies described herein were designed to address the questions: *How well does the latest generation of biodegradable plastics perform relative to the traditional paper leaf bag?* and *How well do the results obtained in standard, laboratory-scale biodegradation tests relate to a product's biodegradability in an actual, full-scale compost environment?* To answer these questions, a field trial was conducted from late March to mid June 1997 at a commercial yard waste composting facility in Westford, MA. In addition, an attempt was made to address the question: *How much does the source material for the compost and composting technology affect the biodegradation of the plastic materials?* This question was addressed by exposing the plastics to a single cycle of composting in an *in-vessel* compost system located at the University of Massachusetts Amherst.

Sample handling protocols. The test and reference materials were cut into 7.5-cm square pieces. Each piece was dried to a constant weight in a convection oven (50°C for 12 to 18 h), cooled to room temperature in a desiccator, and weighed. The individual pieces of plastic were then assigned a sample ID number, placed in a 10-cm square litter bag (made from nylon-coated fiberglass screen with a 1.5-mm square mesh), and stored in polyethylene ziplock bags at 4°C until needed. One day prior to placing the samples in the compost, 2.6 ± 0.1 g of matured compost (collected from the appropriate composting facility and screened to pass a 2-mm sieve) was added to the litter bags on each side of the film. The bags were then tagged, placed in polyethylene ziplock bags, and stored in Styrofoam coolers at 4°C until they were placed in the compost.

Upon retrieval from the composts, the litter bags were placed in zip-lock bags and transported under refrigeration to the Biology Department at UMass Lowell. Upon arrival at UMass, the bags were stored at 4°C until sample cleaning was initiated (usually within 24 h of their arrival). Each sample was carefully removed from its litter bag and, prior to cleaning, a visual rating of sample disintegration (Table III) was recorded. The residual films were then cleaned as described for the Tier I test. Any residual plastic material was dried to a constant weight in a convection oven (50°C for 12 to 18 h), cooled to room temperature in a desiccator, and weighed.

Table III. Rating system used to describe the extent of visible sample disintegration during compost exposures

VDR[a]	*Observed change in test specimen*
0	No apparent change
1	Slight discoloration with little or no evidence of physical disintegration
2	Moderate discoloration and/or physical disintegration
3	Major changes and/or flaws – samples highly fragmented and nearly unrecognizable
4	Samples unretrievable (i.e., complete disintegration)

[a] Visual disintegration rating. Ratings assigned prior to sample cleaning.

Site 1 – Laughton's Garden Center (Westford, MA). Composting was carried out in a large (ca. 4-m × 3-m × 100-m) static windrow that was turned only infrequently (i.e., ca. once every 6 to 9 weeks). Source materials for the compost included yard trimmings from the town of Westford, MA and plant materials from the garden center's commercial horticulture operation (see Table II). At the time the study was initiated, the compost pile was about two months old and small additions of paper sludge had been made to the compost matrix. Due to labor and size limitations, the Westford study was set up using a 2-way randomized complete block design (Figure 2A); each "block" consisted of a PVC frame (8-ft × 4-ft) strung with six rows of nylon rope spaced at 8-in intervals and ten sample bags (one for each test material) attached to the rope in each row (with an 8-in spacing between samples). The blocking frames were laid flat and stacked vertically in the windrow with ca. 18 to 20 cm of compost between each block (Figure 2B); both the base of the compost and the "cap" were about 1.0-m thick. The test materials were placed into the windrow on March 26, 1997 and five replicate samples of each test material (i.e., one row per block) were recovered after 14, 21, 28, 42, 56, and 84 days. The litter bags were returned to the Biology Department at UMass–Lowell where the residual films were assigned a visual disintegration rating (*VDR*; see Table III), cleaned, dried, and analyzed for changes in total dry weight. These parameters were then used to provide quantitative data to assess the degradability of the plastics under field composting conditions and to provide a comparison with the lab-scale tests. In addition, critical composting parameters (i.e., surface and core temperature of the compost, moisture content, and gas analysis of the core atmosphere) were monitored on each sampling date.

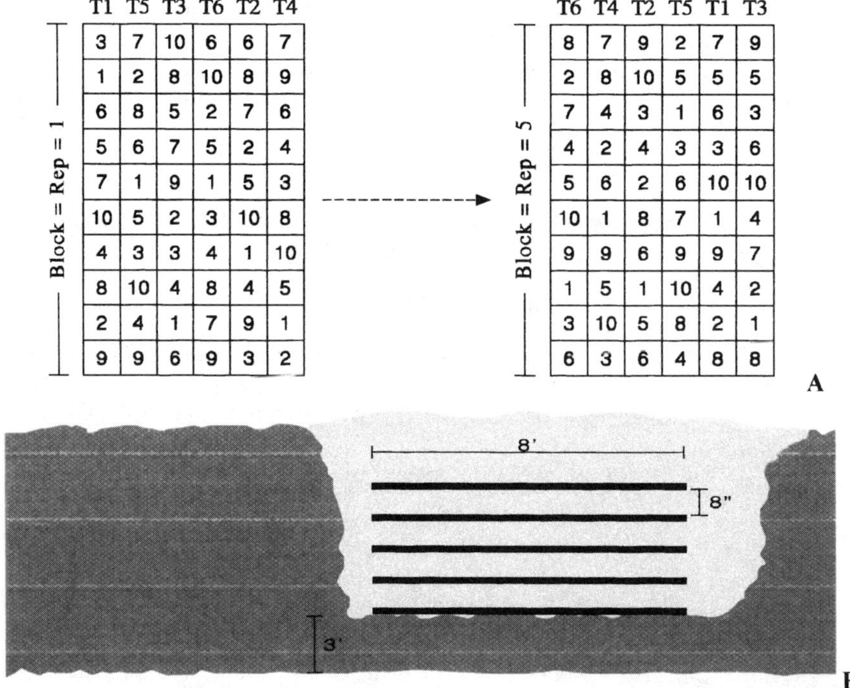

Figure 2. Experimental design employed at the Westford, MA composting site (FIELD SITE 1). [A] Schematic showing the arrangement of the test plastics in a two-way randomized complete block design: 1st factor = polymer type (1 - 10), 2nd factor = time (T1 - T6); block = rep (1 - 5). [B] Cut-away view of the windrow – illustrating the arrangement of the blocking frames in the compost pile. (Not drawn to scale.)

Site 2 – University of Massachusetts (Amherst, MA). Composting was carried out using an *in-vessel* compost system (Wright Environmental Management, Inc., Toronto, ON, Canada) operated by the staff of the UMass Office of Waste Management – Intermediate Processing Facility (IPF). The in-vessel composter (*IVC*) is a self-contained, self-heating, continuous feed composting system in which the entire composting process takes place in a completely enclosed and controlled environment (Figure 3). Waste to be composted is mixed thoroughly and fed onto a series of stainless steel transporters within the unit which then move the material through two temperature controlled zones where the composting takes place. Optimum composting conditions are maintained via a control system operated by the staff of the IPF.

The *IVC* is designed to process 1500 to 3000 lbs of solid waste per day on a 14 to 28 day cycle. Source materials for the compost included food waste, manure and bedding materials from the university farm, yard trimmings, and wood chips (see Table II). Seven replicate samples of each test material, as well as samples of newspaper and glossy paper (included as additional reference materials), were placed in the in-vessel composter on March 18th and were recovered starting April 3rd and continuing through April 10th (average residence time = 20 days). Upon their return to the Biology Department at UMass–Lowell, the residual films were assigned visual disintegration ratings, cleaned, dried, and analyzed for changes in total dry weight. These parameters were then used to provide quantitative data to assess the degradability of the plastics during the shorter, but more intense composting exposure in the *in-vessel* composting unit and to provide a comparison with the lab-scale and other field tests. Critical composting parameters (i.e., air temperature, composting temperatures in the in-vessel unit, and moisture content) were monitored throughout the test exposure.

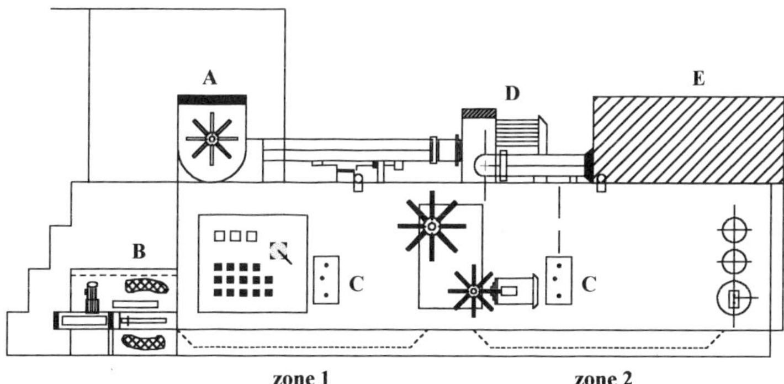

Figure 3. Schematic of the Wright in-vessel composter. A: mixer unit; B: transporter trays; C: temperature probes; D: exhaust fan; E: bio-filter.

Statistical Analyses. Exploratory data analysis (*10*) was used in the initial stage of the statistical analysis to assess the nature of the frequency distribution for each variable and to identify outliers. Outliers (measured value ≥ median value + k) were defined using a k value of 1.5 times the interquartile range and were excluded from the analysis of variance, means separation, and correlation analysis. Analysis of variance (ANOVA) and correlation analysis were conducted using the GLM procedure of CoStat ver. 5.0 (*11*). However, because percent data form a binomial distribution, rather than a normal distribution, all such data (i.e., % weight loss and % mineralization) were transformed using the arcsine transformation (*12*) prior to analysis of variance. The lab-scale weight loss test employed a 2-way randomized complete block design (RCBD) – with polymer type and time as the main factors, weight loss as the variable, and replicates as the blocks. The lab-scale mineralization test employed a one-way completely randomized design (CRD) – with polymer type as the main factor and cumulative net mineralization as the variable. FIELD SITE 1 (Laughton's Garden Center; Westford, MA) was set up as a 2-way RCBD – with polymer type and time as the main factors, weight loss and visual disintegration rating as the variables, and replicates as the blocks. FIELD SITE 2 (UMass Amherst), employed a one-way CRD – with polymer type as the main factor and weight loss and visual disintegration rating as the variables. Correlations between the various degradation variables (i.e., weight loss, net mineralization, and visual disintegration rating) were assessed using the appropriate correlation technique.

Whereas weight loss (or mineralization) values obtained on the final sampling date were used to assess the ultimate degradability of the test materials, the integrated-mean values (i.e., means averaged across time) were used to provide a general *performance index* for each material (i.e, a quantitative index of the relative ease of biodegradation). Because the performance index is a function of both the rate and extent of degradation, materials that degrade rapidly and extensively rank higher than materials that degrade slowly or are only partially degraded. The significance of these ratings is reflected in the *time* × *polymer* interaction determined in the ANOVA.

The least significant differencc (LSD) test was used to detect significant differences between mean values (grouped by test material and time). By their very nature, standard lab-scale tests lend themselves to the precise control of operating parameters, thus differences between mean values were assessed at the 5% level of probability ($P \leq 0.05$). Field studies, on the other hand, are inherently more variable and uncontrollable. Consequently, differences between means for the field tests were assessed using a less rigorous level of significance (i.e., 10%; $P \leq 0.10$). Use of a more rigorous level (e.g., 5%) would have greatly enhanced the probability of failing to detect a difference between two or more samples when in fact one did occur in the field. (Note: Throughout the text, the symbols †, *, **, and *** are used to denote significance at the $P \leq 0.10$, 0.05, 0.01, and 0.001 levels of probability, respectively.)

The consistency of the rankings between test environments was assessed using the Kendall coefficient of concordance, W (*12, 13*), which measures the degree of association (agreement) in the ranking of biodegradation indices (i.e., weight loss or net mineralization) among the test materials. The value of W ranges from 0 (no agreement) to 1.0 (perfect agreement).

Results & Discussion

It is generally agreed that a "compostable" material should exhibit three fundamental characteristics: (i) it must disintegrate during composting so that no significant residues are visible to detract from the value of the compost; (ii) it must biodegrade (mineralize) at a rate equal to or greater than that of the compost itself; and (iii) it must be shown to have no negative impact on the ability of the finished compost to support plant growth. Thus, the present study was carried out to assess the compostability of several environmentally degradable plastic lawn & leaf bags. The bags were chosen to provide a wide range of polymer types and formulations and included plastics that were both commercially available and in developmental stages. Because of the complex nature of the issues involved, we chose a multiple-tiered testing scheme – involving both laboratory and field studies – to address several key questions: (i) Are the test materials biodegradable?, (ii) How well do they perform relative to a conventional paper lawn & leaf bag? and (iii) How well do the results obtained in laboratory-scale (Tier I & II) tests relate to a product's compostability under field (Tier III) conditions?

Tier I – Rapid Screening Test

The Tier I weight-loss test (ASTM D 6003) was used as a rapid screening test to assess the inherent biodegradability of the test plastics under controlled aerobic composting conditions. In addition, the extent to which the plastics degraded abiotically (chemically) was determined by simultaneously exposing test samples to an inactive (i.e., poisoned) compost under environmental conditions identical to those of the active compost. Validation of the lab-scale weight loss test was made by assessing key indicators of compost activity; i.e., a reduction in weight and increase in pH of the compost matrix itself during the test exposure. Weight loss from the active bioreactors averaged about 67% (w/w) while the pH of the compost matrix increased from about 6.7 to 8.1 during the 42-d test exposures. On the other hand, compost in the poisoned-control exhibited only about 5% (w/w) weight loss and there was no significant change in pH during the 42-d test exposures. As a result of these observations, all test exposures were considered valid. In addition, analysis of variance determined that there was no blocking effect ($F = 0.219$ ns); i.e., degradation of the various samples proceeded at essentially the same rate and to the same extent in each of the three active bioreactors.

Results from the Tier I test are presented in Figure 4. In general, two trends emerged from the data analysis: (i) the PE-based plastics (LLDPE, Bio-Solo™, and ECO Bag) remained essentially undegraded throughout the 42-d test exposure and (ii) weight loss from the non-PE based plastics increased steadily throughout the first 28 days of the test exposure, but then slowed considerably with little or no significant weight loss after DAY 35. The one notable exception to these trends was the 80:20 PLA:Bionolle blend (UML1) which, following an initial lag period of about one week, degraded in a near-linear fashion (19.5% per week; $R^2 = 0.922^{***}$) throughout the remainder of the test exposure. These differences in the inherent degradability of the test materials were reflected in the significant *time* × *plastic* interaction (F = 5.12***) determined in the analysis of variance and were quantified using the performance index (*PI*).

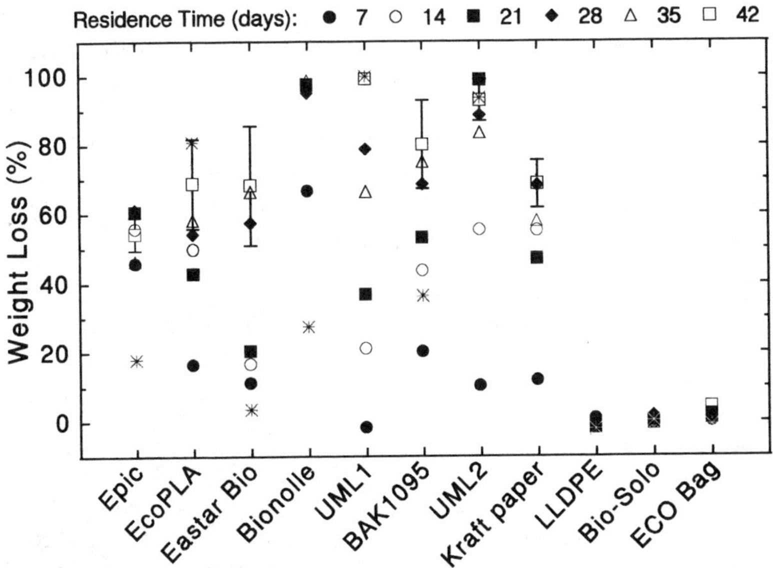

Figure 4. Weight loss from the test plastics degraded in a 4-L, continuously aerated simulated municipal solid waste compost environment. The bioreactors were externally heated and maintained at a core temperature of 52 ± 2°C for 42 days. Note: ✳ = weight loss measured following a 42-day test exposure in the poisoned control reactor.

The performance index is a function of both the rate and extent of degradation, thus it serves as a useful indicator of the relative degradability of the test plastics under a given set of environmental conditions. Performance indices for the test and reference materials are summarized and ranked in Table IV. Whereas the Bionolle® and 50:50 PLA:Bionolle blend outperformed the positive reference (Kraft paper); Epic, EcoPLA®, UML1, and BAK 1095 yielded performance indices that were not significantly different from that of the Kraft paper. Despite exhibiting extensive weight loss at the end the 42-day test exposure, degradation of the Eastar Bio® proceeded slowly at first (Figure 4), resulting in an overall performance index that was marginally lower than that of the Kraft paper. The PE-based plastics, on the other hand, were characterized by performance indices significantly lower than those of the positive reference and non-PE-based plastics – reflecting the fact that they exhibited no significant weight loss during the test exposure.

As a general rule, a performance index at either end of the scale (i.e., $PI \leq 30$ or ≥ 70) is a good indicator of both the rate and total extent of degradation that is achieved during a test exposure. That is, readily degradable materials are characterized by a high *PI*, whereas recalcitrant materials are characterized by a low *PI*. Moderately degradable materials (i.e., those that degrade extensively but slowly or are only partially degraded) are characterized by an intermediate *PI* ($30 < PI < 70$). The performance index is especially useful when comparing materials that degrade to about the same extent. Indeed, the utility of the performance index as an indicator of relative degradability can be illustrated by comparing the *PI*s for Bionolle®, EcoPLA®, and the PLA:Bionolle blends (Table IV). Whereas both blends degraded to about the same extent as the Bionolle® (97 ± 3% after 42 days), they were characterized by *PI*s that more closely reflected the actual composition of the blend. That is, the 80:20 blend (UML1) yielded a *PI* that was similar to that of the EcoPLA®, while the 50:50 blend (UML2) was characterized by a *PI* that was intermediate to that of either homopolymer-based plastic.

Table IV. Results of the least significant difference test for the 42-day weight loss from test and reference samples and overall performance index obtained during composting in the laboratory-scale (4-L) bioreactors (ASTM D 6003)

Sample ID	*Weight Loss (%)*[a,b]	*Sample ID*	*PI*[a,b]
UML1	99.20 a	Bionolle	92.46 a
Bionolle	96.90 a	UML2	70.41 b
UML2	93.75 ab	BAK 1095	56.80 c
BAK 1095	80.12 bc	Epic	54.34 cd
EcoPLA	69.93 cd	UML1	50.09 cd
Kraft paper	68.63 cd	Kraft paper	49.88 cd
Eastar Bio	68.27 cd	EcoPLA	48.56 d
Epic	54.17 d	Eastar Bio	40.37 e
ECO Bag	4.13 e	ECO Bag	1.23 f
Bio-Solo	-0.30 e	Bio-Solo	-0.05 f
LLDPE	-0.40 e	LLDPE	-0.71 f

[a] Within columns, mean values followed by the same letter are not significantly different ($P \leq 0.05$).

[b] Overall performance index = weight loss (%) averaged across time.

Weight loss values obtained on the final sampling date were used to assess the ultimate degradability of the test materials. Six of the test plastics (EcoPLA®, Eastar Bio™, Bionolle®, the two PLA:Bionolle blends, and BAK 1095) exhibited greater than 65% weight loss after 42 days (comparable to the Kraft paper); with three of the plastics (Bionolle® and the two PLA:Bionolle blends) exhibiting weight loss in excess of 90% (Figure 4; Table IV). Moreover, in most cases the residual plastics were so highly fragmented that the majority of the material was recovered in the < 2 mm size fraction of the compost (i.e., by using the solvent extraction procedure). On the other hand, neither of the PE-based test samples (Bio-Solo™ nor ECO Bag) exhibited any significant degree of weight loss (biological or abiotic) during the 42-day test exposure – mirroring the results obtained for the negative reference (LLDPE).

Despite containing only about 15% (w/w) polyethylene, degradation of the PCL-starch-PE blend (Epic) plateaued at about 55 ± 5% weight loss after only 14 days in the active compost (Figure 4). Moreover, upon completion of the 42-day test exposure, most of the residual plastic was recovered in the ≥ 2 mm size fraction of the compost. In contrast, Tosin et al. (*15*) assessed the compostability of Mater-Bi™ ZI101U (a plasticized PCL-starch blend) in a lab-scale (ca. 11-L) reactor and reported complete degradation in only 15–25 days. These authors also reported that the degradation of Mater-Bi™ was carried out primarily by fungi and was characterized by a progressive erosion of the surface of the plastic (*14, 15*). In light of this, our results suggest that some of the PCL and/or starch in the Epic blend had become encapsulated by polyethylene during processing. Thus, essentially isolating the most readily degradable components of the blend from the compost environment and rendering them unavailable for fungal attack.

It was generally observed that weight loss from the "degradable" plastics decreased noticeably during the last two weeks of the test exposure. Given the limited amount of compostable material present in the bioreactors at the start of the test exposure (ca. 300 g in a 4-L bioreactor) and consistently high temperature of the incubation (51–53°C), these results suggest a general depletion of the available carbon and nutrients needed to maintain a high level of microbial activity. This, in turn, may account for the incomplete degradation of the positive control (Kraft paper), as well as that of the EcoPLA® and Eastar Bio™, in the lab-scale compost reactors.

Abiotic Degradation

Though it must be demonstrated that degradation involves a microbial process (be it transformation or assimilation) to support a claim of *bio*degradability or compostability, this does not preclude microbial attack taking place via gratuitous metabolism occurring either concomitantly with an abiotic degradation process (e.g., hydrolysis or oxidation) or at the end of the abiotic process. For example, it is widely known that polylactic acid is sensitive to both enzymatic and chemical hydrolysis (16). Indeed, we found that during test exposures in poisoned-control reactors, abiotic degradation of the EcoPLA® and both PLA:Bionolle blends was equal to, or greater than, that observed in the bioactive reactors (Figure 4). The polyester amide (BAK 1095) also exhibited significant chemical (hydrolytic) degradation (ca. 40% w/w); though abiotic degradation accounted for only about half the total degradation observed in the biologically active compost. Likewise, abiotic degradation of the Bionolle® accounted for only about 25% of the total weight loss observed in the biologically active compost. The PE-based samples, on the other hand, remained essentially undegraded – as did the Kraft paper and Eastar Bio™ copolyester. These results are consistent with those reported by other investigators (*1, 17*).

There was a moderately strong correlation between weight loss in the biologically active and poisoned-control reactors ($r = 0.613$*); however, there was a much stronger correlation ($r_S = 0.918$***) when comparing relative degradation in the two systems (i.e., the rankings derived from the LSD tests). These results clearly support the notion that plastics which are subject to both chemical and microbial attack (regardless of the order in which this attack occurs) have a competitive advantage in terms of their relative degradability.

It must be noted, however, that just because a plastic disintegrates to the point where it cannot be visually distinguished from the other materials within the compost does not mean that it has become fully mineralized. Thus, the ultimate biodegradability (i.e., mineralization) of the test plastics was evaluated using a respirometric method (employing the static compost biometer system) to quantify the conversion of polymer-C into CO_2.

Tier II – Polymer Mineralization Study

Measurements of polymer-derived CO_2-C being produced during biodegradation in an active (fresh) compost are hampered by the high levels of CO_2 produced as a result of the decomposition of native organic matter in the compost matrix itself. Thus, to avoid problems associated with these high background levels of CO_2, the Tier II test employs a matured (finished) compost as the test matrix. In addition to minimizing background levels of CO_2, the matured compost matrix provides a rich source of both thermophilic microorganisms and the inorganic nutrients needed to support these microorganisms. However, because mature compost is inherently less active than fresh (active) compost, the Tier II test should be considered a "conservative" test intended only to elucidate information relating to the ultimate biodegradability (i.e., mineralization) of the test plastic.

Standard (ASTM and ISO) respirometry tests involve the use of continuously aerated composting systems (*1, 6*). However, of the approximately 3500 yard waste composting facilities operating in the United States in 1997 (*2*), it is estimated that 85–90% of these are low-tech (passively managed) windrow systems (*18*). Because of this, we developed a static (passively aerated) system in which to assess the mineralization of polymeric materials. Preliminary tests of the static compost biometer system involved comparing the mineralization of three known biodegradables (glucose, cellulose, and PHB) in both the static and continuously aerated (equivalent to ASTM Standard D 5338) systems. These tests demonstrated that the static system produced results (data not shown) that were highly correlated with, and not significantly different from, those obtained using the standard ASTM method (D 5338).

Results of the mineralization studies were interpreted using the following response parameters (Figure 5): (i) MAX-CO_2, defined as the percent cumulative net CO_2-C evolved during the test exposure; (ii) LAG, defined as the time required for net CO_2-C evolution to reach 10% of the MAX-CO_2 ; (iii) r_{MAX}, defined as the slope of the linear least-squares regression line plotted between the end of the lag period and the start of the plateau region of the net mineralization curve (i.e., the point where net mineralization = $^2/_3$•MAX-CO_2); and (iv) the relative biodegradation index (RBI), defined as the ratio of cumulative net mineralization of the TEST SAMPLE to cumulative net mineralization of the POSITIVE CONTROL.

All mineralization studies were terminated after a 42-day test exposure (i.e., equivalent to that of the Tier I weight loss test); consequently, not all the test plastics yielded net CO_2 evolution curves exhibiting the plateau region that identifies the net mineralization maximum (see Figure 5). Nevertheless, we were able to demonstrate that several of the plastics underwent significant mineralization during the test exposures (Figure 6). Indeed, biodegradation of the Kraft paper produced net CO_2 yields equivalent to 97% (± 7%) mineralization. Despite these indications of near-complete mineralization, however, pieces of paper that were only partially degraded were recovered from each of the triplicate biometer vessels upon completion of the test exposure. This suggests that there may have been a substantial *priming effect* (i.e., substrate-induced mineralization of the compost matrix in response to a change in microbial dynamics) associated with biodegradation of the Kraft paper. Indeed, occurrence of the priming effect under aerobic composting conditions has been demonstrated for readily biodegradable polymers such as cellulose (*19*). Though any "priming" that occurred would introduce a "positive bias" into the test results, there was no evidence that this phenomenon occurred with any of the test plastics.

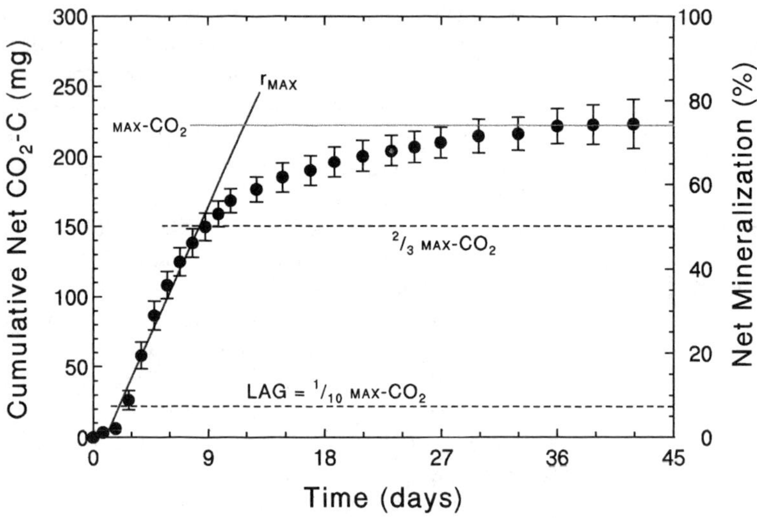

Figure 5. Parameters used to describe a typical net mineralization curve.

The analysis of variance indicated the presence of significant differences in net mineralization of the test polymers (F = 36.34***). Additional data analysis revealed that the plastics could be separated into four distinct groups based on their degradation characteristics (Figure 6; Table V). Group I materials were characterized by a short (2–3 day) lag period, followed by 5–7 days of rapid mineralization (20.2 ± 2.7 mg CO_2-C d^{-1}) and another 5–7 days of declining mineralization, culminating in a plateau by DAY 15 to DAY 18; this group included the Kraft paper, Bionolle®, and Epic. Group II materials exhibited a slightly longer lag period (5–8 days), followed by a near-linear increase in mineralization throughout the remainder of the test exposure; this group included EcoPLA™ and the 50:50 PLA:Bionolle blend (UML2). The plastics in Group III were characterized by a relatively long lag period (10–18

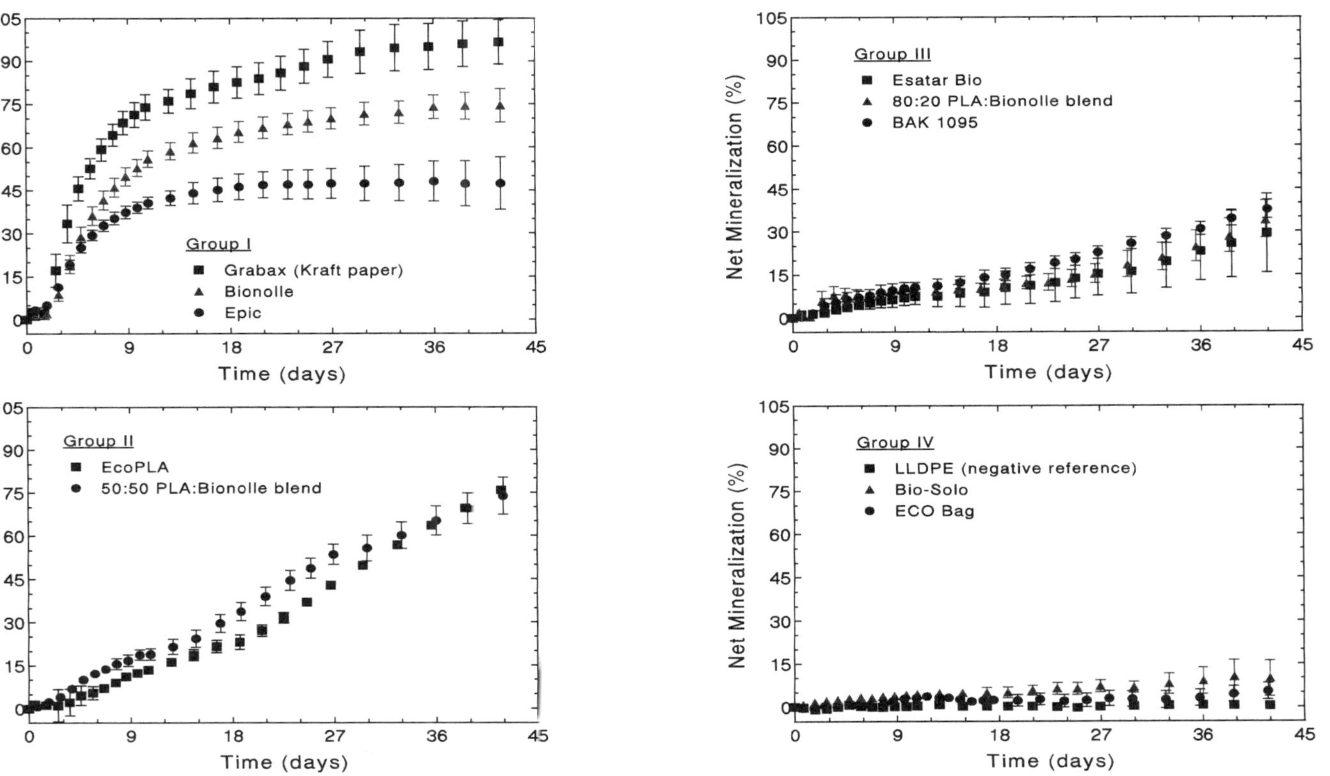

Figure 6. Net mineralization curves obtained for the test plastics and reference materials in the "static" compost biometer system. The test matrix consisted of a matured compost sample from Laughton's Garden Center, Westford, MA (LGC'97) maintained at 60% of water holding capacity. The biometers were externally heated and maintained at a core temperature of 52 ± 1 ℃ for 42 days.

days) followed by a slow, but steady increase in the rate of mineralization throughout the remainder of the test exposure; this group included Eastar Bio®, the 80:20 PLA:Bionolle blend (UML1), and BAK 1095. The final group (Group IV) was characterized by a lag period of 39–42 days and exhibited little or no net mineralization; this group consisted of the three PE-based plastics (LLDPE, Bio-Solo™, and ECO Bag). These groupings are reflected in the LSD ranking of the overall performance indices of the test materials (Table VI).

Table V. Mineralization of plastic films during a 42-day test exposure in compost at 51 ± 1°C and a moisture content of 60% WHC. Films were used as provided by the manufacturer, except that they were cut into small (10 mm × 10 mm) pieces prior to exposure in the compost. Sample loading ≈ 10 mg polymer g^{-1} compost (oven dry weight basis)

Sample ID	*Lag[a] (days)*	*Plateau*	*Cumulative Net Mineralization (mg C)*	*(%)*	*RBI[b]*	*r_{MAX}[c] (mg C d^{-1})*	*t_{60}[d] (days)*
Kraft paper	1.9	yes	201.8	97.0	1.00	22.0	7
EcoPLA	8.1	no	201.4	76.0	0.78	5.34	35
Bionolle	2.8	yes	224.4	74.6	0.77	21.4	14
UML2	4.8	no	195.0	72.8	0.75	4.76	33
Epic	2.8	yes	141.9	47.5	0.49	17.1	IND
BAK 1095	9.9	no	113.7	36.5	0.38	2.68	60–70[e]
UML1	17.6	no	88.5	33.8	0.35	2.56	55–70[e]
Eastar Bio	18.8	no	86.1	29.4	0.30	2.57	70–80[e]
Bio-Solo	38.8	yes	39.5	9.8	0.10	0.83	IND
ECO Bag	> 42	yes	22.0	5.5	0.06	*ns*	IND
LLDPE	> 42	yes	1.7	0.7	0.01	*ns*	IND

[a] Time required to reach 10% of the net CO_2-C maximum. Note: In those cases where the net mineralization curve did not reach a plateau, the LAG was defined as the time required for net CO_2-C evolution to reach 10% of its theoretical maximum.

[b] Relative biodegradation index = (% net mineralization of the TEST SAMPLE ÷ % net mineralization of the POSITIVE CONTROL); positive control = Kraft paper (Grabax).

[c] Maximum rate of mineralization: defined as the slope of the linear least-squares regression line plotted between the end of the lag period and start of the plateau region (i.e., the point where net mineralization = two-thirds the net CO_2 maximum). Note: In those cases where the net mineralization curve did not reach a plateau, the r_{MAX} was calculated over the region from the end of the lag period to the net CO_2-C maximum.

[d] Time (days) required to achieve 60% theoretical net mineralization. IND = indeterminate.

[e] t_{60} value estimated by extrapolating the net mineralization curves (Figure 6) using nonlinear regression techniques.

[ns] Not significantly different from zero.

As in the Tier I test, degradation of the Epic PCL-starch-PE blend proceeded rapidly at first, but then plateaued with a CO_2 yield equivalent to about 48% mineralization of the plastic (Figure 6 – Group I). In addition, much of the starting material was recovered in the >2-mm size fraction. Indeed, the residual films were recovered mostly intact – exhibiting varying degrees of surface pitting, with significant degradation along the edges only. These results suggest that the initial phase of rapid mineralization (ca. 17 mg CO_2-C d^{-1}) reflects preferential biodegradation of the starch and PCL components of the film. Indeed, a number of investigators have reported that PCL-starch blends undergo complete biodegradation (both primary degradation and mineralization) during composting (*1, 14, 15*). These results also indicate that the incorporation of even a relatively small amount of polyethylene in a blend containing truly biodegradable components can significantly retard the biodegradation of these components.

Table VI. Results of the least significant difference test for the cumulative (42-day) net mineralization of the test and reference samples and overall performance index obtained during test exposures in the static compost biometer system

Sample ID	*Cumulative Net Mineralization (%)*[a]	*Sample ID*	*PI*[ab]
Kraft paper	97.0 a	Kraft paper	83.5 a
EcoPLA	76.0 b	Bionolle	64.9 b
Bionolle	74.6 b	Epic	44.5 c
UML2	72.8 b	UML2	44.3 cd
Epic	47.5 c	EcoPLA	39.2 cd
BAK 1095	36.5 cd	BAK 1095	20.7 cd
UML1	33.8 cd	UML1	17.1 d
Eastar Bio	29.4 d	Eastar Bio	15.5 e
Bio-Solo	9.8 e	Bio-Solo	6.56 f
ECO Bag	5.5 e	ECO Bag	2.91 f
LLDPE	0.7 e	LLDPE	0.06 f

[a] Within columns, mean values followed by the same letter are not significantly different ($P \leq 0.05$).

[b] Overall performance index = net mineralization (%) averaged across time.

Standards currently being drafted by organizations such as ASTM, CEN, and DIN require that, at the very minimum, plastics claiming to be biodegradable should mineralize to an extent ≥ 60% for a single polymer (≥ 90% for polymer blends and copolymers) during a test exposure of 180 days (or less). Whereas our data demonstrated that the Group I and Group II plastics (Figure 6) generally exceeded this minimum value, the 42-day test exposure was too short to allow determination of the t_{60} for the Group III and Group IV plastics. Consequently, non-linear regression techniques were used to estimate the t_{60} for these plastics. Using these techniques, it was estimated that test exposures of 60 to 80 days would be required for the Group III plastics to produce CO_2 yields equivalent to 60% mineralization (Table VI). In the case of the Eastar Bio™ (the only one of the plastic bags for which mineralization data was available), both the r_{MAX} (2.57 mg CO_2-C d^{-1}) and estimated t_{60} (70–80 days) were similar to those found at the Eastman Biodegradation Testing Laboratory (*17*) and somewhat faster than those calculated from data presented in the 1996 ISR report (*1*).

Each of the Group III plastics yielded a significant amount of residual plastic in the compost upon completion of the test exposure. However, whereas most of the residual Eastar Bio™ and BAK 1095 were recovered in the >2-mm size fraction; residual pieces of the PLA:Bionolle blend were highly fragmented – passing through the 2-mm sieve. These data are consistent with those obtained in the Tier I test and suggest that microbial attack is the primary mode of degradation for both the Eastar Bio™ and BAK 1095. Moreover, the data suggests that although there was significant abiotic degradation (disintegration) of the PLA:Bionolle blend, the primary degradation products were mineralized at a relatively slow rate.

Rates of mineralization for the Group IV (PE-based) plastics were generally too low (< 1.0 mg CO_2-C d^{-1}) for estimates of the t_{60} value to be calculated. Indeed, the negative reference (LLDPE) exhibited no sign of biodegradation (i.e., produced no net increase in CO_2 production) during the test exposure (Table V). Conversely, both the Bio-Solo™ and ECO Bag produced small amounts of net CO_2 (Table V; Figure 6 – Group IV). Only in the case of the Bio Solo™, however, was this increase significant (i.e., relative to CO_2 production in the control – compost only – system). It also was observed that the Bio Solo™ bag became brittle and faded (changing from brown to a light tan) during the test exposure. This suggests that the net CO_2 produced during the test exposure was most likely a result of biodegradation of one or more of the additives used in manufacturing the plastic (i.e., a plasticizer or colorant). Similar results have been observed for plasticized PLA films during test exposures in soil (*21*).

It is to be expected that the level of biological activity in the compost will affect both the rate and extent of biodegradation (*15*). Thus, given that the Tier I test used an active compost as the test matrix, whereas the Tier II test used a matured compost, it was not surprising to find that degradation measured as cumulative net mineralization after a 42-day test exposure was substantially lower (by an average of about 23%) than that measured as weight loss in the Tier I test. Nevertheless, there was a moderately strong correlation ($r = 0.662^*$) between net mineralization of the test materials and degradation measured as weight loss. Furthermore, when the ranked data were compared, weight loss and mineralization were more strongly correlated ($r_S = 0.736^{**}$), indicating that the Tier I and II tests were equally good at distinguishing differences in the relative biodegradability of the test plastics.

In general, the combined results of the Tier I and Tier II tests indicated that (i) Bionolle®, EcoPLA®, and the PLA:Bionolle blends were readily biodegradable; (ii) primary biodegradation (i.e., disintegration) of the Eastar Bio™ and BAK 1095 was comparable to that of the Kraft paper, though mineralization of these plastics was considerably slower than that of the positive reference; (iii) following a short period of rapid degradation (both primary biodegradation and mineralization), biodegradation of the PCL-starch-PE blend (Epic) was ultimately retarded by the presence of polyethylene in the blend; and (iv) there was no evidence to support claims of biodegradability for the PE-based plastics (LLDPE, Bio-Solo™, and ECO Bag).

It is generally accepted that laboratory–scale testing provides an excellent platform for rapidly assessing the bio-environmental degradability of plastic materials under simulated disposal conditions, as well as facilitating inter-laboratory comparisons. However, predicting the long-term fate and effects of a plastic from results obtained during test exposures carried out under carefully controlled and optimized conditions is, at best, extremely difficult. Thus, while the Tier I and Tier II tests have been demonstrated to provide a fair assessment of the relative biodegradability of test plastics under standard conditions, rates of degradation determined from these tests should not be considered representative of those that would occur under actual field composting conditions. Indeed, positive results from the Tier I and II tests should be interpreted as demonstrating only that the test plastic is *inherently* biodegradable. Nevertheless, a positive result can be perceived as an indication that there is a good probability the test plastic will biodegrade under field conditions. Negative test results, on the other hand, do not necessarily mean that the test plastic is not biodegradable. Instead, a negative result may indicate that the test plastic is biodegraded at very slow rate (e.g., the Group III plastics) or is only partially degraded (e.g., the Epic sample) under the conditions of the test exposure.

Tier III – Field Tests

Full-scale (Tier III) testing was carried out to confirm the results obtained in the Tier I and Tier II laboratory-scale tests; i.e., to demonstrate that plastics which biodegraded under controlled laboratory conditions also would biodegrade under field conditions. In addition, results from the field studies allowed us to assess the reliability of the lab-scale tests in predicting the *in situ* biodegradability of the test plastics.

Process monitoring at both the Westford (FIELD SITE 1) and Amherst (FIELD SITE 2) sites was carried out by personnel from the individual composting facilities. Likewise, control of the composting process itself was left to the discretion of each facility's manager. This decision not to interfere with the normal operation of the composting facilities (beyond what was necessary to recover the test samples) was intended to enhance the relevance of the test results to *real world* conditions.

Site 1 – Laughton's Garden Center (Westford, MA)

The composting facility at Laughton's Garden Center was chosen as being representative of the overwhelming majority of yard-trimmings compost facilities in the United States. That is, the facility accepts only grass clippings, leaves, small brush, and horticultural waste as its source material and composting is carried out in a passively managed windrow compost (*18*). The facility is approved by the Massachusetts Department of Environmental Protection–Office of Solid Waste Management and, at the time of this study, had been in operation for about nine years (using the finished compost primarily in the Center's commercial horticulture operation) and had been producing compost for sale for about three years. Source materials, dropped off on-site by residents of the town of Westford, are mixed and stacked into windrows using a front-end loader. Normal maintenance of the facility is such that the temperature and water content of the compost is monitored weekly and the windrows turned every six to eight weeks. However, the turning frequency is varied as necessary to maintain adequate composting conditions. On average, active composting takes place for eight to ten months; this is followed by six to nine months of curing before the finished compost is ready for market.

Composting Conditions. Test samples were placed in the windrow on March 26, 1997 and exposed to active composting conditions for 12 weeks. At the time of insertion, active composting had been underway for about two months and temperatures in the windrow were in the 48–54°C range. Whereas initial air samples collected from the core of the windrow yielded O_2 concentrations of only 8–10%, samples collected 24-h after opening and restacking the windrow were in the 18–20% range. However, the site received about 60 cm of snow on March 31st, which upon melting saturated the outer layer of the windrow with water – thus reducing gaseous exchange between the compost and atmosphere. Consequently, on the first sampling date (April 9th), O_2 concentrations in the windrow were again relatively low (9–11%). Continued warming throughout the spring resulted in the water content of the compost decreasing from a high of about 70% (immediately after the snow melt) to about 50–55%; thus, allowing for greater gaseous exchange between the compost and atmosphere. As a result, O_2 concentrations in gas samples collected from the compost were generally in the 12–16% range throughout the remainder of the test exposure. The lone exception occurred on the third sampling date (April 23rd; DAY 28); on this date, O_2 concentrations near the center of the windrow were only in the 4–5% range – increasing to about 14% in the outer layers of the compost. This was attributed to restricted air flow resulting from compaction of the compost during sample recovery and the restacking of the windrow, and was alleviated by "raking out" the compost as the blocking frames were placed back into the windrow for each subsequent exposure period. Core temperatures in the compost varied from 40°C to 62°C (with a median 54°C and an interquartile range of 48–56°C) throughout the 12-week test exposure. Temperatures such as these generally indicate that microbial activity in the windrow was maintained at a moderate level.

Biodegradation (Weight Loss). Samples retrieved from the compost were inspected and assigned a visual disintegration rating (Figure 7A). In general, the visual disintegration ratings followed the same trend as actual weight loss from the test samples (Figure 7B). Indeed, there was a very strong correlation between weight loss and the visual disintegration rating ($r_S = 0.924$***).

The ANOVA for the overall performance indices revealed that blocking effects were not significant (F = 1.53ns); indicating that the composting process itself was proceeding in a uniform manner. A closer look at the data, however, revealed that there was a significant blocking effect (F = 2.95, P = 0.032*) on DAY 28 of the test exposure, with block No. 5 (located at the bottom of the windrow) yielding weight loss values that were 50–75% lower than those recorded for the overlying blocks. Field observations suggested that this was most likely a result of poor aeration in the layers of compost immediately above and below the 5th block. That is, the O_2 content of gas samples collected at depth from the compost was only 4–5% (v/v) – the lowest of the entire 84-d test exposure, and the only date on which the O_2 concentration fell below the threshold level for the onset of anaerobiosis (i.e., 6% v/v). Consequently, starting on DAY 28, the compost below the 5th block was "raked out" to a depth of about 30-cm immediately prior to placing the blocking frames back into the windrow for the next exposure period. Likewise, additional care was taken to ensure that there was minimal compaction of the layers above the 5th block. As a result of these procedures, the O_2 content of the compost air increased to about 14 ± 2% and remained there throughout the remainder of test exposure. No further blocking effects were observed.

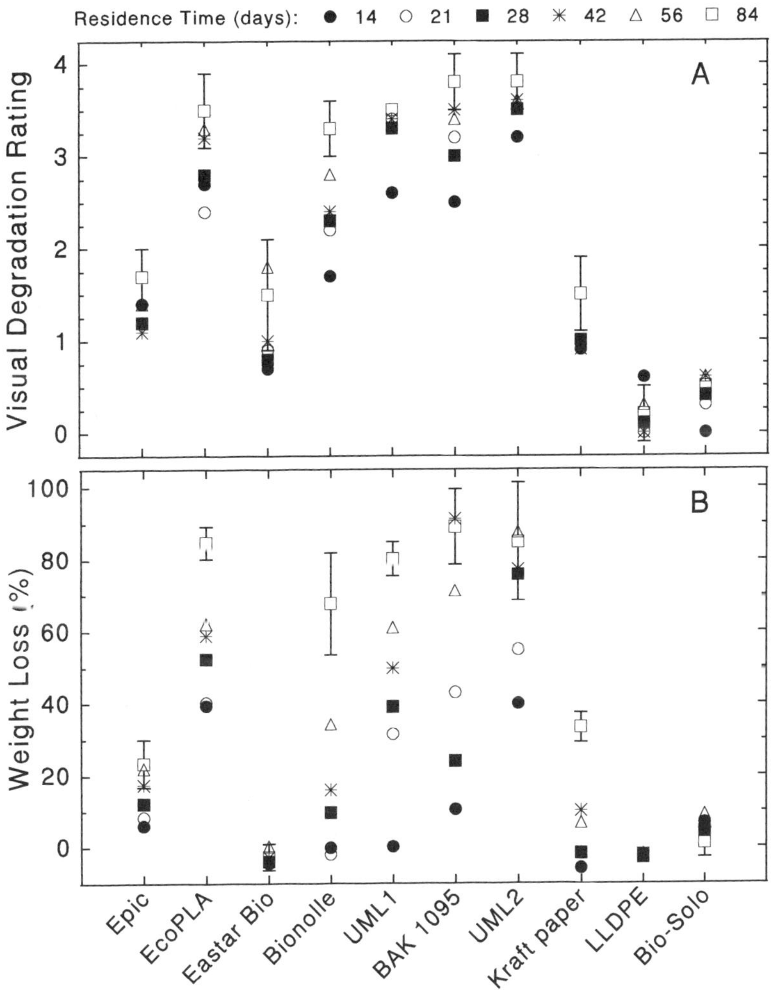

Figure 7. Degradation of test plastics exposed to windrow composting at FIELD SITE No. 1 (Laughton's garden center, Westford, MA). [A] Visual disintegration ratings; [B] Weight loss (%).

Analysis of variance determined that there were significant differences in both the performance index ($F = 4.05$*** for the *time × test material* interaction) and total degradability of the test plastics ($F = 96.31$***). Furthermore, test materials that exhibited little or no abiotic degradation in the Tier I test were characterized by *PI*s (Table VII) that were substantially lower in the windrow compost than in the lab-scale test. Conversely, test materials whose degradation pathway included a significant abiotic component (i.e., EcoPLA®, the PLA:Bionolle blends, and BAK 1095) exhibited *PI*s in the windrow compost that were comparable to those obtained in the lab-scale weight loss test. As a result, the overall performance indices generated in the two tests were only weakly correlated ($r = 0.518^{\dagger}$). However, when comparing the relative performance of the test materials (i.e., the ranked *PI*s), there was a much stronger correlation between the performance indices ($r_S = 0.673$*) – indicating that the comparative biodegradability of the test plastics was relatively unaffected by differences in the test environments. Nevertheless, as with the process control data, these results suggest that the level of microbial activity in the windrow compost was only moderate and was lower than that obtained in the lab-scale composting system.

Table VII. Results of the least significant difference test for the overall performance indices and visual disintegration ratings for test and reference samples exposed to windrow composting at FIELD SITE NO. 1 (Laughton's Garden Center, Westford, MA)

Sample ID	*PI*[a,b]	*Sample ID*	*VDR*[b,c]
UML2	70.08 a	UML2	3.53 a
EcoPLA	56.37 b	UML1	3.25 b
BAK 1095	53.62 b	BAK 1095	3.23 b
UML1	43.70 c	EcoPLA	2.98 c
Bionolle	21.01 d	Bionolle	2.45 d
Epic	14.90 d	Epic	1.37 e
Kraft paper	6.87 e	Eastar Bio	1.11 f
Bio-Solo	4.90 e	Kraft paper	1.05 f
LLDPE	-2.27 f	Bio-Solo	0.40 g
Eastar Bio	-2.88 f	LLDPE	0.20 h

[a] Overall performance index = weight loss (%) averaged across time.

[b] Within columns, mean values followed by the same letter are not significantly different ($P \leq 0.10$).

[c] Visual disintegration rating; integrated mean value (i.e., individual VDR values averaged across time).

Additional data analysis suggested that the test materials could be classed into one of four groups based on their overall performance characteristics. The materials in Group A (EcoPLA®, BAK 1095, and UML2) did not exhibit an appreciable lag period and yielded *PIs* >50. Group B was characterized by a lag period of 14–21 days and a *PI* between 25 and 50; this group consisted of a single test material

(UML1). Group C (Epic, Bionolle®, Kraft paper, and Bio-Solo™) was characterized by a relatively long lag period (≥ 28 days) and a *PI* between 0 and 25. The materials in the final group (Group D: Eastar Bio™ and LLDPE) exhibited little or no degradation during the 84-day test exposure and were the only samples to yield negative *PIs*.

Whereas the Group A and Group B plastics generally out-performed the Kraft paper at all stages of the test, the positive reference itself exhibited little or no degradation during the first 56 days of the test exposure. This result was somewhat surprising, given the generally good performance of the Kraft paper in both the Tier I and Tier II tests. The same was true of the Eastar Bio™, which exhibited poor performance characteristics in the windrow compost relative to the lab-scale compost tests. Indeed, despite the fact that the Eastar Bio™ received a much higher visual disintegration rating than either of the PE-based plastics (Table VII), its overall performance rating at FIELD SITE 1 was essentially the same as that of the negative reference (LLDPE). As in the lab-scale tests, the performance index proved to be a good indicator of relative degradability, allowing us to distinguish between materials that exhibited equivalent amounts of total degradation (i.e., weight loss).

Five of the eight test plastics (EcoPLA®, Bionolle®, BAK 1095, and the two PLA:Bionolle blends) exhibited weight loss in excess of 60% by DAY 84 of the composting exposure (Figure 7B; Table VIII). Conversely, neither the Eastar Bio™ nor Bio-Solo™ samples exhibited any significant weight loss during the 84-day test exposure. Again, the Epic bag (PCL-starch-PE blend) was only partially degraded – yielding about 30% less weight loss than the positive control and exhibiting no appreciable change in total weight loss (ca. 20 ± 3%) after DAY 42 of the test exposure. These results again demonstrate that the incorporation of polyethylene in a blend with readily biodegradable components (in this case, PCL and starch) can have a significant negative impact on the biodegradability of the resulting plastic material. By comparison, Itävaara et al. (*20*) demonstrated that Mater-Bi ZFO3U (a PCL-starch based plastic) underwent near complete degradation (measured as weight loss) during a 16 week test exposure in a windrow compost.

Biodeterioration (Sample Disintegration). In addition to the samples placed in nylon mesh bags and recovered for the determination of weight loss, larger (15-cm × 15-cm) samples of each test material were mounted in Plexiglas window-pane frames and placed in the windrow along with the blocking frames. These "inspection" frames were recovered on each of the first five sampling dates (the frames themselves being too badly damaged to be of use beyond DAY 56 of the test exposure) and visual disintegration ratings assigned to each of four replicate samples (one rep per frame). Newspaper and glossy magazine paper were included in the inspection frames as additional positive reference materials. Five of the test plastics (EcoPLA®, Bionolle®, the two PLA:Bionolle blends, and BAK 1095) showed signs of extensive disintegration (mean *VDR* ≥ 3.3) by DAY 14 of the test exposure. Extensive disintegration of the newspaper and glossy paper also was evident after 21–28 days (mean *VDR* ≥ 3). Samples from the Epic bags exhibited signs of moderate degradation after only 14 days (mean *VDR* = 2.5); however, further disintegration of the sample proceeded slowly – with no significant increase in the mean *VDR* until DAY 42. Disintegration ratings for the Eastar Bio™ generally were higher than those for the Kraft paper; nevertheless, disintegration of the two materials proceeded in essentially the same fashion. That is, there were no visible signs of significant degradation during the first 21 days of the test exposure (mean *VDR* ≤ 1.5), the rate of disintegration for both materials increased with time, and both were completely disintegrated by DAY 56. Conversely, both PE-based plastics (LLDPE and Bio-Solo™) remained essentially intact throughout the test exposure. However, whereas the negative reference material (LLDPE) showed no signs of degradation whatsoever, the Bio-Solo™ samples became faded and less flexible as the test exposure proceeded. This result presumably reflected the loss of both colorant and plasticizer from the plastic matrix and was similar to that observed in the Tier II mineralization study.

Table VIII. Results of the least significant difference test for the total (84 day) weight loss and final visual disintegration rating for test and reference samples exposed to windrow composting at FIELD SITE NO. 1 (Laughton's Garden Center, Westford, MA)

Sample ID	*Weight Loss (%)*[a]	*Sample ID*	*VDR*[a,b]
BAK 1095	89.06 a	BAK 1095	3.80 a
UML2	84.86 a	UML2	3.80 a
EcoPLA	84.82 a	UML1	3.50 ab
UML1	80.32 a	EcoPLA	3.50 ab
Bionolle	67.82 b	Bionolle	3.30 b
Kraft paper	33.36 c	Epic	1.70 c
Epic	23.36 d	Eastar Bio	1.50 c
Bio-Solo	1.14 e	Kraft paper	1.50 c
LLDPE	-2.22 e	Bio-Solo	0.50 d
Eastar Bio	-2.62 e	LLDPE	0.20 d

[a] Within columns, mean values followed by the same letter are not significantly different ($P \leq 0.10$).

[b] Visual disintegration rating.

In general, the *VDR*s for samples mounted in the inspection frames paralleled those for samples recovered from the nylon mesh bags. That is, despite the fact that the ratings assigned to the inspection frames were always higher (by an average of more than one rating point), there was a strong correlation between the two sets of *VDR*s ($r_S = 0.905$***). It is important to note, however, that extensive disintegration of materials from the inspection frames indicates only that these materials had become visually indistinguishable from the compost matrix itself, and does not imply complete biodegradation of the materials. Indeed, visual evidence of disintegration was always indicative of more extensive degradation than the actual weight loss data – especially in the case of the Eastar Bio™ and Kraft paper (see Tables VII and VIII). Thus, while the nylon bags facilitated the recovery of materials (especially those that degraded extensively), our results suggest that the weight loss data obtained using the bags should be considered conservative. Similar results were reported by Itäraava et al. (*20*).

Site 2 – University of Massachusetts (Amherst, MA)

The Wright in-vessel composter (*IVC*) at the University of Massachusetts Amherst (FIELD SITE 2) is an example of a conveyer-type, horizontal-flow static bed reactor. It is considered a "high-end", aggressive composting system and is representative of a technology that continues to gain popularity for both institutional and urban composting. At the time of this study, in-vessel composting was the initial step in a two-phase process aimed at assessing the feasibility of composting the food, plant, and animal wastes generated by the university community – and had been in operation for about nine months. Source materials (see Table II) were mixed thoroughly (using a food waste–to–non-food waste ratio of approximately 3:2) and fed onto a series of stainless steel trays within the *IVC* unit (see Figure 3). As waste was added to the *IVC*, the trays advanced through temperature controlled zones where the composting took place. Normal operation of the *IVC* revolved around a 5-day work week, thus yielding an average composting time of 18 days. Temperatures in the *IVC* were controlled by fans linked to temperature probes that could be activated to move air through the unit as needed. Likewise, appropriate levels of oxygen in the compost were maintained by intermittent operation of the circulating fans and by "turning" the compost about half-way through the composting cycle – using a spinner to agitate the compost and reduce compaction. The moisture content of the initial waste mix was maintained in the 50–60% range by making small adjustments to the daily mix ratio. Thus, given the level of process control provided by the *IVC*, it was expected that the very active and species-rich microbial populations associated with the feedstock would take advantage of the enhanced environmental conditions within the *IVC* to produce an extremely active composting environment.

Composting Conditions. Test samples (seven replicates of each test material) were placed into the *IVC* by mixing with approximately 1.2 tons of waste. Thereafter, the *IVC* received an average load of about 1.3 tons of waste per day and was operated on a 14-day cycle (based on a 5-day work week). During the test exposure, self-heating of the solid waste matrix produced average daily temperatures ranging from 64 to 78°C in ZONE 1 of the *IVC* (with a median temperature of 69°C). As the composting process proceeded, there was a significant reduction in the mass of the solid waste compost – resulting in lower composting temperatures in ZONE 2. Indeed, during the test exposure, average daily temperatures in ZONE 2 ranged from 46 to 63°C (with a median temperature of 52°C). Temperatures such as these are indicative of a high level of biological activity in the *IVC* compost. Additional indications of the high degree of activity within the *IVC* were provided by ancillary testing of the pre- and post-*IVC* solid waste compost. For example, representative changes in the solid waste mix itself during composting included: (i) a significant increase in pH – from 4.4 to 7.2, (ii) a 40-45% reduction in dry weight – with only a 5-10% increase in water content, (iii) a 48-50% reduction in soluble salts and total dissolved solids, and (iv) a 40-50% reduction in C:N ratio. Changes of this magnitude clearly indicate that the composting environment was produced and maintained within the *IVC* was more intense than that produced in either the lab-scale or windrow composts.

In all, only 80% of the samples placed in the *IVC* were recovered – with a minimum of five samples recovered (over a three to five day period) for each test material. However, only about 20% of the bags were recovered with the compost collected on DAY 18; i.e., the day on which trays entering the *IVC* on March 17th (DAY 1) exited the in-vessel unit. Indeed, more than 70% of the bags were recovered with composted wastes that entered the *IVC* 1-4 days after the test samples. Despite variations in total residence (exposure) time, there was no apparent relationship between weight loss and the time an individual sample spent in the *IVC*. In addition, whereas most of the litter bags were recovered intact, more than one-third of the bags were either torn or partially shredded and, in several instances, the test samples themselves had become knotted or balled during the test exposure. Thus, it was apparent that samples placed in the *IVC* were subjected to a greater range of physical processing than had been anticipated. Indeed, it was obvious that some redistribution of compost within the *IVC* occurred during the turning cycle and that the solid waste compost (including the test plastics) was subjected to forces that were both intense and non-uniform.

Biodegradation (Weight Loss). A minimum of five replicate samples were recovered for each test material and, prior to cleaning, were inspected and assigned a visual disintegration rating (Figure 8A). As was the case at FIELD SITE 1, the visual disintegration ratings followed the same general trend as actual weight loss from the test samples (Figure 8B; Table IX) and there was a strong correlation between weight loss and the visual disintegration rating ($r_S = 0.916$***).

In general, samples processed through the *IVC* were more highly degraded than samples placed in the windrow compost for an equivalent period of time (Figure 9), which suggests that the *IVC* provided a more robust composting environment. Four of the test materials (EcoPLA®, Bionolle®, and the two PLA:Bionolle blends) degraded to an extent of greater than 60% during the short test exposure in the *IVC*, with the 50:50 PLA:Bionolle blend (UML2) yielding more than 90% weight loss. Three of the other test plastics (Epic, BAK 1095, and Eastar Bio™) also degraded extensively (ca. 40%). Despite little or no visual evidence of degradation (other than some color fading), the Bio-Solo™ also yielded a significant weight loss. On the other hand, both the Kraft paper and LLDPE exhibited a net gain in weight, reflecting both the adsorption of soluble organics by the paper and the difficulty involved in cleaning the residual materials.

Enhanced hydrolytic degradation – brought on by the relatively high moisture content of the MSW compost and high temperatures attained during the test exposure – undoubtedly contributed to the overall weight loss observed for some samples (e.g., EcoPLA®, the two PLA:Bionolle blends, and BAK 1095). At the same time, plastics that degraded primarily (or entirely) via microbial attack (e.g., Eastar Bio™ and Epic) also exhibited more extensive degradation in the *IVC* than in either the lab-scale or windrow composts. Again, these results are a clear indication that there was a much higher level of microbial activity in the *IVC* than in either the lab-scale or in-vessel compost systems.

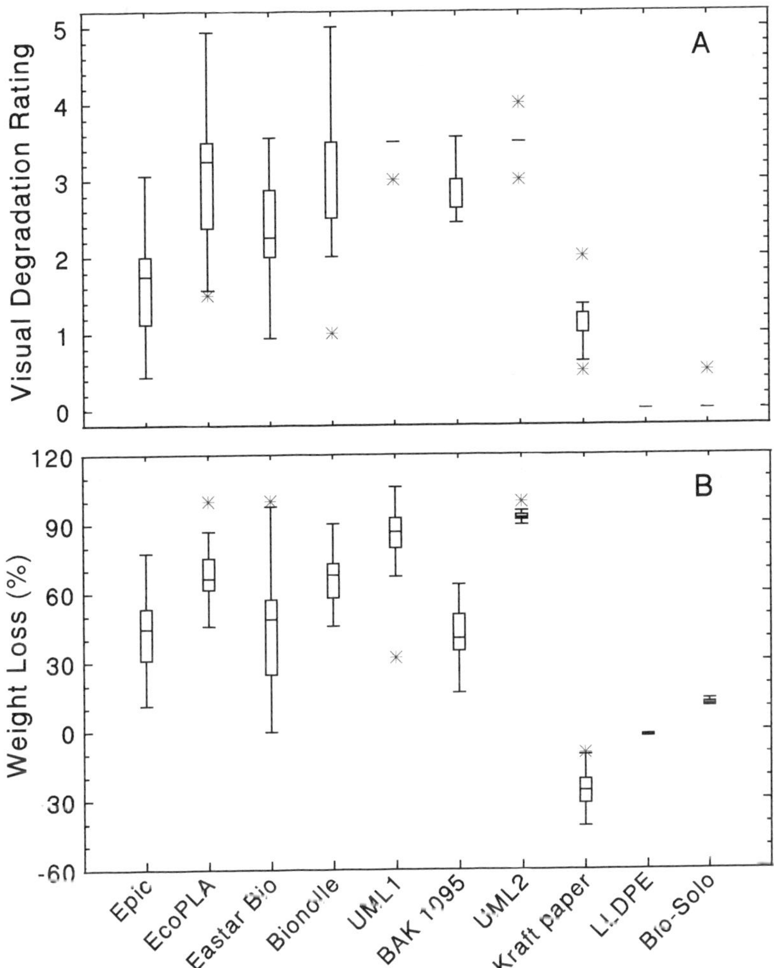

Figure 8. Degradation of test plastics exposed to in-vessel composting at FIELD SITE NO. *2 (University of Massachusetts Amherst, Intermediate Processing Facility, Amherst, MA). [A] Visual disintegration ratings; [B] Weight loss (%). Average residence time = 20 days. ✱ = statistical outlier.*

Weight loss from test samples degraded in the *IVC* averaged about 55% greater than that from samples recovered after a similar exposure in the windrow, and the two were only weakly correlated (r = 0.642[†]). However, data from the windrow study demonstrated that as the duration of the test exposure increased, total weight loss in this less active system approached that in the *IVC* (Figure 9). For example, average total weight loss from samples degraded during an 84-day exposure in the windrow compost was only about 12% lower than that from samples degraded in the *IVC*.

Likewise, average total weight loss from samples degraded for 21 days in the lab-scale (Tier I) compost was about 14% lower than that from samples degraded in the *IVC*. Samples degraded in the lab-scale reactors for the full 42-days, however, yielded an average total weight loss that was about 10% greater than that obtained during the *IVC* test exposure. Thus it can be concluded that, in addition to environmental conditions, the duration of the test exposure is a key factor in determining the final results of any biodegradation (compostability) test. A similar conclusion was reached during the ISR study (1). Furthermore, comparison of the ranked *PI*s revealed that there was a fairly strong correlation between compostability in the Tier III-*IVC* test and compostability in both the Tier I ($r_S = 0.685$*) and Tier III-windrow ($r_S = 0.735$*) tests. Again, indicating that differences in the test environment and exposure time had relatively minor effects on the *relative* biodegradability of the test plastics.

Table IX. Results of the least significant difference test for the 20-day weight loss and visual disintegration ratings for test and reference samples exposed to in-vessel composting at FIELD SITE NO. 2 (University of Massachusetts Amherst, Intermediate Processing Facility, Amherst, MA)

Sample ID	*Weight Loss (%)*[a]	*Sample ID*	*VDR*[ab]
UML2	92.34 a	UML2	3.50 a
UML1	88.32 a	UML1	3.42 ab
Bionolle	66.20 b	EcoPLA	3.10 abc
EcoPLA	64.80 b	BAK 1095	2.92 bc
Epic	42.42 c	Bionolle	2.90 bc
BAK 1095	40.43 c	Eastar Bio	2.25 d
Eastar Bio	37.76 c	Epic	1.58 ef
Bio-Solo	11.36 d	Kraft paper	1.08 f
LLDPE	-2.34 e	Bio-Solo	0.10 g
Kraft paper	-27.99 f	LLDPE	0.00 g

[a] Within columns, mean values followed by the same letter are not significantly different ($P \leq 0.10$).

[b] Visual disintegration rating.

Consistency of the Biodegradation Test Rankings

The consistency of the biodegradation test results was assessed using Kendall's coefficient of concordance, *W*, which measures the degree of association in the ranking of the polymers following the different biodegradation tests. First, the mean weight loss (%) value for each test/reference material was ranked for the individual biodegradation tests (Table X). The average rank for each plastic was then calculated and compared with the average rank for all the test materials. The calculated coefficient of concordance for the ranking of test materials between the different tests was 0.652** for the overall performance index and 0.724** for the total weight loss recorded upon completion of the various test exposures. Both values indicate a high degree of association between the different tiers of testing; thus, we concluded that the results obtained from lab-scale (Tier I and Tier II) tests can be used as reliable indicators of the *relative biodegradability* of test plastics in the field (i.e., at the Tier III level).

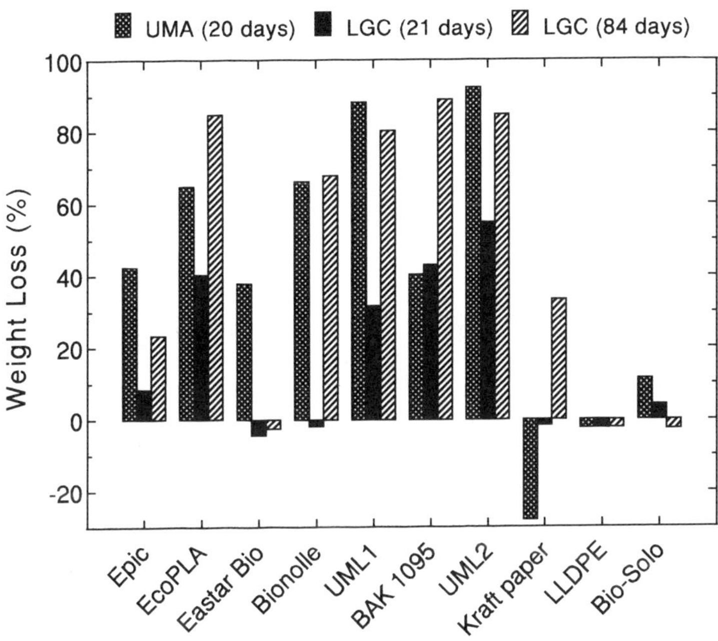

Figure 9. Comparison of the total weight loss from test plastics exposed to in-vessel composting at FIELD SITE No. 2 (UMA) with that from test samples exposed to windrow composting at FIELD SITE No. 1 (LGC).

Summary

Compostable plastic bags continue to make inroads with both consumers and composters; however, confusion surrounding the claims of biodegradability for bags of vastly different composition and properties remains a major obstacle to the growth and development of this market. For now, bags made from Kraft paper are viewed as the most trusted vehicle for the collection of yard wastes. Thus, it seems only logical that any "compostable" plastic bag must exhibit biodegradation characteristics comparable to (if not better than) paper collection bags. Accordingly, we evaluated the relative compostability of nine "environmentally degradable" plastic lawn & leaf bags relative to that of a standard Kraft paper bag. Studies were conducted at both the laboratory and field scales and were intended to address such issues as

Table X. Summary of the individual test rankings used to determine Kendall's coefficient of concordance (W)

Tier	Scale	Parameter	Sample										
			Epic	EcoPLA	Eastar Bio	Bionolle	UML1	BAK 1095	UML2	Kraft paper	LLDPE	Bio-Solo	ECO Bag
I	Lab[a]	weight loss	8.0	5.0	5.0	1.5	1.5	3.0	2.0	5.0	10.0	10.0	10.0
II	Lab[b]	CO_2 evolution	5.0	3.0	8.0	3.0	6.5	6.5	3.0	1.0	10.0	10.0	10.0
III	field: (LGC)[c]	weight loss	7.0	2.5	9.0	5.0	2.5	2.5	2.5	6.0	9.0	9.0	- - -
III	field: (UMA)[d]	weight loss	6.0	3.5	6.0	3.5	1.5	6.0	1.5	10.0	9.0	8.0	- - -
	Average rank		***6.50***	***3.50***	***7.00***	***3.25***	***3.00***	***4.50***	***2.25***	***5.50***	***9.50***	***9.25***	- - -

[a] ASTM D 6003: Standard test method for determining weight loss from plastic materials exposed to a simulated municipal solid-waste (MSW) aerobic compost environment (ASTM, 1997).

[b] Static compost biometer test; based on the methods of Bartha and Pramer (1965) and ASTM standards D 5338 and D 5988.

[c] Static windrow compost (primarily yard trimmings and horticultural waste); operated and maintained by Laughton's Garden Center, Westford, MA.

[d] In-vessel compost system (Wright Environmental Management, Inc., Toronto, ON) operated by the University of Massachusetts (Amherst), Office of Waste Management–Intermediate Processing Facility.

biodegradability, relative performance under a range of environmental conditions, and predictability.

Are the test plastics truly biodegradable? Results of the rapid screening (Tier I) weight loss test demonstrated that six of the test plastics (EcoPLA®, Eastar Bio™, Bionolle®, the two PLA:Bionolle blends, and BAK 1095) were degraded extensively (>65% weight loss) during a 42 day test exposure. In general, the plastics were so highly fragmented that most of the residual material passed through a 2-mm sieve; the end result being that these materials met or exceeded the "≤10% recoverable material" criterion proposed for claiming biodegradability. Likewise, results of the mineralization (Tier II) tests demonstrated that, along with the positive reference (2-ply, 50# Kraft paper), both the EcoPLA® and Bionolle® exceeded the 60% mineralization threshold during a 42-day test exposure. However, given that both homopolymers used in the PLA:Bionolle blends could be classified as biodegradable, it was surprising that only the 50:50 blend actually mineralized to an extent >60% during the test exposure. Nevertheless, based on analysis of the mineralization vs. time curves (see Figure 6), it was estimated that the 80:20 PLA:Bionolle blend, Eastar Bio™, and BAK 1095 bags would have reached the 60% threshold well within the mandated (180 day) time limit (see Table V) and could be considered biodegradable.

Degradation measurements obtained at all levels of testing yielded a wide range of standard deviations, but were generally smaller at the lab-scale (Tier I & II) than at the field-scale (Tier III). Examination of the standard deviations themselves, however, revealed that there were three fairly distinct ranges which reflected variations in both the overall degradability of the test/reference materials and the mode of degradation. The PE-based plastics were generally found to resist degradation and, as a result yielded data characterized by a narrow range of relatively small standard deviations (i.e., <5% with a median value of 1.2% and an inter-quartile range of 1.0–3.9%). Plastics exhibiting extensive degradation (EcoPLA®, Bionolle®, the PLA:Bionolle blends, and BAK 1095) were characterized by a much larger standard deviation – with a median value of 6.4% and an inter-quartile range of 4.7–9.5%. Materials exhibiting only moderate degradation, on the other hand (Epic, Eastar Bio™, and Kraft paper), generally yielded both the largest and widest range of standard deviations – with a median value of 8.7% and an inter-quartile range of 5.8–15.6%. These results reflect the fact that plastics whose degradation pathway included a potentially significant abiotic contribution (i.e., EcoPLA®, Bionolle®, the PLA:Bionolle blends, and BAK 1095) appeared to be somewhat less affected by variations in the biological activity of the various compost environments. The plastics whose degradation was primarily a result of microbial attack, on the other hand (i.e., Epic, Eastar Bio™, and Kraft paper), tended to exhibit considerably more variation both among and between tests. Nevertheless, our results confirm those of earlier studies (*1, 15, 20*) which demonstrated that the second- and third-generation of environmentally degradable plastic lawn & leaf bags are indeed biodegradable – meeting or exceeding the minimum requirements needed to claim compostability.

Whereas field-scale studies are useful as a means of demonstrating and confirming the compostability of a test material under actual composting conditions, in and of themselves, they offer little proof of a material's true (inherent) biodegradability. Laboratory- or bench-scale studies, on the other hand, are essential for evaluating and validating a material's biodegradability. However, due caution must be exercised when extrapolating the results of a short-term test exposure – carried out under carefully controlled and optimized conditions – to the more uncontrolled and long-term exposure that can be anticipated during composting exposures in the natural environment. Consequently, we also addressed the question: *How well did the results obtained in standard, laboratory-scale biodegradation tests relate to a product's compostability under actual field conditions?*

The concept of correlation between degradation measured at one tier with that measured at another tier is confounded by variations inherent in the different test methods (e.g., weight loss vs. CO_2 production, duration of the test exposure, level of experimental control, and intensity of biological activity). Consequently, the Spearman rank-order correlation (r_S) was used to express the degree of association between any two degradation variables measured in, or transformed to, ranks. Kendall's coefficient of concordance (W) takes this one step further, allowing us to express the degree of concordance (association) between k sets of variables (rankings) (*12, 13*). Thus, using Kendall's coefficient of concordance, we determined that there was significant agreement among the different compost tests both for the overall performance indices ($W = 0.652$**) and total degradation ($W = 0.724$**). From this, we can conclude that agreement among the various tiers of testing was higher than it would have been by chance; i.e., had the rankings been random or independent. Furthermore, we can interpret these results as meaning that although variations in composting conditions can have a significant effect on the rate and extent of degradation, the relative biodegradability of the test/reference materials was essentially unaffected by these variations. That is, only the *relative* results obtained from laboratory-scale (Tier I & II) tests can be expected to carry over to field exposures.

A significant conclusion of the ISR study (*1*) was that laboratory- and pilot-scale tests were more conservative than full-scale (field) studies. However, we found that the Tier I rapid screening test yielded overall performance indices that were generally greater than those obtained from the field studies. The Tier II (mineralization) test, on the other hand, was more conservative than the field tests and clearly reflected differences in substrate (matrix) degradability and the level of microbial activity attained in the different test environments. Our results suggest that the Tier I and II tests should be interpreted only in terms of a material's inherent and/or relative biodegradability and that due care must be taken to avoid false positive results.

Paper bags are the most common vehicle for the collection and disposal of yard wastes and Kraft papers are commonly used as standards for developing relevant time frames for the compostability of biodegradable plastic bags (i.e., the time that composters anticipate is necessary to produce a finished compost). Thus, it is reasonable to expect that, under identical disposal conditions, a compostable plastic bag will biodegrade at a rate equal to (if not faster than) conventional paper lawn & leaf bags. Hence, we asked the question: *How well did the test plastics perform relative to a commercially available paper lawn & leaf bag?* Given that the coefficient of concordance (W) was significant for both the overall performance index and total degradability, we concluded that the best estimate of the "true" ranking of the compostability of the test and reference materials was given by the average rank (see Table X). Thus our data indicate that the overall compostability of the various test and reference materials increased in the order:

LLDPE ≈ ECO Bag ≈ Bio-Solo™ < Eastar Bio™ ≈ Epic ≈ **Kraft paper**
≤ BAK 1095 ≤ EcoPLA® ≈ Bionolle® ≈ the 80:20 PLA:Bionolle blend (UML1)
≤ the 50:50 PLA:Bionolle blend (UML2)

Three of the commercially available test plastics (EcoPLA®, Bionolle®, and BAK 1095) and two of the developmental plastics (the two PLA:Bionolle blends) generally outperformed the Kraft paper bag (Garbax™). In addition, the Epic and Eastar Bio™ plastics yielded average compostability rankings that were marginally higher, but not significantly different, than the Kraft paper. Indeed, only the three PE-based plastics (LLDPE, Bio-Solo™, and ECO Bag) yielded average rankings that were significantly higher than those of the Kraft paper.

Our results suggest that both primary biodegradation and mineralization of the Eastar Bio™ were particularly susceptible to variations in composting conditions. Presumably, this reflects the fact that degradation of the copolyester was predominantly (if not entirely) a function of microbial activity. Indeed, this plastic performed as well or better than the Kraft paper when composting activity was high – as in the lab-scale (Tier I) and in-vessel (Tier III, field site 2) composters. When composting conditions were sub-optimal, however, biodegradation of the Eastar Bio™ was slowed significantly.

The recalcitrance of polyethylene to microbial attack is well established (*22*). For example, Albertsson and Karlsson (*23*) reported only minimal (< 10%) mineralization of a ^{14}C-labelled low-density polyethylene during a 10 year exposure in a humid, composted soil. Likewise, Palmisano and Pettigrew (*24*) reported that there was no evidence to indicate that incorporation of starch in blends with polyethylene resulted in an increase in susceptibility of the polyethylene to microbial attack. Moreover, they suggested that the starch may fail to degrade when "embedded in a hydrophobic matrix such as polyethylene". Our results also indicate that incorporation of a relatively small amount of polyethylene (12–15%, w/w) in a blend containing inherently biodegradable components (i.e., Epic: a PCL-starch-PE blend) resulted in incomplete biodegradation of these components. In addition, we found no evidence for biodegradation of the PE-based plastics during test exposures in either the lab-scale or field composts.

An unexpected result of this study was the relatively slow degradation of the positive control (i.e., Kraft paper). Indeed, the Garbax™ paper bag (a 2-ply, 50# wet strength Kraft paper) failed to disintegrate completely in any of the test exposures. Moreover, its disintegration in both the windrow and in-vessel composts was considerably slower than that of either newspaper or glossy paper. These results were attributed to the much greater weight and thickness of the paper bag relative to the test plastics and other papers. That is, the Garbax™ paper bag was about 2.5–4 times thicker and 2–3 times heavier than the various test plastics. Nevertheless, results such as those obtained for the Garbax™ paper bag serve to highlight the need to validate claims of biodegradability or compostability over a wide range of environmental conditions at both the lab- and field-scale.

Composting continues to gain momentum as a cost-effective and environmentally acceptable alternative to the landfilling of organic wastes, as well as the preferred method for disposing/recycling of yard wastes. However, composting is generally considered a viable option only as long as collection and handling are convenient and the end-product is of a high quality (e.g., free of plastic residues). Clearly, collection bags made from "truly" biodegradable plastics could make a major contribution to the acceptance of composting as an important component of modern solid waste management strategies. Despite their obvious potential, however, there is still much debate concerning the role that biodegradable plastics can/will play in yard waste composting and a number of scientific issues remain to be explored before the skepticism surrounding the compostability of biodegradable plastics is itself degraded away.

Acknowledgments

This study was made possible by funding from the NSF–Biodegradable Polymer Research Center at the University of Massachusetts Lowell and, in part, by grants from the Chelsea Center for Recycling & Economic Development and the Massachusetts Department of Environmental Protection. The authors would like to thank the following people for their assistance and cooperation during the performance of this study: Sumner Martenson, Massachusetts Department of Environmental Protection; Illya Michelson, Chelsea Center for Recycling & Economic Development; Chris Laughton, Laughton's Garden Center, Westford, MA; Amy Baranowski & Cheryl Chaves, University of Massachusetts Amherst, Office of Waste Management–Intermediate Processing Facility. Saskatchewan Centre for Soil Research Publication no. R846.

References

1. American Society for Testing and Materials. *ISR degradable polymeric materials program: Degradability of polymeric materials in a commercial, full-scale composting environment.* ASTM Institute for Standards Research: West Conshoshocken, PA; 1996.
2. Glenn, J. *BioCycle* **1998,** April, pp. 32-43.
3. Glenn, J. *BioCycle* **1998,** May, pp. 48-51.
4. Ma, T.S. and R.C. Ritter. *Modern Organic Elemental Analysis.* Marcel Dekker: New York, 1979; Chapter 2, pp. 51-58.
5. American Society for Testing and Materials. *Annual Book of ASTM Standards.* ASTM: West Conshoshocken, PA; 1997, Vol. 08-03; Standard D 6003.
6. American Society for Testing and Materials. *Annual Book of ASTM Standards.* ASTM: West Conshoshocken, PA; 1995, Vol. 08-03; Standard D 5338.
7. American Society for Testing and Materials. *Annual Book of ASTM Standards.* ASTM: West Conshoshocken, PA; 1997, Vol. 08-03; Standard D 5988.
8. Bartha R. and Pramer, D. *Environ. Lett.* **1972,** 2(4), 217-224.

9. Bartha R. and Pramer, D. *Soil Sci.* **1965,** 100, 68-70.
10. Tukey, J.W. *Exploratory data analysis.* Addison-Wesley: Reading, MA; 1977.
11. CoHort. *CoStat Statistical Software, Version 5.0.* CoHort Software: Minneapolis, MN; 1995.
12. Zar, J.H. *Biostatistical analysis.* 2nd Edition. McGraw-Hill: New York; 1984.
13. Siegel, F. and N.J. Castellan Jr. *Nonparametric Statistics for the Behavioral Sciences.* McGraw-Hill: New York, 1988.
14. Bastioli, C., Cerutti, A., Guanella, I., Romano, G.C., and Tosin, M. *J. Environ. Polym. Degrad.* **1995,** 3, 81-95.
15. Tosin, M., F. Degli-Innocenti, and C. Bastioli. *J. Environ. Polym. Degrad.* **1996,** 4, 55-63.
16. Kaplan, D.L., J.M. Mayer, D. Ball, J. McCassie, A.L. Allen, and P. Stenhouse. In *Biodegradable Polymers and Packaging*; Ching, C., Kaplan, D., and Thomas, N. Eds.; Technomic: Lancaster, PA; 1991; Chapter 1.
17. White, A. Eastman Chemical Co., Kingsport, TN, personal communication, 1998.
18. Block, D. <BioCycle@igpress.com> 29 June 1998. Composting facilities [Personal e-mail]. Accessed 30 June 1998.
19. Farrell, R.E., T.J. Adamczyk, R.A. Gross, D.T. Eberiel, and S.P. McCarthy. In *ISR Degradable Polymeric Materials Research Program: Compilation of ISR Contractor Compost Test Reports* PCN: 33-000018-19. ASTM: West Conshoshocken, PA; 1996.
20. Itäraava, M., M. Vikman, and O. Venelampi. *Compost Sci. & Utilization* **1997,** 5(2), 84-92.
21. Kouloungis, N.R. M.Sc. Thesis, University of Massachusetts Lowell, June 1996.
22. Gould, J.M, Gordon, S.H., Dexter, L.B., and Swanson, C.L. In *Agricultural and Synthetic Polymers*; Glass, J.E. and Swift, G. Eds.; American Chemical Society: Washington, DC; 1990; Pp. 65-75.
23. Albertsson, A.-C. and S. Karlsson. 1988. *J. Appl. Polym. Sci.* **1988,** 35, 1289-1302.
24. Palmisano, A.C. and C.A. Pettigrew. *BioScience* **1992,** 42(9), 680-685.

Chapter 22

The Status of the Biodegradable Plastics Industry in Japan

Hideo Sawada

Biodegradable Plastics Society, 3–2534 Sayama, Osaka-Sayama, Osaka 589–0005, Japan

Several dozens of companies in Japan seem to be either developing or selling biodegradable plastics. Although biodegradable plastics are rather expensive compared to commodity plastics, it is expected to be one of the fastest growing niches in the plastics industry in situation where recycling is difficult or expensive. Biodegradable plastics are new and still developing. If a cost reduction to 200-300 ¥/kg is realized with favorable market conditions, biodegradable plastics could replace commodity plastics in some applications.

Introduction

Japan is the world's second largest plastics producer and produced more than 15 million tons of plastics in 1997 (Figure 1) (1). The disposal of 9 million tons of plastics per year in Japan has raised the question of biodegradable plastics as a means of reducing the environmental impact related to the waste management of plastics. Biodegradable plastics are considered one possible option to solve the plastics industry's solid waste problem. In 1995, Green-Pla, which refers to environmentally friendly plastics, was selected as a generic name for the biodegradable plastics in Japan after a public survey. Throughout this paper, biodegradable plastics are referred to as "Green-Pla".

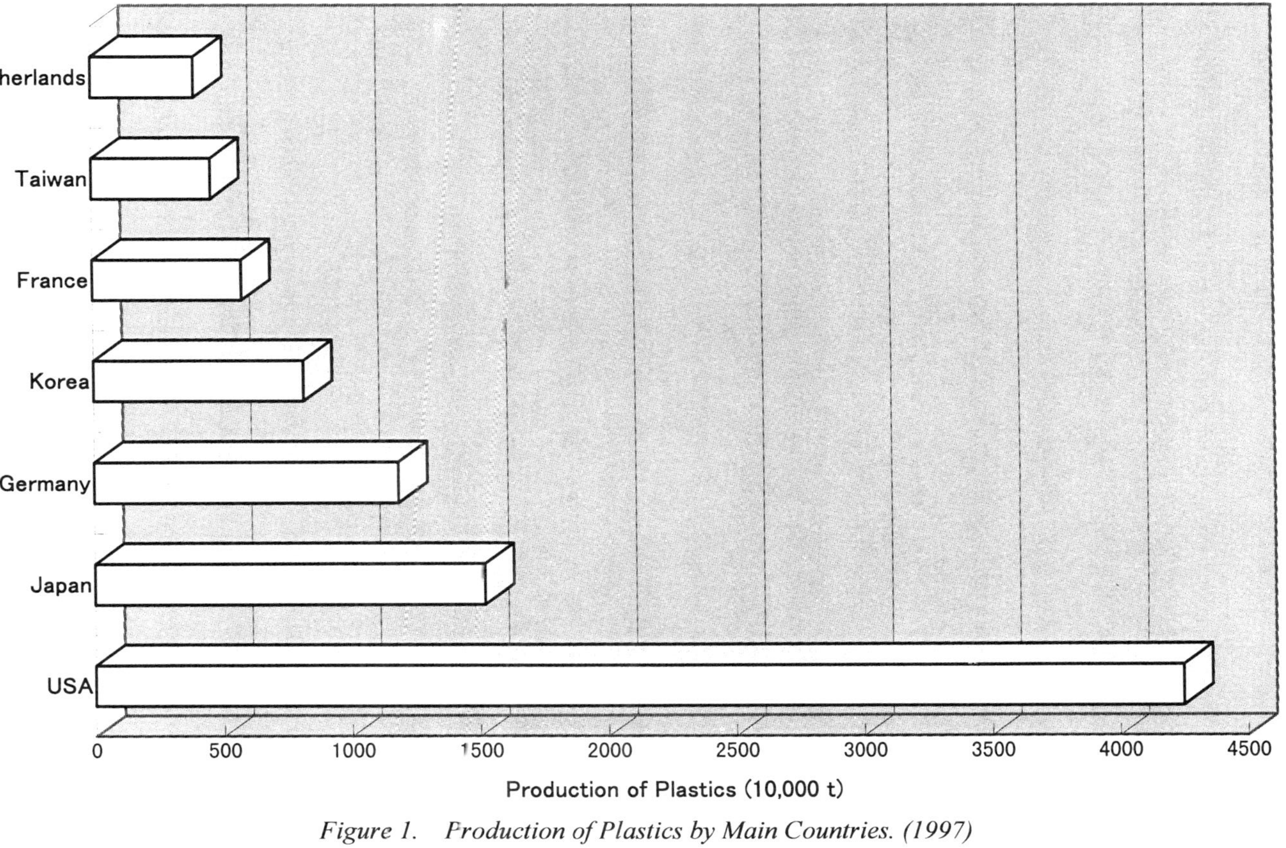

Figure 1. Production of Plastics by Main Countries. (1997)

Current Status of Green-Pla

Currently, plastic waste is considered one of the most serious environmental problems. Plastic waste accounts for approximately 30-40 % of the total waste volume. At present, Japan is trying to reduce its amount of plastic waste, mainly by promoting recycling and providing as many disposal facilities as necessary.

As far as solid waste is concerned, a combination of recycling, source reduction, landfill and waste-to-energy incineration is the reasonable route to take. Generally, plastics will tend to leak from a closed infrastructure into the natural environment. Examples of these cases can be found in the disposal of fishing tackle, such as fishing rods, lines, hooks or agricultural mulch films. These plastic wastes are difficult to recycle due to high cost and the difficulty of collecting and sorting them.

Green-Pla are especially useful as a solution to such plastic waste, as well as post-consumer waste which is difficult or too expensive to recycle. Green-Pla are still rather expensive compared with commodity plastics and are not widely mass-produced at present. However, Green-Pla are expected to be one of the fastest growing niches of the plastics industry for applications where recycling is difficult or too expensive.

Japanese Plastics Industry and Plastic Waste Disposal

The production of plastic products in 1997 was estimated to be 6,463,781 tons, a 4 % increase over the previous year. Containers and packaging materials such as films, sheets and containers accounted for 40 % of the entire plastics market as shown in Figure 2 (2). The amount of plastic waste in 1996 was estimated to be 9,088,000 tons: 4,553,000 tons (50%) of which was municipal solid waste and 4,535,000 tons (50%) of industrial waste (Figure 3). The percent of plastic waste accounted for by application in 1996 is shown in Figure 4 (3).

As far as waste plastics are concerned, a combination of reuse, landfill and incineration is currently the reasonable route to take in Japan. Among the 9,088,000 tons of total plastic waste, nearly 1,030,000 tons (11%) was estimated to be reused, 2,500,000 tons (27%) was burned to produce combustion energy, and 2,140,000 tons (24%) was burned without energy recovery. The other 3,370,000 tons (37%) was disposed of in landfills (4). The incineration of plastics to recover energy is an effective process, however, there is always concern about possible air pollution and hazardous waste by-products such as dioxin from incinerated plastics, particularly polyvinyl chloride (PVC).

Japan has fewer composting plants than found in the US and Europe. The infrastructure and technology for composting is still not strong enough to significantly reduce the amount of waste. Japanese electric equipment manufacturers (Hitachi, Sanyo Electric and Matsushita Electric Works) have recently begun to produce home composting equipment (5). In a batchwise

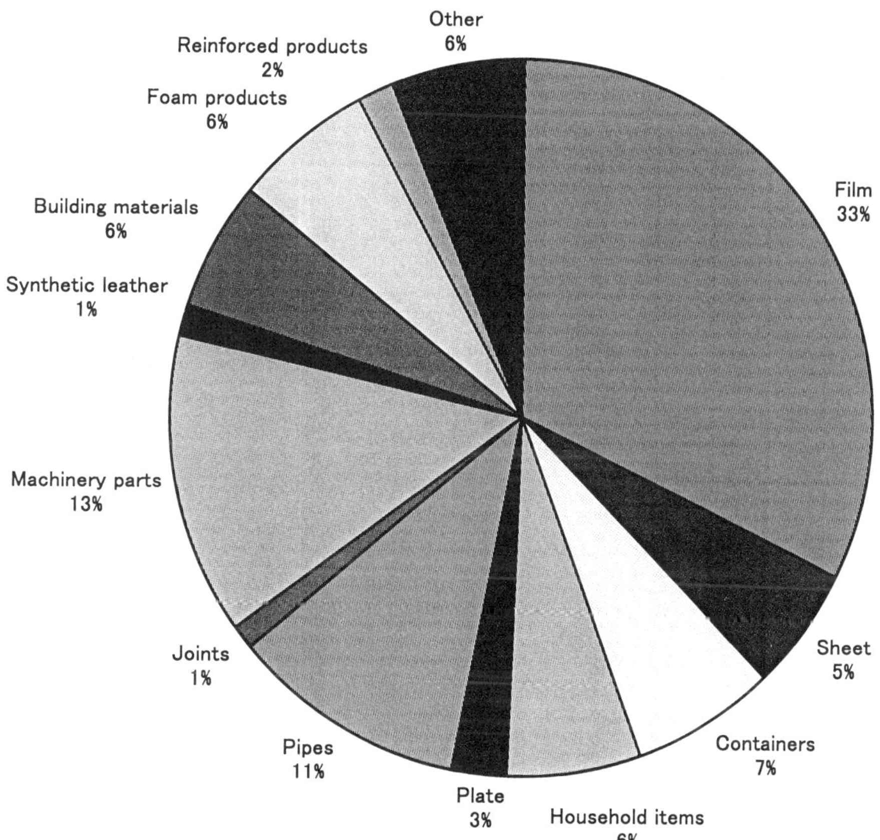

Figure 2. Production of Plastic Products in Japan. (1997)

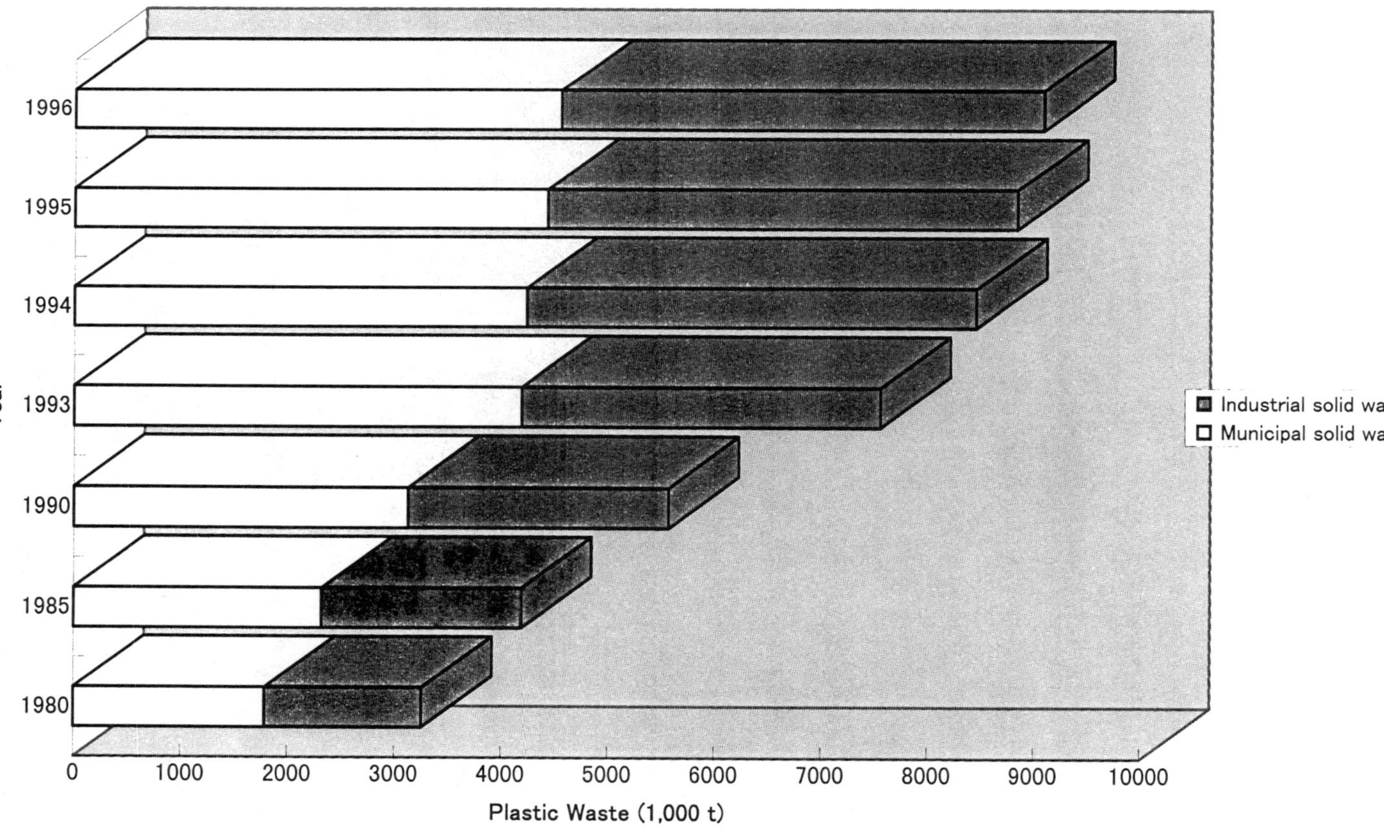

Figure 3. Plastic Waste in Japan.

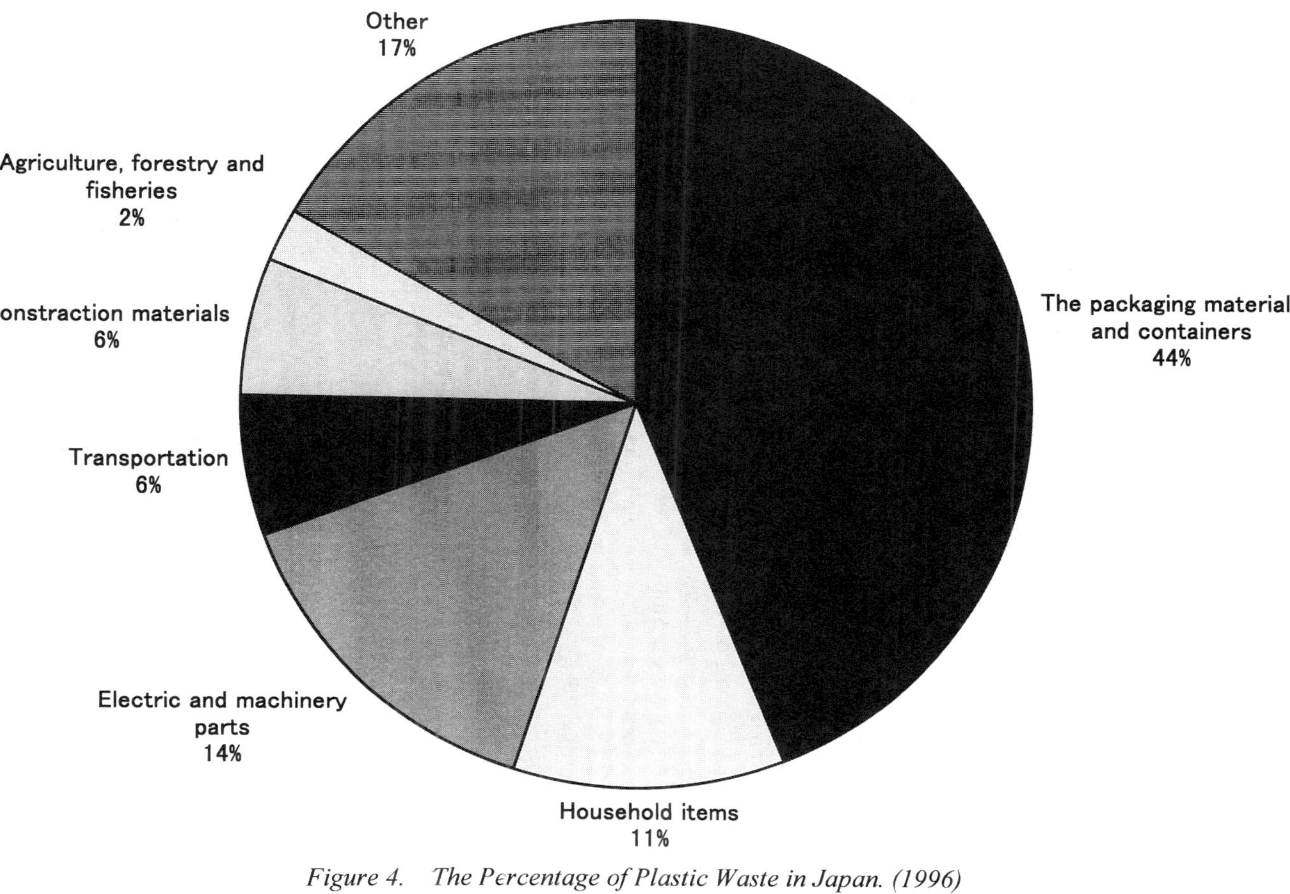

Figure 4. The Percentage of Plastic Waste in Japan. (1996)

operation, garbage is simply intermittently stirred with a cultivation media such as sawdust, porous wood and wood chips in a composting vessel. The surplus water formed is removed through an outlet. The temperature of the contents is kept constant at room temperature. Biodegradation occurs in the presence of aerobic thermophilic microorganisms and garbage from the kitchen is converted into water, carbon dioxide and a homogeneous humus-like substance within a week. Such a device disposes of 1 kg of garbage every day which, after processing, is reduced to 1/5 of a kg. The device is selling for about ￥50,000-60,000 in Japan.

Administrative Actions on Waste Disposal

In April 1997, the Container and Packaging Recycling Law was introduced to promote solid waste reduction and the use of recycled packaging materials and containers. The outlines of this law are as follows: 1) Consumers should sort their waste. 2) Local governments should collect refuse according to type. The methods of collecting refuse are determined by the local governments. 3) Manufacturers (producers of containers and producers of the container contents) should try to reuse materials and may defray expenses.

This law described the duties of entrepreneurs, consumers, national and regional public organizations, and the roles they should play in boosting the recycling of resources. However, at best, this law only stipulates that effort should be expended towards this end. Since the recovery rate of polyethylene terephtalate (PET) bottles is low, only PET bottles among plastic products are included in this law to promote their recovery. Due to a great deal of time and labor required to classify and collect wastes, the recycling of PET bottles is not as advanced as expected. The law will be applied to all plastics in April 2000. If Green-Pla wastes are recognized as organic wastes due to their biodegradability, the law could be applied to all plastics except Green-Pla.

As previously mentioned, the infrastructure and technology for composting is still not good in Japan. The Japanese Government plans to provide low interest funds from the Japan Development Bank if local governments and private companies form joint ventures to build composting facilities. In addition, the government plans to provide subsidies to such joint ventures.

Activities of Biodegradable Plastics Society (BPS)

In 1989, BPS was established to promote the development of technology related to Green-Pla and their commercial use under the sponsorship of the Ministry of International Trade and Industries (MITI). BPS is a non-profit organization whose activities are supported by membership fees and voluntary contributions from member companies (6). The actual activities are conducted through three committees: the Planning, Research and Technical Committees. The Society has played an important role in the development of test methods of

Green-Pla. BPS has been involved in the activities of International Organization for Standardization (ISO), especially, the working group (WG22) on the biodegradability of plastics. Three drafts of International Standards on biodegradability are now available (7).

Environmental labeling will make it very easy to recognize and identify compostable and biodegradable products. Since it is convenient for consumers to classify their wastes, environmental labeling will contribute to a reduction in environmental impact. In Japan, the ecology mark is authorized by the Environmental Agency. This agency advocates strict enforcement of the ecology mark on the basis of Life Cycle Assessment (LCA). Since the Japanese ecology mark presently covers a wide range, a special ecology mark for Green-Pla will be needed. The Japanese ecology mark on Green-Pla will be independent third-party certification. BPS is still studying the details.

Commercially Available Green-Pla

Present commercially available Green-Pla are shown in Table 1. Several dozen companies in Japan seem to be either developing or selling Green-Pla. Chemosynthetic polymers are promising because these polymers can be produced in large volumes by conventional processes. For example, PCL (Daicel Chemical), PVA (Kuraray), PLA (Shimadzu, Mitsui Chemical and Cargill Dow Polymers) and polyalkylene succinate and related copolyesters (Showa Highpolymer), are all gaining acceptance. Among the microbial polymers, poly(3-hydroxybutyrate-co-3-hydroxyvalerate) (PHB/V) (Monsanto Japan) is one thermoplastic which can be processed using conventional techniques. If transgenic plants can be developed, which produce PHB/V agro-plastic on a large-scale and a cost reduction to 200 ¥/kg is realized, PHB/V has a bright future. Starch-based polymers (Nippon Synthetic Chemical and Chisso) are one of the most important Green-Plas in Japan. Its main applications will be packaging foams and compost bags. The prospects of cellulose acetate-based Green-Pla (Daicel Chemical and Nippon Shokubai) are promising because cellulose acetate is mass-produced, mainly for molding clear, tough and flexible sheets.

Market Forecast of Green-Pla

More than one million tons a year of big-volume plastics (commodity plastics) such as PVC, polypropylene (PP), polystyrene (PS), high density polyethylene (HDPE) and low density polyethylene (LDPE) are produced in Japan. Prices of these plastics are 200 ¥/kg or less. The production of plastics, such as PET and acrylonitrile-butadiene-styrene copolymer (ABS) is nearly 500,000 tons. These plastics are sold at the price of 300 ¥/kg. The production

Table 1. Commercially Available Green-Pla

Type	*Composition*	*Trade name*	*Company*
Microbial	PHB/V	Biopol	Monsanto/Monsanto Japan
	Bacteria cellulose		Ajinomoto
Synthetic	Polycaprolactone (PCL)	CELGREEN P-H	Daicel Chemical
		Tone	Union Carbide/Nippon Unica
	Modified PCL	CELGREEN P-HB	Daicel Chemical
	Poly(ethylene succinate)	Bionelle #1000	Showa Highpolymer
	Poly(ethylene succinate/adipate)	Bionolle #3000	Showa Highpolymer
	Poly(buthylene succinate/adipate)	Lunare SE	NIPPON SHOKUBAI
	Poly(lactic acid) (PLA)	Lacty	Shimadzu
		Lacea	Mitsui Chemical
		EcoPLA	Cargill Dow Polymer/ Cargill Japan
	Polyvinylalcohol (PVA)	Poval CP	Kuraray
Natural	Modified Cellulose Acetate	CELGREEN P-CA	Daicel Chemical
		Lunare ZT	NIPPON SHOKUBAI
	Starch based	Eco-ware	NISSEI
		EverCorn	Japan Corn Starch
	Starch and modified PVA or PCL	Mater-Bi	Novamont/Nippon Synthetic Chemical
		Novon	Novon International/ Chisso
	Chitosan/Cellulose/ Starch	Dolon CC	AICELLO

of engineering plastics, such as polycarbonate (PC) and polymethyl methacrylate (PMMA) is nearly 200,000 tons. Sales prices of these plastics are around 500 ¥/kg.

The growth of every plastic has shown the same trend. Several years after its introduction, the production of plastics rapidly increased while plastic prices rapidly declined. Generally, the growth of knowledge in any discipline occurs over time in the shape of an "S" curve. At first, progress is painstakingly slow, followed by a period of rapid growth, and then a leveling off, as the field matures. We view the discovery and development of Green-Pla as the beginning of the "S" curve. Let's call this "S" curve the "foundations" period of Green-Pla. Green-Pla is still in its infancy.

Table Ⅱ. The Prices of Green-Pla

Type	*Prices*
Microbial polymers	1,500~2,000 ¥/kg
Synthetic polymers	500~1,500 ¥/kg
Natural polymers	500 ~ 800 ¥/kg

The prices of Green-Pla are shown in Table 2. To compete with engineering plastics, the target price for Green-Pla is estimated at 500 ¥/kg. Green-Pla occupies an important position among such polymers and seems to be one of the fastest growing niches in the Japanese plastics industry. At present there are several plants producing various Green-Plas at annual production rates of 500~5,000 tons. Although the market for Green-Pla in Japan is roughly doubling every year, the sales of Green-Pla in 1997 was still only 1,500~1,600 tons (8). Due to the relatively low sales of Green-Pla, its fixed cost is a relatively high percent of the manufacturing cost. Therefore, the prices of Green-Pla are high due to high manufacturing costs, and producers hope to increase production volume. If the demand for Green-Pla increases with favorable market conditions, the annual production of Green-Pla will reach 100,000 tons, and resin prices will fall to an estimated 200~300 ¥/kg.

To compete with commodity plastics, the target price for Green-Pla is estimated at less than 300 ¥/kg. If a cost reduction to 200~300 ¥/kg is realized, Green-Pla could be applied to some fields where commodity plastics are used. Japanese producers will try to lower the cost of Green-Pla through technological innovations and by using low cost materials. The demand for Green-Pla is expected to be 3 million tons in Japan (9). The main application will be films, sheets, household items and containers as shown in Figure 2.

Conclusions

Green-Pla is expected to be one of the fastest growing niches in the plastics industry in situations where recycling is difficult or expensive. Green-Pla is particularly applicable to high value-added products, such as higher profit margin, high performance specialty polymers. The target price for the expansion of Green-Pla is probably 200-300 ￥/kg to compete with commodity plastics. If a cost reduction to 200~300 ￥/kg is realized, Green-Pla could replace commodity plastics in some applications. Potential applications are films, sheets, household items and containers, and the maximum demand for Green-Pla is expected to be 3 million tons per year in Japan. In Japan, the composting infrastructure and technology are insufficient to recover organic wastes. If possible future government regulations require municipalities to establish collecting and composting systems for organic wastes, compost bags made of Green-Pla will be used for collecting and composting. At that time, the prospects of Green-Pla will be promising. Several dozens of companies in Japan seem to be either developing or selling Green-Pla. The Green-Pla industry is new and still developing. New industries always take a while, but the long-term prospects for the Green-Pla industry look bright.

References

1. JPIF, *Japan Plastics* **1998**, *49*, 18.
2. *Asahi Shimbun Japan Almanac 1998,* Asahi Shimbun: Tokyo, Japan, 1998: p.163.
3. *Plastic Waste*, Plastic Waste Management Institute, Japan, June 2, 1998.
4. *Plastic Waste*, Plastic Waste Management Institute, Japan, May 28, 1998.
5. *Asahi Shimbun*, March 27, 1998.
6. Sawada, H., *BEDPS News* **1994**, *3,* 6.
7. Sawada, H., Polymer *Degradation and Stability* **1998**, *59*, 365-370.
8. *Sekiyukagaku Shimbun*, January 30, 1998.
9. *The Age of New Plastics*, BPS: Tokyo, Japan, 1995; p.28.

Chapter 23

Worldwide Composting Technologies with Special Reference to Biodegradable Plastics

Eliot Epstein

E&A Environmental Consultants, Inc., 95 Washington Street, Suite 218, Canton, MA 02021

There are more than 50 different types of composting technologies operating worldwide. These can be classified generically into static, turned, or combined systems. In the U.S., static systems are predominantly used for biosolids composting, whereas turned systems are primarily used for municipal solid waste and yard materials composting. In Europe, many static systems are used for biowaste and agitated systems for solid waste. There are more than 274 biosolids, 3,484 yard materials, and 142 food/institutional composting facilities operating in the U.S. today; there are only 15 municipal solid waste composting facilities. The potential for utilization of biodegradable plastics in composting operations is in the use of bags for source-separated solid waste and yard material, coated papers, and packaging materials. There are several factors that could affect the rate and extent of biodegradation of polymers/plastics. The efficiently of biodegradation is dependent upon the feedstock, the system, and maximizing the microbial system.

Composting is the biological decomposition of organic matter under controlled, aerobic conditions. It can involve both mesophilic and thermophilic temperatures. Over 80 different microorganisms have been identified in composting[1].

There are many factors which affect the composting process:

- Oxygen
- Moisture
- Temperature
- pH
- Nutrients, especially carbon, nitrogen, and phosphorus

- Particle size and surface area
- Chemical and physical nature of the feedstock material

These factors greatly affect the rate and extent of microbial decomposition of organic compounds. The ultimate goal in composting is to produce a humus-like material, which is an excellent soil conditioner and can be used for numerous horticultural and agricultural applications. As a soil conditioner, this stable organic material improves soil moisture content, water retention, infiltration and permeability of water, soil structure, soil temperature, cation exchange capacity, and other soil chemical properties. Constraints to product use have been related to heavy metal content exceeding regulatory levels; pathogens as a result of improper composting; unstable material, resulting in odors; rapid decomposition in soil causing nutrient deficiency in plants; presence of undesirable organic compounds (e.g., PCBs); contamination with inerts, such as glass, metals, and plastics; lack of product standards, resulting in inconsistent product quality and consistency; and lack of educational programs to target the market.

Many components of the waste stream can be composted. The materials listed below are some of the feedstocks that have been composted.

- Municipal solid waste, unsorted or sorted
- Biosolids (sewage sludge)
- Septage or night soil
- Animal wastes
- Leaves/yard materials
- Food wastes (restaurants) and food processing waste
- Industrial waste
  - pulp and paper mill sludge
  - organic polymer sludge
  - petroleum waste, tank bottoms
  - munitions waste (TNT, HMX, RDX)
  - pharmaceutical wastes
- Coated paper and cardboard
- Biodegradable polymers
- Artificial blood

In the U.S., there are over 247 biosolids, 3,484 yard material, 147 food/institutional, and 15 municipal solid waste composting facilities.

The extent of composting in North America has primarily been a function of the economics of disposal for a particular feedstock and regulatory restrictions on various management options. There are three major feedstocks that are composted: municipal solids waste, biosolids (sewage sludge), and yard materials. Traditionally, municipal solid waste has been managed by the private sector, while biosolids management has been the responsibility of public entities (wastewater treatment plants). Yard material is managed by both private and public organizations.

Until the mid-1980s, there was little composting of unseparated or source-separated municipal solid waste in North America or Europe. In the 1930s and 1940s, several attempts were made to compost municipal solid waste. One proprietary method, Fairfield-Hardy, was used at several plants, the largest of which was in Altoona,

Pennsylvania. In the 1960s, the U.S. Public Health Service funded two projects in Johnson City, Tennessee and Gainesville, Florida.

Two municipal solid waste composting facilities were constructed and operational, but both of these facilities were subsequently closed. However, by 1971, 14 of the 18 large-scale composting facilities constructed in the U.S. were closed.[2] In 1991, there were 18 municipal solid waste composting facilities, and there has been virtually no growth in the number of facilities in recent years. This can be attributed to the low cost of landfilling in the U.S. as well as the poor performance of existing municipal solid waste composting facilities. Table I lists the solid waste composting facilities in the U.S.

Three large facilities in Portland, Oregon; Dade County, Florida; and Pembroke Pines, Florida have closed within the past four years. The facilities were closed primarily because of odor problems caused by poor design. This has given municipal solid waste composting a bad reputation.

Currently, the largest co-composting (municipal solid waste and biosolids) facility is being designed in Edmonton, Alberta, Canada. This facility will compost approximately 1,000 tonnes of municipal solid waste and 90 dry tonnes of biosolids per day.

Prior to 1975, there was an insignificant amount of biosolids composting. In 1973, the U.S. Department of Agriculture embarked on a major research program at the Beltsville Agricultural Research Station in order to find a biosolids management option for Washington, D.C. In 1974, the Aerated Static Pile composting method was developed.[3] This method can be used for the composting of either undigested (raw) or digested biosolids. From that period forward, biosolids composting increased.

This continued increase was the result of several factors:

- The economics of biosolids composting is very competitive with other technologies, such as land application, heat drying, incineration, and landfilling.
- Regulations in the U.S. favor beneficial use, such as composting, heat drying, and land application.
- The public generally favors composting over other management options.
- Federal funds appropriated through the Clean Water Act of 1978 provided funds for composting facilities and other technologies.

It is expected that this growth will continue, even though no more Federal funds for project development are available. There has been an increased involvement by the private sector in financing and operating biosolids composting facilities.

The greatest growth of composting has been in yard material management. It is estimated that there are over 3,484 yard material composting facilities, both private and public, currently operating in the U.S. There were very few yard material composting facilities until the late 1980s because landfilling was inexpensive and convenient. When local and state governments became concerned about ground water contamination from landfills and diminished landfill capacity, states began restricting landfill disposal of yard materials. The U.S. EPA imposed more strict regulations on landfill construction and monitoring, making this technology more expensive.

At this point, it appears that the growth in yard material composting has peaked. Many private companies, such as Browning Ferris Industries; Waste Management, Inc.; and Scotts Environmental entered the yard waste composting business for profit, but

Table I. Some Solid Waste Composting Facilities in the United States

Location	*Year*	*Capacity (tonnes/day)*	*Technology*	*Status*
New Castle, DE	1984	205	Fairfield	Closed
Pennington Co., MN	1985	22	Lundell	Closed
Fillmore Co., MN	1987	5	Engineered/ aerated windrow	Operating
St. Cloud, MN	1988	228	Recomp	Closed
Sumter Co., FL	1988	46	Retrofitted to Bedminster	Operating
Ashland, KY	--	114	Addington	Closed
Mora, MN	1992	270	Static pile	Restarting
Pinetop-Lakeside, AZ	1991	8	Bedminster	Operating
Cobb Co., GA	1995	273	Bedminster	Restarted after fires
Buena Vista Co., IA	--	35	Engineered/ windrow	Operating
Swift Co., MN	1992	5	Engineered/ windrow	Operational
Truman, MN	1992	91	OTVD	Operational
Lexington, KY	--	91	Agranom	Operational
East Hampton, NY	--	27	Engineered/ IPS	Operational
Sevierville, TN	1993	209	Bedminster	Operational
Wright Co., MN	1992	150	Buhler	Closed
Pembroke Pines, FL	1992	546	Buhler	Closed
Toronto, CN	1978	41	Fairfield-Hardy	Closed
Altoona, PA	1951	23	Fairfield-Hardy	Closed
Portland, OR	1991	546	DANO	Closed
Columbia Co., WI	--	64	Engineered	Operational

recently, several of these large waste corporations have discontinued their organics operations, citing low or negative profit margins.

Collection costs are an important factor. Early attempts at biodegradable plastic bags failed, and paper bags are costly. Removal of plastics from the waste stream has also been problematic and costly. Some facilities have solved this problem by restricting bagging of yard material; others have restricted bagging to paper only. Odor has also been a problem for yard material facilities, especially those that process large volumes of grass.

Factors Affecting the Growth of Composting

The major factors affecting growth and development of the composting industry are:

- Economics of waste management
- Public attitudes towards recycling/composting
- Incentives
- Regulations
- Product markets
- Past experience
- Environmental and public health issues

Economics of Waste Management

In most of the U.S., the least expensive method of waste management has always been landfilling. Landfill costs have increased in recent years as a result of regulations that require leachate and gas emissions control. In several parts of the U.S., particularly the northeast where landfill costs can exceed $60 per ton, municipal solid waste composting can compete and succeed. Composting can generally compete in cost with waste-to-energy facilities.

The economics of biosolids composting is competitive with most biosolids management options, as landfilling of low-solids material is often not allowed. Incineration or heat drying is more expensive; direct land application is often less expensive.

Public Attitude Towards Recycling

Composting is recycling. The U.S. EPA and many state agencies have established a hierarchy of solid waste management options. Reuse, recycling, and composting are rated as the most desirable, and waste-to-energy, incineration, and landfilling are ranked as least desirable. Recycling goals and mandates have had much influence on the development of composting facilities, particularly yard materials facilities. Many states have instituted recycling goals of up to 50 percent diversion from landfills. Composting plays an essential role in attaining these goals.

The public generally favors composting over other waste management options. However, many individuals, as well as communities, do not want composting facilities or other waste handling industries in their neighborhoods. This has mostly been a result of odor problems experienced at some facilities.

Incentives

Financial incentives from both the U.S. Federal and state governments have encouraged composting. The U.S. Federal government has provided money through Clean Water Act funding for the development of biosolids composting facilities. Several states have provided financial incentives in the form of grants for composting development and facilities. For example, the state of Minnesota has provided up to $2 million for construction of municipal solid waste composting facilities. Thus, this state has the highest number of municipal solid waste composting facilities. Several states tax waste going into landfills and use the moneys to provide grants for composting, recycling, and other activities.

Regulations

Regulations can impact the economics of waste disposal as well as the availability of options. Regulations regarding air emissions from waste-to-energy facilities increase the cost of combustion of wastes. Regulations on leachate control and gas emissions from landfills increase the cost of landfill construction and operation. The U.S. EPA has promulgated regulations that have significantly increased the cost of incineration and landfilling. The Clean Water Act, which specifically banned ocean dumping of biosolids, left many wastewater treatment plants searching for options. The U.S. EPA 40 CFR Part 503 regulations, encouraging and regulating the beneficial reuse of biosolids, make composting a more attractive alternative than many other options.

Canadian regulations of trace elements can discourage composting by limiting its use and distribution. In the Netherlands, composting has shifted from solid waste composting to composting of yard material and biowaste. Although landfilling has decreased significantly, incineration of solid waste has increased. Improved technology for effective separation and refining of products could have resulted in less incineration and more composting.

Product Markets

The availability of markets for compost encourages composting. Before financing a private facility, lending institutions want assurance that a market for the finished product exists. Past experience has shown that a high quality product is easily marketed. Prices vary from $7 to over $20 per cubic meter. The principal paying markets in the U.S. are landscapers, nurseries, greenhouses, turf grass growers, and soil blenders.

Past Experience

A community or industry evaluating different waste management strategies can only base its judgment on previously built facilities. Unfortunately, the failure of facilities reflects badly on the industry and can make siting new facilities difficult. This has been especially true with respect to municipal solid waste facilities.

The principal reason for the failure of several large municipal solid waste composting facilities was the poor design of materials handling systems and odor control. Facilities were sited near residential communities, and odorous emissions caused complaints and concern about public health.

Most of the biosolids composting systems that have failed have been vertical systems. The failures were due to mechanical problems, the inability to process the waste properly, creation of odors, and inadequate odor control. It is doubtful that new vertical systems will be designed and accepted in the U.S.

Environmental and Public Health Issues

Environmental and public health aspects have often restricted the development of composting facilities. The most important issues regarding facilities are odors and bioaerosols. The presence of odors suggests to the public that the air being emitted may contain unhealthy compounds or material.

The potential for odors exists, regardless of the technology or system. The key to odor control is proper design of the facility and proper operation. Enclosing facilities can result in better odor management. However, this raises the cost of the facility. Odors have also been the major cause of failure at yard material facilities. Although brush and leaves are relatively easy to handle, grass, which arrives at facilities in concentrated batches, can quickly create an odor problem if left unmanaged.

The major potential use of biodegradable polymer bags is in source-separated municipal solid waste, such as dry/wet or blue bag systems or the collection of yard materials.

Composting Technology

Basic Concepts

Composting systems fall into one of several categories: static, turned, or a combination of both. Within these categories, there are several different basic designs:

- Static systems
  - aerated static pile
  - vertical silos or bins
  - tunnels

- Turned systems
  - windrow
  - agitated bed

Different systems are more appropriate for different feedstocks. Depending on the feedstock, more structures, odor control, or special material handling equipment may be required. For example, biosolids are best composted in an aerated static pile or agitated bed system. Both of these systems can be enclosed with effective odor control. The aerated static piles can often be built outdoors with negative forced air that is piped into a biofilter for treatment. Turned systems can be more effective in reducing particle size and are generally more suitable for municipal solid waste or yard materials.

Table II lists numerous composting system that we have evaluated. These systems are based on static or turned technologies and would be very different in their ability to biodegrade feedstocks.

Table II. Various Composting Systems

Composting Systems	
Aerated Static Pile (non-proprietary)	Windrow (non-proprietary)
International Processing System (IPS)	Bedminster
Longwood	OTVD
DANO	Buhler
Taulman-Weiss	ABV-Purac
BAV	Daneco
Arus-Ruthner	PLM Selbergs
EBARA	Japan Steel Works
Paygro	AmeriCycle
American BioTech	Ashbrook Tunnel
Enadisma	Gicom Tunnel
Seerdrum	Fairfield
Agripost	Heidelberg Silo
VAM	Sorain Cecchini
Weser Bio-waste	Koch Bio-waste

Composting of organic material involves three basic processes:

- Pre-processing
- Composting
- Post-processing

Pre-processing involves the preparation of the feedstock for composting. It may involve shredding, removal of non-desirable materials (plastics, metals), or blending of amendments or bulking agents.

Table II lists the various systems that have been investigated.

The first stage of composting involves a high rate of decomposition, which results in high temperatures and rapid breakdown of the organic matter. This stage has a higher potential for odor generation. The length of this stage depends on the feedstock. The second stage involves curing. Curing is a continuation of composting, but since materials are partially stabilized, heat generation is lower and there is less oxygen demand in the composting medium. Curing produces a stable and mature compost product.

Post-processing involves the production of a material for marketing. This stage can involve screening, destoning, and other refining processes.

Figures 1, 2, and 3 show general process flow diagrams for municipal solid waste, biosolids, and yard materials composting facilities.

Composting as Related to Biodegradable Polymers/Plastics

There is considerable potential for the use of biodegradable plastics, especially in yard material, food waste, and solid waste composting. The different systems could greatly affect the rate and extent of the biodegradation of polymers and plastics. Composting conditions will vary within each system. For example, anaerobic conditions are much more prevalent in windrow or unaerated static systems. This can be overcome by installing aeration systems. Aerobic conditions accelerate composting and reduce the potential for odors. Moisture content will vary. Windrow and agitated bed systems will tend to dry unless water is applied at frequent intervals. Temperatures will generally be higher in static systems, often approaching high thermophilic temperatures exceeding 60°C. The rate of biodegradation is a function of particle size. Different systems affect physical properties, especially particle size and surface area. All these factors affect the microbial population both, in types and numbers, and they, in turn, influence the rate and extent of biodegradation.

There is a need for improved technology in debagging, separation of inerts, and other product refining systems in order to produce high quality products.

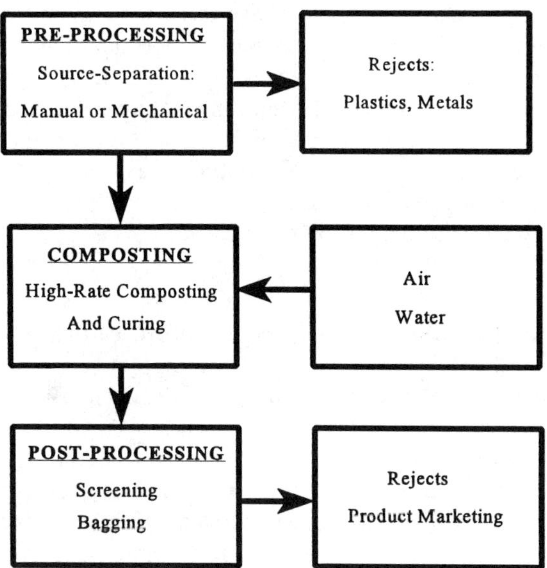

Figure 1. Process Flow for Municipal Solid Waste Composting.

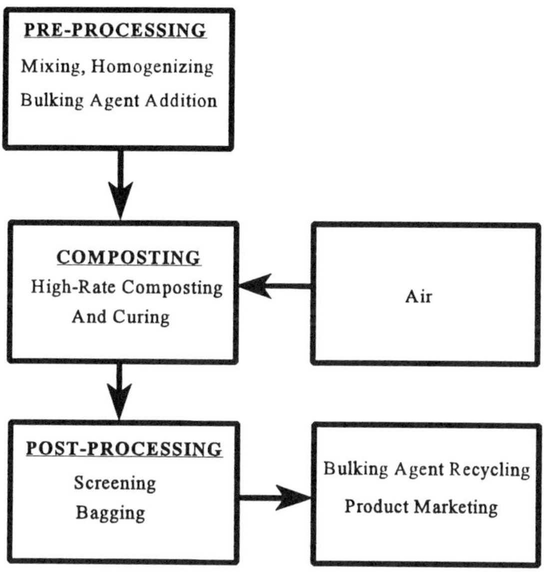

Figure 2. Process Flow for Biosolids Composting.

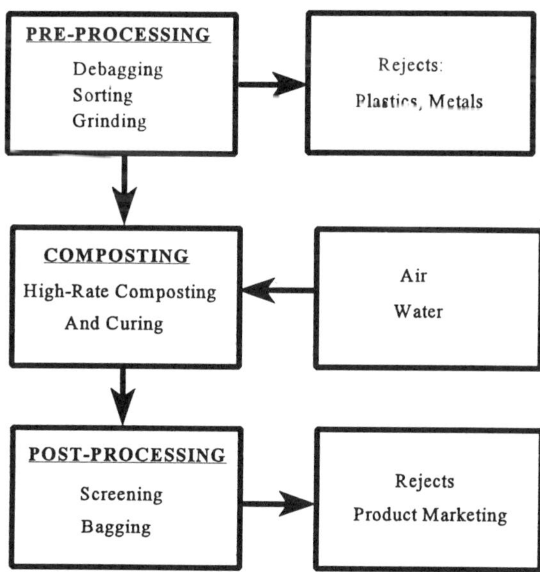

Figure 3. Process Flow for Yard Materials Composting.

Literature Cited

1. Epstein, E. *The Science of Composting*; Technomic Publishing Co., Inc.: Lancaster, PA, 1997.
2. Stickelberger, D. "Survey of City Refuse Composting;" WHO International Reference Centre, 1974.
3. Epstein, E.; Willson, G.B.; Burge, W.D.; Mullen, D.C.; Enkiri, N.K. *J. Water Pollut. Control Fed.* **1976**, *48*,688-694.

Author Index

Subject Index

C

H

I

J

K

L

N

R

S

T

U

V

W